Student Solutions Manual an...
for
Serway & Jewett's

Principles of Physics

Volume 2

FIFTH EDITION

Prepared by

Vahé Peroomian
University of California at Los Angeles

John R. Gordon
Emeritus, James Madison University

Raymond A. Serway
Emeritus, James Madison University

John W. Jewett
Emeritus, California State Polytechnic University, Pomona

BROOKS/COLE
CENGAGE Learning

Australia • Brazil • Japan • Korea • Mexico • Singapore • Spain • United Kingdom • United States

For product information and technology assistance, contact us at **Cengage Learning Customer & Sales Support, 1-800-354-9706**

For permission to use material from this text or product, submit all requests online at **www.cengage.com/permissions** Further permissions questions can be emailed to **permissionrequest@cengage.com**

ISBN-13: 978-1-133-11075-0
ISBN-10: 1-133-11075-4

Brooks/Cole
20 Channel Center Street
Boston, MA 02210
USA

Cengage Learning is a leading provider of customized learning solutions with office locations around the globe, including Singapore, the United Kingdom, Australia, Mexico, Brazil, and Japan. Locate your local office at: **www.cengage.com/global**

Cengage Learning products are represented in Canada by Nelson Education, Ltd.

To learn more about Brooks/Cole, visit **www.cengage.com/brookscole**

Purchase any of our products at your local college store or at our preferred online store **www.cengagebrain.com**

Printed in the United States of America
1 2 3 4 5 6 7 16 15 14 13 12

Table of Contents

PREFACE

This *Student Solutions Manual and Study Guide* has been written to accompany the textbook *Principles of Physics, Fifth Edition, Volume 2,* by Raymond A. Serway and John W. Jewett, Jr. The purpose of this *Student Solutions Manual and Study Guide* is to provide students with a convenient review of the basic concepts and applications presented in the textbook, together with solutions to selected end-of-chapter problems. This is not an attempt to rewrite the textbook in a condensed fashion. Rather, emphasis is placed upon clarifying typical troublesome points, and providing further practice in methods of problem solving.

Every textbook chapter has a matching chapter in this book and each chapter is divided into several parts. Very often, reference is made to specific equations or figures in the textbook. Each feature of this Study Guide has been included to ensure that it serves as a useful supplement to the textbook. Most chapters contain the following components:

- **Notes From Selected Chapter Sections:** This is a summary of important concepts, newly defined physical quantities, and rules governing their behavior.

- **Equations and Concepts:** This is a review of the chapter, with emphasis on highlighting important concepts and describing important equations and formalisms.

- **Suggestions, Skills, and Strategies:** This offers hints and strategies for solving typical problems that the student will often encounter in the course. In some sections, suggestions are made concerning mathematical skills that are necessary in the analysis of problems.

- **Review Checklist:** This is a list of topics and techniques the student should master after reading the chapter and working the assigned problems.

- **Answers to Selected Objective and Conceptual Questions:** Suggested answers are provided for selected Objective and Conceptual Questions.

- **Solutions to Selected End-of-Chapter Problems:** Solutions are given for approximately 20 percent of the problems in each textbook chapter. Problems were selected to illustrate important concepts in each chapter. The solutions follow the *Conceptualize—Categorize—Analyze—Finalize* strategy presented in the text.

An important note concerning significant figures: When the statement of a problem gives data to three significant figures, we state the answer to three significant figures. The last digit is uncertain; it can, for example, depend on the precision of the values assumed for physical constants and properties. When a calculation involves several steps, we carry out intermediate steps to many digits, but we write down only three. We "round off" only at the end of any chain of calculations, never anywhere in the middle. We also note that integers (typically as components of a vector) are assumed to have three significant figures.

ACKNOWLEDGMENTS

We take this opportunity to thank everyone who contributed to this edition of *Student Solutions Manual and Study Guide to accompany Principles of Physics, Fifth Edition, Volume 2*.

Special thanks for managing and directing this project go to Publisher, Physics and Astronomy, Charles Hartford; Development Editor Ed Dodd, Associate Development Editor Brandi Kirksey; and Editorial Assistant Brendan Killion.

Our appreciation goes to our reviewer, Susan English. Her careful reading of the manuscript and checking the accuracy of the problem solutions contributed in an important way to the quality of the final product. Any errors remaining in the manual are the responsibility of the authors.

Finally, we express our appreciation to our families for their inspiration, patience, and encouragement.

We sincerely hope that this *Student Solutions Manual and Study Guide* will be useful to you in reviewing the material presented in the text, and in improving your ability to solve problems and score well on exams. We welcome any comments or suggestions that could help improve the content of this study guide in future editions; and we wish you success in your study of physics.

Vahé Peroomian
Los Angeles, CA

John R. Gordon
Harrisonburg, VA

Raymond A. Serway
Leesburg, VA

John W. Jewett, Jr.
Anaheim, CA

SUGGESTIONS FOR STUDY

We have seen a lot of successful physics students. The question, "How should I study this subject?" has no single answer, but we offer some suggestions that may be useful to you.

1. Work to understand the basic concepts and principles before attempting to solve assigned problems. Carefully read the textbook before attending your lecture on that material. Jot down points that are not clear to you, take careful notes in class, and ask questions. Reduce memorization to a minimum. Memorizing sections of a text or derivations does not necessarily mean you understand the material.

2. After reading a chapter, you should be able to define any new quantities that were introduced and discuss the first principles that were used to derive fundamental equations. A review is provided in each chapter of the Study Guide for this purpose, and the marginal notes in the textbook (or the index) will help you locate these topics. You should be able to correctly associate with each *physical quantity* the *symbol* used to represent that quantity (including vector notation, if appropriate) and the SI *unit* in which the quantity is specified. Furthermore, you should be able to express each important principle or equation in a concise and accurate prose statement. Perhaps the best test of your understanding of the material will be your ability to answer questions and solve problems in the text, or those given on exams.

3. Try to solve plenty of the problems at the end of the chapter. The worked examples in the text will serve as a basis for your study. This Study Guide contains detailed solutions to about 20% of the problems at the end of each chapter. You will be able to check the accuracy of your calculations for any odd-numbered problem, since the answers to these are given at the back of the text.

4.   Besides what you might expect to learn about physics concepts, a very valuable skill you can take away from your physics course is the ability to solve complicated problems. The way physicists approach complex situations and break them down into manageable pieces is widely useful. Starting in Section 1.10, the textbook develops a general problem-solving strategy that guides you through the steps. To help you remember the steps of the strategy, they are called *Conceptualize, Categorize, Analyze,* and *Finalize.*

General Problem-Solving Strategy

Conceptualize

- The first thing to do when approaching a problem is to *think about* and *understand* the situation. Read the problem several times until you are confident you understand what is being asked. Study carefully any diagrams, graphs, tables, or photographs that accompany the problem. Imagine a movie, running in your mind, of what happens in the problem.

- If a diagram is not provided, you should almost always make a quick drawing of the situation. Indicate any known values, perhaps in a table or directly on your sketch.

- Now focus on what algebraic or numerical information is given in the problem. In the problem statement, look for key phrases such as "starts from at rest" ($v_i = 0$), "stops" ($v_f = 0$), or "freely falls" ($a_y = -g = -9.80 \text{ m/s}^2$). Key words can help simplify the problem.

- Next focus on the expected result of solving the problem. Exactly what is the question asking? Will the final result be numerical or algebraic? If it is numerical, what units will it have? If it is algebraic, what symbols will appear in it?

- Incorporate information from your own experiences and common sense. What should a reasonable answer look like? What should its order of magnitude be? You wouldn't expect to calculate the speed of an automobile to be 5×10^6 m/s.

Categorize

- Once you have a really good idea of what the problem is about, you need to *simplify* the problem. Remove the details that are not important to the solution. For example, you can often model a moving object as a particle. Key words should tell you whether you can ignore air resistance or friction between a sliding object and a surface.

- Once the problem is simplified, it is important to *categorize* the problem. How does it fit into a framework of ideas that you construct to understand the world? Is it a simple *plug-in problem*, such that numbers can be simply substituted into a definition? If so, the problem is likely to be finished when this substitution is done. If not, you face what we can call an *analysis problem*—the situation must be analyzed more deeply to reach a solution.

- If it is an analysis problem, it needs to be categorized further. Have you seen this type of problem before? Does it fall into the growing list of types of problems that you have solved previously? Being able to classify a problem can make it much easier to lay out a plan to solve it. For example, if your simplification shows that the problem can be treated as a particle moving under constant acceleration and you have already solved such a problem (such as the examples in Section 2.6), the solution to the new problem follows a similar pattern.

Analyze

- Now, you need to analyze the problem and strive for a mathematical solution. Because you have already categorized the problem, it should not be too difficult to select relevant equations that apply to the type of situation in the problem. For example, if your categorization shows that the problem involves a particle moving under constant acceleration, Equations 2.10 to 2.14 are relevant.

- Use algebra (and calculus, if necessary) to solve symbolically for the unknown variable in terms of what is given. Substitute in the appropriate numbers, calculate the result, and round it to the proper number of significant figures. Substituting numbers into the algebraic solution should ideally be the very last step, in order to avoid numerical and rounding errors.

Finalize

- This final step is the most important part. Examine your numerical answer. Does it have the correct units? Does it meet your expectations from your conceptualization of the problem? What about the algebraic form of the result—before you substituted numerical values? Does it make sense? Try looking at the variables in it to see whether the answer would change in a physically meaningful way if they were drastically increased or decreased or even became zero. Looking at limiting cases to see whether they yield expected values is a very useful way to make sure that you are obtaining reasonable results.

- Think about how this problem compares with others you have solved. How was it similar? In what critical ways did it differ? Why was this problem assigned? You should have learned something by doing it. Can you figure out what? Can you use your solution to expand, strengthen, or otherwise improve your framework of ideas? If it is a new category of problem, be sure you understand it so that you can use it as a model for solving future problems in the same category.

When solving complex problems, you may need to identify a series of sub-problems and apply the problem-solving strategy to each. For very simple problems, you probably don't need this whole strategy. But when you are looking at a problem and you don't know what to do next, remember the steps in the strategy and use them as a guide.

Work on problems in this Study Guide yourself and compare your solutions with ours. Your solution does not have to look just like the one presented here. A problem can sometimes be solved in different ways, starting from different principles. If you wonder about the validity of an alternative approach, ask your instructor.

5.   We suggest that you use this Study Guide to review the material covered in the text, and as a guide in preparing for exams. You can use the sections Chapter Review, Notes From Selected Chapter Sections, and Equations and Concepts to focus in on any points which require further study. The main purpose of this Study Guide is to improve upon the efficiency and effectiveness of your study hours and your overall understanding of physical concepts. However, it should not be regarded as a substitute for your textbook or for individual study and practice in problem solving.

Chapter 16
Temperature and the Kinetic Theory of Gases

NOTES FROM SELECTED CHAPTER SECTIONS

Section 16.1   Temperature and the Zeroth Law of Thermodynamics

Thermal physics is the study of the behavior of solids, liquids and gases, using the concepts of heat and temperature. Two approaches are commonly used:

- macroscopic approach, called thermodynamics, in which one explains the bulk thermal properties of matter;

- microscopic approach, called statistical mechanics, in which properties of matter are explained on an atomic scale.

All thermal phenomena are manifestations of the laws of mechanics as we have learned them. For example, internal energy is actually a consequence of the random motions of a large number of particles making up the system.

The concept of the temperature of a system can be understood in connection with a measurement, such as the reading of a thermometer. Temperature, a scalar quantity, is a property that can only be defined when the system is in thermal equilibrium with another system. Thermal equilibrium implies that two (or more) systems are at the same temperature.

The **zeroth law of thermodynamics** states that if two systems are in thermal equilibrium with a third system, they must be in thermal equilibrium with each other. The third system can be a calibrated thermometer whose reading determines whether or not the systems are in thermal equilibrium.

Section 16.2   Thermometers and Temperature Scales

Thermometers are devices used to define and measure the temperature of a system. All thermometers make use of a change in some physical property with temperature.

Some of the physical properties used to establish a temperature scale are:

- change in volume of a liquid

- change in length of a solid

- change in pressure of a gas held at constant volume

- change in volume of a gas held at constant pressure

- change in electric resistance of a conductor

- change in color of a very hot object

The **gas thermometer** is a standard device for defining temperature. In the constant-volume gas thermometer, a low-density gas is placed in a flask, and its volume is kept constant while it is heated or cooled. The pressure is measured as the temperature of the gas is changed. Experimentally, one finds that the temperature is proportional to the absolute pressure.

The **thermodynamic temperature scale** is based on a scale for which the reference temperature is taken to be the **triple point of water**; that is, the temperature and pressure at which water, water vapor, and ice coexist in equilibrium. On this scale, the SI unit of temperature is the kelvin, defined as the fraction 1/273.16 of the temperature of the triple point of water.

Section 16.5   The Kinetic Theory of Gases

A **microscopic model of an ideal gas** is based on the following assumptions:

- **The number of molecules is large, and the average separation between them is large compared with their dimensions.**

Therefore, the molecules occupy a negligible volume compared with the volume of the container.

- **The molecules obey Newton's laws of motion, but the individual molecules move in a random fashion** – the molecules move in all directions with equal probability and with various speeds. *This distribution of velocities does not change in time, despite the collisions between molecules.*

- **The molecules undergo elastic collisions with each other.** The molecules do not have structure, and for any pair of colliding molecules considered as a system both kinetic energy and momentum are conserved.

- **The forces between molecules are negligible except during a collision.** The forces between molecules are short-range, so that the only time the molecules interact with each other is during a collision.

- **The gas is in thermal equilibrium with the walls of the container.** The collisions of molecules with the walls are perfectly elastic.

- **The gas under consideration is a pure gas.** That is, all molecules are identical.

EQUATIONS AND CONCEPTS

Pressure is force per unit area and has units of N/m^2. The SI unit of pressure is the pascal (Pa). One atmosphere of pressure is atmospheric pressure at sea level.

$$1\,Pa = 1\,N/m^2$$

$$1\,atm = 1.013 \times 10^5\,Pa$$

The **Celsius temperature** T_C is related to the Kelvin temperature (sometimes called the absolute temperature) T according to Equation 16.1. *The size of a degree on the Kelvin scale equals the size of a degree on the Celsius scale.*

$$T_C = T - 273.15 \tag{16.1}$$

$$0°C = 273.15\,K$$

Conversion between **Fahrenheit and Celsius scales** is based on the freezing and boiling point temperatures set for each scale.
Freezing point: $0°C = 32°F$
Boiling point: $100°C = 212°F$

$$T_F = \tfrac{9}{5}T_C + 32°F \qquad (16.2)$$

$$T_C = \tfrac{5}{9}(T_F - 32)°C$$

$$\Delta T_F = \tfrac{9}{5}\Delta T_C \qquad (16.3)$$

Thermal expansion results in changes in length, area and volume. In each case the *fractional change* is proportional to the change in temperature and a thermal coefficient characteristic of the particular type of material.

$\alpha = $ average coefficient of linear expansion.

$$\Delta L = \alpha L_i \Delta T \qquad (16.4)$$

$\beta = $ average coefficient of volume expansion.

$$\Delta V = \beta V_i \Delta T \qquad (16.6)$$
$$\beta \approx 3\alpha$$

$\gamma = $ average coefficient of area expansion.

$$\Delta A = \gamma A_i \Delta T \qquad (16.7)$$
$$\gamma \approx 2\alpha$$

The **number of moles** (n) in a sample of any substance can be stated in terms of:
(1) the sample mass (m) and molar mass (M), or
(2) the number of molecules in the sample (N) and Avogadro's number (N_A).

$$n = \frac{m}{M} \qquad (16.8)$$

$$n = \frac{N}{N_A}$$

The **equation of state of an ideal gas**, relates the variables P, V, and T at equilibrium and is shown in terms of:
(1) the number of moles and the universal gas constant (Equation 16.9), or
(2) the number of molecules and Boltzmann's constant (Equation 16.11).

In terms of number of moles (n)
$$PV = nRT \qquad (16.9)$$

In terms of number of molecules (N)
$$PV = N k_B T \qquad (16.11)$$

In Equations 16.9 and 16.11, T is in kelvins and R is in units consistent with those used for P and V.

The **pressure of an ideal gas** is proportional to the number of molecules per unit volume and the average translational kinetic energy per molecule.

$$P = \frac{2}{3}\left(\frac{N}{V}\right)\left(\frac{1}{2}m_0\overline{v^2}\right) \qquad (16.13)$$

m_0 = mass of a single molecule.

The **average molecular kinetic energy** of the molecules of an ideal gas and the **total translational kinetic energy** of N molecules of the gas are proportional to the absolute temperature.

$$\frac{1}{2}m_0\overline{v^2} = \frac{3}{2}k_BT \qquad (16.15)$$

$$E_{total} = \frac{3}{2}Nk_BT = \frac{3}{2}nRT \qquad (16.17)$$

The expression for the **root-mean-square speed** of molecules shows that at a given temperature, lighter molecules move faster on the average than heavier molecules.

$$v_{rms} = \sqrt{\overline{v^2}} = \sqrt{\frac{3k_BT}{m_0}} = \sqrt{\frac{3RT}{M}}$$

$$(16.19)$$

The **Maxwell-Boltzmann distribution** describes the distribution of speeds of gas molecules as a function of temperature.

$$N_v = 4\pi N\left(\frac{m_0}{2\pi k_BT}\right)^{3/2} v^2 e^{-m_0v^2/2k_BT}$$

$$(16.20)$$

On a speed distribution curve for gas molecules at a given temperature, the root-mean-square speed, the average speed, and the most probable speed values are ranked as:

$v_{rms} > v_{avg} > v_{mp}$.

$$v_{rms} = 1.73\sqrt{\frac{k_BT}{m_0}} \qquad (16.21)$$

$$v_{avg} = 1.60\sqrt{\frac{k_BT}{m_0}} \qquad (16.22)$$

$$v_{mp} = 1.41\sqrt{\frac{k_BT}{m_0}} \qquad (16.23)$$

REVIEW CHECKLIST

- Understand the concepts of thermal equilibrium and thermal contact between two bodies, and state the zeroth law of thermodynamics.

- Discuss some physical properties of substances, which change with temperature, and the manner in which these properties are used to construct thermometers.

- Describe the operation of the constant-volume gas thermometer and how it is used to define the ideal gas temperature scale.

- Convert between the various temperature scales, especially the conversion from degrees Celsius into kelvins, degrees Fahrenheit into kelvins, and degrees Celsius into degrees Fahrenheit.

- Provide a qualitative description of the origin of thermal expansion of solids and liquids; define the linear expansion coefficient and volume expansion coefficient for an isotropic solid, and learn how to deal with these coefficients in practical situations involving expansion or contraction.

- State and understand the assumptions made in developing the molecular model of an ideal gas.

- Recognize that the temperature of an ideal gas is proportional to the average molecular kinetic energy.

ANSWER TO AN OBJECTIVE QUESTION

12. A rubber balloon is filled with 1 L of air at 1 atm and 300 K and is then put into a cryogenic refrigerator at 100 K. The rubber remains flexible as it cools. **(i)** What happens to the volume of the balloon? (a) It decreases to $\frac{1}{3}$ L. (b) It decreases to $1/\sqrt{3}$ L. (c) It is constant. (d) It increases to $\sqrt{3}$ L. (e) It increases to 3 L. **(ii)** What happens to the pressure of the air in the balloon? (a) It decreases to $\frac{1}{3}$ atm. (b) It decreases to $1/\sqrt{3}$ atm. (c) It is constant. (d) It increases to $\sqrt{3}$ atm. (e) It increases to 3 atm.

Answer   **(i)** (a) and **(ii)** (c). The tiny quantities of carbon dioxide and water in the air solidify on the way down to 100 K, but the oxygen, nitrogen, and argon remain gases, and it is reasonable to assume that their densities are low enough to permit use of the ideal gas equation of state. To qualify as a balloon, the container cannot have a fast leak, so in $PV = nRT$ we take the quantity of gas n, as well as the universal constant R, as constant in the contraction process. A refrigerator is not airtight, so the pressure inside the box stays equal to the constant atmospheric pressure outside. The flexible wall of the balloon moves to equalize the pressure of the air inside with the pressure outside. The rubber is in mechanical equilibrium throughout the process, stationary at the start and finish and moving with negligible acceleration in between, while the air inside is coming to thermal equilibrium. An ordinary person "blowing up" a balloon does not stretch the rubber enough to make the interior pressure significantly different from atmospheric pressure; as the air cools, the rubber will go slack and the interior pressure will become exactly equal to atmospheric pressure. Then the process for the sample within is isobaric and the answer to question **(ii)** is (c). In the equation of state the only quantities that change are V and T. When T falls to one-third of its initial value, $V = nRT/P$ falls to 1/3 L. The answer to **(i)** is (a).

ANSWERS TO SELECTED CONCEPTUAL QUESTIONS

4. A piece of copper is dropped into a beaker of water. (a) If the water's temperature rises, what happens to the temperature of the copper? (b) Under what conditions are the water and copper in thermal equilibrium?

Answer (a) If the water's temperature increases, that means that energy is being transferred by heat to the water. This can *happen* either if the copper changes phase from liquid to solid, or if the temperature of the copper is above that of the water, and falling.

In this case, the copper is referred to as a "piece" of copper, so it is already in the solid phase; therefore, its temperature must be falling. (b) When the temperature of the copper reaches that of the water, the copper and water will reach equilibrium, and the subsequent net energy transfer by heat between the two will be zero.

□   □   □   □

6. What happens to a helium-filled latex balloon released into the air? Does it expand or contract? Does it stop rising at some height?

Answer As the balloon rises it encounters air at lower and lower pressure, because the bottom layers of the atmosphere support the weight of all the air above. The flexible wall of the balloon moves and then stops moving to make the pressures equal on both sides of the wall. Therefore, as the balloon rises, the pressure inside it will decrease. At the same time, the temperature decreases slightly, but not enough to overcome the effects of the pressure.

By the ideal gas law, $PV = nRT$. Since the decrease in temperature has little effect, the decreasing pressure results in an increasing volume; thus the balloon expands quite dramatically. By the time the balloon reaches an altitude of 10 miles, its volume will be 90 times larger.

Long before that happens, one of two events will occur. The first possibility, of course, is that the balloon will break, and the pieces will fall to the Earth. The other possibility is that the balloon will come to a spot where the average density of the balloon and its payload is equal to the average density of the air. At that point, buoyancy will cause the balloon to "float" at that altitude, until it loses helium and descends.

People who remember releasing balloons with pen-pal notes will perhaps also remember that replies most often came back when the balloons were low on helium, and were barely able to take off. Such balloons found a fairly low flotation altitude, and were the most likely to be found intact at the end of their journey.

□   □   □   □

12. When the metal ring and metal sphere in Figure CQ16.12 are both at room temperature, the sphere can barely be passed through the ring. (a) After the sphere is warmed in a flame, it cannot be passed through the ring. Explain. (b) **What if?** What if the ring is warmed and the sphere is left at room temperature? Does the sphere pass through the ring?

Figure CQ16.12

Answer (a) The hot sphere has expanded, so it no longer fits through the ring. (b) When the ring is warmed, the cool sphere fits through more easily.

Suppose a cool sphere is put through a cool ring and then warmed so that it does not come back out. With the sphere still hot, you can separate the sphere and ring by warming the ring as shown in the drawing here. This more surprising result occurs because the thermal expansion of the ring is not like the inflation of a blood-pressure cuff. Rather, it is like a photographic enlargement; every linear dimension, including the hole diameter, increases by the same factor. The reason for this is that neighboring atoms everywhere, including those around the inner circumference, push away from each other. The only way that the atoms can accommodate the greater distances is for the circumference—and corresponding diameter—to grow. This property was once used to fit metal rims to wooden wagon and horse-buggy wheels.

□   □   □   □

SOLUTIONS TO SELECTED END-OF-CHAPTER PROBLEMS

4. In a student experiment, a constant-volume gas thermometer is calibrated in dry ice (–78.5°C) and in boiling ethyl alcohol (78.0°C). The separate pressures are 0.900 atm and 1.635 atm. (a) What value of absolute zero in degrees Celsius does the calibration yield? What pressures would be found at the (b) freezing and (c) boiling points of water? Hint: Use the linear relationship $P = A + BT$, where A and B are constants.

Solution

Conceptualize: Visualize a straight-line graph. The temperature at which the extrapolated pressure drops to zero will be far below –100°C. The pressure will be around 1.2 atm at 0°C and 1.8 atm at 100°C.

Categorize: We need not actually draw the graph. We can find the equation of the graph line and solve all parts of the problem from it.

Analyze: To represent a linear graph, the pressure is related to the temperature as $P = A + BT$, where A and B are constants. To find A and B, we use the given data:

$$0.900 \text{ atm} = A + (-78.5°C)B \quad \text{and} \quad 1.635 \text{ atm} = A + (78.0°C)B$$

Solving these simultaneously, we find

$$0.735 \text{ atm} = (156.5°C)B$$

so $\quad B = 4.696 \times 10^{-3} \text{ atm/°C} \quad$ and $\quad A = 1.269$ atm.

Therefore, for all temperatures

$$P = 1.269 \text{ atm} + (4.696 \times 10^{-3} \text{ atm/°C})T$$

(a) At absolute zero, $P = 0 = 1.269 \text{ atm} + (4.696 \times 10^{-3} \text{ atm/°C})T$

which gives $\quad T = -270°C$. ∎

(b) At the freezing point of water, $P = 1.269 \text{ atm} + 0 = 1.27$ atm. ∎

(c) And at the boiling point,

$$P = 1.269 \text{ atm} + (4.696 \times 10^{-3} \text{ atm/°C})(100°C) = 1.74 \text{ atm}$$ ∎

Finalize: Our estimates were good. In a typical student experiment the extrapolated value of absolute zero would have high uncertainty.

5.   Liquid nitrogen has a boiling point of –195.81°C at atmospheric pressure. Express this temperature (a) in degrees Fahrenheit and (b) in kelvins.

Solution

Conceptualize: Estimate minus several hundred degrees Fahrenheit and positive eighty-something kelvins.

Categorize: We can puzzle through the conversion to Fahrenheit. Expressing the temperature on the absolute scale will be easy.

Analyze:

(a) By Equation 16.2, $T_F = \frac{9}{5}T_c + 32°F = \frac{9}{5}(-195.81) + 32 = -320°F.$ ∎

(b) Applying Equation 16.1, $273.15\text{ K} - 195.81\text{ K} = 77.3\text{ K}.$ ∎

Finalize: A convenient way to remember Equations 16.1 and 16.2 is to remember the freezing and boiling points of water, in each form:

$$T_{freeze} = 32.0°F = 0°C = 273.15\text{ K}$$

$$T_{boil} = 212°F = 100°C$$

To convert from Fahrenheit to Celsius, subtract 32 (the freezing point), and then adjust the scale by the liquid range of water.

$$\text{Scale factor} = \frac{(100-0)°C}{(212-32)°F} = \frac{5°C}{9°F}$$

A kelvin is the same size change as a degree Celsius, but the Kelvin scale takes its zero point at **absolute zero**, instead of the freezing point of water. Therefore, to convert from the Kelvin scale to the Celsius scale, subtract 273.15 K.

—————————————

15. The active element of a certain laser is made of a glass rod 30.0 cm long and 1.50 cm in diameter. Assume the average coefficient of linear expansion of the glass is equal to $9.00 \times 10^{-6}(°C)^{-1}$. If the temperature of the rod increases by 65.0°C, what is the increase in (a) its length, (b) its diameter, and (c) its volume?

Solution

Conceptualize: We expect a change on the order of a part in a thousand. That would be a fraction of a millimeter for the length, a much smaller distance for the diameter, and maybe some cubic millimeters for the volume.

Categorize: It's all based on the definition of the coefficient of expansion.

Analyze:

(a) $\Delta L = \alpha L_i \Delta T = (9.00 \times 10^{-6}\ °C^{-1})(0.300\ m)(65.0°C) = 1.76 \times 10^{-4}\ m$ ∎

(b)   The diameter is a linear dimension, so the same equation applies:

$$\Delta D = \alpha D_i \Delta T = (9.00 \times 10^{-6}\ °C^{-1})(0.015\ 0\ m)(65.0°C) = 8.78 \times 10^{-6}\ m$$ ∎

(c)   The original volume is

$$V = \pi r^2 L = \frac{\pi}{4}(0.015\ 0\ m)^2(0.300\ m) = 5.30 \times 10^{-5}\ m^3$$

Using the volumetric coefficient of expansion β,

$$\Delta V = \beta V_i \Delta T \approx 3\alpha V \Delta T$$

$$\Delta V \approx 3(9.00 \times 10^{-6}\ °C^{-1})(5.30 \times 10^5 m^3)(65.0°C) = 93.0 \times 10^{-9}\ m^3$$ ∎

Finalize: The above calculation of ΔV ignores ΔL^2 and ΔL^3 terms. Let us calculate the change in volume more precisely, and compare the answer with the approximate solution above.

The volume will increase by a factor of

$$\frac{\Delta V}{V_i} = \left(1 + \frac{\Delta D}{D_i}\right)^2\left(1 + \frac{\Delta L}{L_i}\right) - 1$$

$$\frac{\Delta V}{V_i} = \left(1 + \frac{8.78 \times 10^{-6}\ m}{0.015\ 0\ m}\right)^2\left(1 + \frac{1.76 \times 10^{-4}\ m}{0.300\ m}\right) - 1 = 1.76 \times 10^{-3}$$

The amount of volume change is $\Delta V = \dfrac{\Delta V}{V_i}V_i$:

$$\Delta V = (1.76 \times 10^{-3})(5.30 \times 10^{-5}\ m^3) = 93.1 \times 10^{-9}\ m^3 = 93.1\ mm^3$$ ∎

The answer is virtually identical; the approximation $\beta \approx 3\alpha$ is a good one.

═══════════════

25. An automobile tire is inflated with air originally at 10.0°C and normal atmospheric pressure. During the process, the air is compressed to 28.0% of its original volume and the temperature is increased to 40.0°C. (a) What is the tire pressure? (b) After the car is driven at high speed, the tire's air temperature rises to 85.0°C and the tire's interior volume increases by 2.00%. What is the new tire pressure (absolute)?

Solution

Conceptualize: In part (a), both temperature increase and volume decrease contribute to increasing the pressure. In part (b) the temperature increase will dominate the volume increase to produce still higher pressure of some hundreds of kilopascals.

Categorize: The ideal gas equation of state describes each condition of the air. We can set up ratios to describe the changes.

Analyze:

(a) Taking $PV = nRT$ in the initial (i) and final (f) states, and dividing, we have

$$P_iV_i = nRT_i \qquad \text{and} \qquad P_fV_f = nRT_f \qquad \text{yielding} \qquad \frac{P_fV_f}{P_iV_i} = \frac{T_f}{T_i}.$$

$$\text{So} \quad P_f = P_i\frac{V_iT_f}{V_fT_i} = \left(1.013 \times 10^5 \text{ Pa}\right)\left(\frac{V_i}{0.280V_i}\right)\left(\frac{273 \text{ K} + 40.0 \text{ K}}{273 \text{ K} + 10.0 \text{ K}}\right)$$

$$= 4.00 \times 10^5 \text{ Pa} \qquad\qquad \blacksquare$$

(b) Introducing the hot (h) state, we have $\dfrac{P_hV_h}{P_fV_f} = \dfrac{T_h}{T_f}$:

$$P_h = P_f\left(\frac{V_f}{V_h}\right)\left(\frac{T_h}{T_f}\right) = \left(4.00 \times 10^5 \text{ Pa}\right)\left(\frac{V_f(358 \text{ K})}{1.02V_f(313 \text{ K})}\right) = 4.49 \times 10^5 \text{ Pa} \qquad \blacksquare$$

Finalize: These pressures are a bit higher than usual for a car tire. An absolute pressure of 400 kPa means a gauge pressure of about 300 kPa or three atmospheres. Around two atmospheres is more typical.

29. Review. The mass of a hot-air balloon and its cargo (not including the air inside) is 200 kg. The air outside is at 10.0°C and 101 kPa. The volume of the balloon is 400 m^3. To what temperature must the air in the balloon be warmed before the balloon will lift off? (Air density at 10.0°C is 1.244 kg/m^3.)

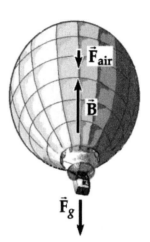

Solution

Conceptualize: The air inside the balloon must be significantly hotter than the outside air in order for the balloon to feel a net upward force, but the temperature must also be less than the melting point of the nylon used for the balloon's envelope. (Rip-stop nylon melts around 200°C.) Otherwise the results could be disastrous!

Categorize: We model the balloon with load and interior hot air as a particle in equilibrium. When it is just ready to lift off, the density of the air inside the balloon must be sufficiently low so that the buoyant force is equal to the weight of the balloon, its cargo, and the air inside. The temperature of the air required to achieve this density can be found from the equation of state of an ideal gas.

Analyze: The density of the air inside the balloon, ρ_{in}, must be reduced until the buoyant force of the outside air is at least equal to the weight of the balloon plus the weight of the air inside it:

$$\sum F_y = 0: B - W_{air\ inside} - W_{balloon} = 0$$

Plugging in for each term gives

$$\rho_{out}gV - \rho_{in}gV - m_bg = 0 \quad \rightarrow \quad (\rho_{out} - \rho_{in})V = m_b \qquad [1]$$

where $\rho_{out} = 1.244$ kg/m^3, $V = 400$ m^3, and $m_b = 200$ kg.

From $PV = nRT$, $\dfrac{n}{V} = \dfrac{P}{RT}$. This equation means that at constant pressure the density is inversely proportional to the temperature. Thus, the density of the hot air inside the balloon is

$$\rho_{in} = \rho_{out}\left(\frac{283\ \text{K}}{T_{in}}\right)$$

Substituting this result into equation [1] gives

$$\rho_{out}\left(1 - \frac{283 \text{ K}}{T_{in}}\right) = \frac{m_b}{V} \quad \rightarrow \quad \frac{283 \text{ K}}{T_{in}} = 1 - \frac{m_b}{\rho_{out} V}$$

Solving for T_{in}, we obtain

$$T_{in} = \frac{283 \text{ K}}{\left(1 - \dfrac{m_b}{\rho_{out} V}\right)}$$

$$= \frac{283 \text{ K}}{\left(1 - \dfrac{200 \text{ kg}}{(1.244 \text{ kg/m}^3)(400 \text{ m}^3)}\right)} = 473 \text{ K} = 198°\text{C} \qquad \blacksquare$$

Finalize: The average temperature of the air inside the balloon required for lift off appears to be close to the melting point of the nylon fabric, so this seems like a dangerous situation! A larger balloon would be better suited for the given load on the balloon. (Many sport balloons have volumes of about 3 000 m^3 or 10 times larger than the one in this problem.)

---

36. The pressure gauge on a tank registers the gauge pressure, which is the difference between the interior pressure and exterior pressure. When the tank is full of oxygen (O_2), it contains 12.0 kg of the gas at a gauge pressure of 40.0 atm. Determine the mass of oxygen that has been withdrawn from the tank when the pressure reading is 25.0 atm. Assume that the temperature of the tank remains constant.

Solution

Conceptualize: We estimate a bit less than half of the 12 kg, or about 5 kg.

Categorize: We will use the ideal gas law and relate the mass to the number of moles.

Analyze: The equation of state for gases far from liquefaction is $PV = nRT$.

At constant volume and temperature, $\dfrac{P}{n}$ = constant, or

$$\frac{P_1}{n_1} = \frac{P_2}{n_2} \quad \text{and} \quad n_2 = \frac{P_2}{P_1} n_1$$

Further, n is proportional to m, so

$$m_2 = \left(\frac{P_2}{P_1}\right) m_1 = \left(\frac{26.0 \text{ atm}}{41.0 \text{ atm}}\right)(12.0 \text{ kg}) = 7.61 \text{ kg}$$

The mass removed is $|\Delta m| = 12.0 \text{ kg} - 7.61 \text{ kg} = 4.39 \text{ kg}$. ∎

Finalize: Our estimate was pretty good. Most of the mass of a high-pressure cylinder of gas is the mass of the metal cylinder.

---

40. A cylinder contains a mixture of helium and argon gas in equilibrium at 150°C. (a) What is the average kinetic energy for each type of gas molecule? (b) What is the rms speed of each type of molecule?

Solution

Conceptualize: Molecules are so small that their energies are tiny fractions of a joule. The speeds will be many hundreds of meters per second.

Categorize: The kinetic-theory identification of the meaning of temperature, $\frac{1}{2} m_0 \overline{v^2} = \frac{3}{2} k_B T$, will give us all the answers.

Analyze:

(a) Both kinds of molecules have the same average kinetic energy. It is

$$\overline{K} = \frac{1}{2} m_0 \overline{v^2} = \frac{3}{2} k_B T = \frac{3}{2}\left(1.38 \times 10^{23} \text{ J/K}\right)(273 \text{ K} + 150 \text{ K})$$

$$\overline{K} = 8.76 \times 10^{-21} \text{ J}$$ ∎

(b) The root-mean square velocity can be calculated from the kinetic energy:

$$v_{\text{rms}} = \sqrt{\overline{v^2}} = \sqrt{\frac{2\overline{K}}{m_0}}$$

These two gases are noble, and therefore monatomic. The masses of the molecules are:

$$m_{\text{He}} = \frac{(4.00 \text{ g/mol})(10^{-3} \text{ kg/g})}{6.02 \times 10^{23} \text{ atoms/mol}} = 6.64 \times 10^{-27} \text{ kg}$$

$$m_{\text{Ar}} = \frac{(39.9 \text{ g/mol})(10^{-3} \text{ kg/g})}{6.02 \times 10^{23} \text{ atoms/mol}} = 6.63 \times 10^{-26} \text{ kg}$$

Substituting these values,

$$v_{\text{rms,He}} = \sqrt{\frac{2\left(8.76\times10^{-27}\text{ J}\right)}{6.64\times10^{-27}\text{ kg}}} = 1.62\times10^{3}\text{ m/s}$$ ■

and $$v_{\text{rms,Ar}} = \sqrt{\frac{2\left(8.76\times10^{-21}\text{ J}\right)}{6.63\times10^{-26}\text{ kg}}} = 514\text{ m/s.}$$ ■

Finalize: Observe that argon has molecules ten times more massive than helium and that their rms speed is smaller by the square root of ten. A dust particle or a floating paper airplane is bombarded with molecules and participates in the random motion with the same average kinetic energy but with vastly smaller rms speeds.

---

44. (a) How many atoms of helium gas fill a spherical balloon of diameter 30.0 cm at 20.0°C and 1.00 atm? (b) What is the average kinetic energy of the helium atoms? (c) What is the rms speed of the helium atoms?

Solution

Conceptualize: The balloon will have a volume of a few liters, so the number of atoms is of the same order of magnitude as Avogadro's number, $\sim 10^{24}$. The speed will be some kilometers per second, faster than any reasonable airplane.

Categorize: The equation of state will tell us the quantity of gas as a number of moles, so multiplying by Avogadro's number gives us the number of molecules. $(1/2)m_{\text{o}}(v^{2})_{\text{avg}} = (3/2)k_{B}T$ will tell us both the average kinetic energy and the rms speed.

Analyze:

(a) The volume is $V = \dfrac{4}{3}\pi r^{3} = \dfrac{4}{3}\pi(0.150\text{ m})^{3} = 1.41\times10^{-2}\text{ m}^{3}.$

The quantity of gas can be obtained from $PV = nRT$:

$$n = \frac{PV}{RT} = \frac{\left(1.013\times10^{5}\text{ N/m}^{2}\right)\left(1.41\times10^{-2}\text{ m}^{3}\right)}{\left(8.314\text{ N}\cdot\text{m/mol}\cdot\text{K}\right)(293\text{ K})} = 0.588\text{ mol}$$

The number of molecules is

$$N = nN_A = (0.588 \text{ mol})(6.02 \times 10^{23} \text{ molecules/mol})$$

$$N = 3.54 \times 10^{23} \text{ helium atoms} \qquad \blacksquare$$

(b) The kinetic energy is given by $\bar{K} = \frac{1}{2} m_0 \overline{v^2} = \frac{3}{2} k_B T$:

$$\bar{K} = \frac{3}{2} (1.38 \times 10^{23} \text{ J/K})(293 \text{ K}) = 6.07 \times 10^{-21} \text{ J} \qquad \blacksquare$$

(c) An atom of He has mass

$$m_0 = \frac{M}{N_A} = \frac{4.002\ 6 \text{ g/mol}}{6.02 \times 10^{23} \text{ molecules/mol}} = 6.65 \times 10^{-24} \text{ g} = 6.65 \times 10^{-27} \text{ kg}$$

So the root-mean-square speed is given by

$$v_{rms} = \sqrt{\overline{v^2}} = \sqrt{\frac{2\bar{K}}{m_0}} = \sqrt{\frac{2 \times 6.07 \times 10^{-27} \text{ J}}{6.65 \times 10^{-27} \text{ kg}}} = 1.35 \text{ km/s} \qquad \blacksquare$$

Finalize: The quantity of internal energy in the sample is much more than two grams of material usually possesses as mechanical energy. The fact that helium is chemically inert and can be said to have no chemical energy makes the contrast sharper. To tap into the internal energy to use it for human purposes would mean to cool off the helium. If the helium is warmer than its environment, a heat engine can systematically extract some but not all of its internal energy, as we will study in Chapter 18.

═══════════════

45. Fifteen identical particles have various speeds: one has a speed of 2.00 m/s, two have speeds of 3.00 m/s, three have speeds of 5.00 m/s, four have speeds of 7.00 m/s, three have speeds of 9.00 m/s, and two have speeds of 12.0 m/s. Find (a) the average speed, (b) the rms speed, and (c) the most probable speed of these particles.

Solution

Conceptualize: All of the measures of 'central tendency' will be around 5 m/s, but all will be different because of the scatter in speeds.

Categorize: We find each kind of average from its definition.

Analyze:

(a) The average is

$$\bar{v} = \frac{\sum n_i v_i}{\sum n_i} = \frac{1(2.00) + 2(3.00) + 3(5.00) + 4(7.00) + 3(9.00) + 2(12.0)}{1 + 2 + 3 + 4 + 3 + 2} \text{ m/s}$$

$$\bar{v} = 6.80 \text{ m/s} \qquad \blacksquare$$

(b) To find the average squared speed we work out

$$\overline{v^2} = \frac{\sum n_i v_i^2}{\sum n_i}$$

$$\overline{v^2} = \frac{1(2.00^2) + 2(3.00^2) + 3(5.00^2) + 4(7.00^2) + 3(9.00^2) + 2(12.0^2)}{15} \text{ m}^2/\text{s}^2$$

$$\overline{v^2} = 54.9 \text{ m}^2/\text{s}^2$$

Then the rms speed is $\quad v_{rms} = \sqrt{\overline{v^2}} = \sqrt{54.9 \text{ m}^2/\text{s}^2} = 7.41 \text{ m/s}.$ $\qquad \blacksquare$

(c) More particles have $v_{mp} = 7.00$ m/s than any other speed. $\qquad \blacksquare$

Finalize: The average in (a) is called the mean. The most probable value in (c) is called the mode. We could also find the median, as the speed of the seventh of the fifteen molecules ranked by speed. It is 7 m/s.

════════════════

46. From the Maxwell–Boltzmann speed distribution, show that the most probable speed of a gas molecule is given by Equation 16.23. *Note:* The most probable speed corresponds to the point at which the slope of the speed distribution curve dN_v/dv is zero.

Solution

Conceptualize: We are determining the coordinate of one point on the Maxwell–Boltzmann graph of the distribution of the number of molecules with different speeds.

Categorize: From the Maxwell speed distribution function,

$$N_v = 4\pi N \left(\frac{m_o}{2\pi k_B T} \right)^{3/2} v^2 e^{-m_o v^2 / 2 k_B T}$$

we locate the peak in the graph of N_v versus v by evaluating dN_v/dv and setting it equal to zero, to solve for the most probable speed.

Analyze: The derivative is

$$\frac{dN_v}{dv} = 4\pi N\left(\frac{m_o}{2\pi k_B T}\right)^{3/2}\left(\frac{-2m_o v}{2k_B T}\right)v^2 e^{-m_o v^2/2k_B T}$$

$$+ 4\pi N\left(\frac{m_o}{2\pi k_B T}\right)^{3/2} 2v e^{-m_o v^2/2k_B T} = 0$$

This equation has solutions $v = 0$ and $v \to \infty$, but those correspond to minimum-probability speeds. Then we divide by v and by the exponential function, obtaining

$$v_{mp}\left(-\frac{m_o\left(2v_{mp}\right)}{2k_B T}\right) + 2 = 0 \qquad \text{or} \qquad v_{mp} = \sqrt{\frac{2k_B T}{m_o}}$$

This is Equation 16.23. ∎

Finalize: It might help you to keep a table of symbols. k_B, T, N, and m_o are known quantities. N_v is the variable function of the variable v. As soon as we set its variable derivative dN_v/dv equal to zero, we start looking for the unknown v that is the most probable speed.

---

53. A mercury thermometer is constructed as shown in Figure P16.53. The Pyrex glass capillary tube has a diameter of 0.004 00 cm, and the bulb has a diameter of 0.250 cm. Find the change in height of the mercury column that occurs with a temperature change of 30.0°C.

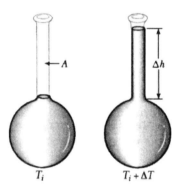

Figure P16.53

Solution

Conceptualize: For an easy-to-read thermometer, the column should rise by a few centimeters.

Categorize: We use the definition of the coefficient of expansion.

Analyze: The volume of the liquid increases as $\Delta V_l = V_i\beta\Delta T$, where V_i is the original interior volume of the bulb. Expansion of the glass makes the volume of the bulb increase as $\Delta V_g = 3\alpha V_i \Delta T$. Therefore, the overflow into the capillary is $V_c = V_i \Delta T (\beta - 3\alpha)$. For mercury $\beta = 1.82 \times 10^{-4} \, {}^\circ\text{C}^{-1}$ and for Pyrex glass $\alpha = 3.20 \times 10^{-6} \, {}^\circ\text{C}^{-1}$. In the capillary the mercury overflow fills a cylinder according to $V_c = A \Delta h$.

Then we have

$$\Delta h = \frac{V_i}{A}(\beta - 3\alpha)\Delta T = \frac{\frac{4}{3}\pi R_{bulb}^3}{\pi R_{cap}^2}(\beta - 3\alpha)\Delta T$$

$$= \frac{4(0.125 \text{ cm})^3}{3(0.002\ 00 \text{ cm})^2}[1.82 \times 10^{-4} - 3(3.20 \times 10^{-6})](^\circ\text{C})^{-1}(30.0^\circ\text{C})$$

$$= 3.37 \text{ cm} \qquad\qquad\blacksquare$$

Finalize: This is a practical thermometer. If we assumed that all of the glass undergoes the temperature change, not just the bulb, then we would include an extra factor of $[1 + 2(3.2 \times 10^{-6})(30)]$ in the bottom of the fraction that we used to compute h. It would account for expansion of the area of the capillary. But it does not affect the answer to three significant digits.

---

56. A vertical cylinder of cross-sectional area A is fitted with a tight-fitting, frictionless piston of mass m (Fig. P16.56). The piston is not restricted in its motion in any way and is supported by the gas at pressure P below it. Atmospheric pressure is P_0. We wish to find the height h in Figure P16.56. (a) What analysis model is appropriate to describe the piston? (b) Write an appropriate force equation for the piston from this analysis model in terms of P, P_0, m, A, and g. (c) Suppose n moles of an ideal gas are in the cylinder at a temperature of T. Substitute for P in your answer to (b) to find the height h of the piston above the bottom of the cylinder.

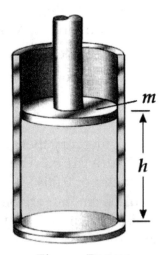

Figure P16.56

Solution

Conceptualize: The "spring of air," as Robert Boyle called it, does not follow the same equation as Hooke's law, but the weight of the piston compresses the gas just enough to raise its pressure enough to support the weight.

Categorize: We will use the definition of pressure and the ideal gas law.

Analyze:

(a) The air above the piston remains at atmospheric pressure, P_0. Model the piston as a particle in equilibrium. ∎

(b) $\sum F_y = ma_y$ yields $-P_0 A - mg + PA = 0$ ∎

where P is the pressure exerted by the gas contained. Noting that $V = Ah$, and that n, T, m, g, A, and P_0 are given,

(c) $PV = nRT$ becomes $P = \dfrac{nRT}{Ah}$,

so we substitute to eliminate P:

$$-P_o A - mg + \frac{nRT}{Ah}A = 0 \quad \text{and} \quad h = \frac{nRT}{P_0 A + mg}$$ ∎

Finalize: It is reasonable that n is on top in the fraction in the derived equation, because more gas will occupy more volume. Expansion makes it reasonable that T is on top. A heavier piston would compress the gas, so it is reasonable that the piston mass is on the bottom.

---

60. The rectangular plate shown in Figure P16.60 has an area A_i equal to ℓw. If the temperature increases by ΔT, each dimension increases according to Equation 16.4, where α is the average coefficient of linear expansion. (a) Show that the increase in area is $\Delta A = 2\alpha A_i \Delta T$. (b) What approximation does this expression assume?

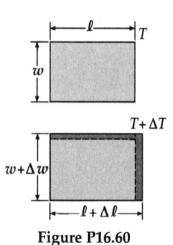

Figure P16.60

Solution

Conceptualize: We expect the area to increase in thermal expansion. It is neat that the coefficient of area expansion is just twice the coefficient of linear expansion.

Categorize: We will use the definitions of coefficients of linear and area expansion.

Analyze: From the diagram in Figure P16.60, we see that the **change** in area is

$$\Delta A = \ell \Delta w + w \Delta \ell + \Delta w \Delta \ell$$

Since $\Delta \ell$ and Δw are each small quantities, the product $\Delta w \Delta \ell$ will be very small compared to the original or final area.

Therefore, we assume $\Delta w \Delta \ell \approx 0.$ ■

Since $\Delta w = w \alpha \Delta T$ and $\Delta \ell = \ell \alpha \Delta T,$

we then have $\Delta A = \ell w \alpha \Delta T + w \ell \alpha \Delta T.$

(a) Finally, since $A = w \ell$ we have $\Delta A = 2 \alpha A \Delta T.$ ■

(b) A clearer way to state the approximation we have made is that $\alpha \Delta T$ is much less than 1.   ■

Then the change in area is small compared to the original area and $\Delta w \Delta \ell$ is negligibly small.

Finalize: Increase in length and increase in width both contribute to the increase in area, through the narrow ribbons of extra area in the diagram. Their contributions accumulate to give the 2 in $\Delta A = 2 \alpha A \Delta T$. The bit we ignored is the little rectangle off in the corner of the diagram. It is truly small if the temperature change is small.

---

62. A liquid has a density ρ. (a) Show that the fractional change in density for a change in temperature ΔT is $\Delta \rho / \rho = -\beta \Delta T$. (b) What does the negative sign signify? (c) Fresh water has a maximum density of 1.000 0 g/cm^3 at 4.0°C. At 10.0°C, its density is 0.999 7 g/cm^3. What is β for water over this temperature interval? (d) At 0°C, the density of water is 0.999 9 g/cm^3. What is the value for β over the temperature range 0°C to 4.00°C?

Solution

Conceptualize: Mass does not change with temperature, but density does because of thermal expansion in volume.

Categorize: We will use the definitions of density and of expansion coefficient.

Analyze: We start with the two equations

$$\rho = \frac{m}{V} \quad \text{and} \quad \frac{\Delta V}{V} = \beta \Delta T$$

(a) Differentiating the first equation gives $d\rho = -\dfrac{m}{V^2} dV.$

For very small changes in V and ρ, this can be written

$$\Delta\rho = -\frac{m}{V}\frac{\Delta V}{V} = -\left(\frac{m}{V}\right)\frac{\Delta V}{V}$$

Substituting both of our initial equations, we find that

$$\Delta\rho = -\rho\beta\Delta T \qquad \text{so} \qquad \frac{\Delta\rho}{\rho} = -\beta\Delta T \qquad\blacksquare$$

(b) The negative sign means that if β is positive, any increase in temperature causes the density to decrease and vice versa. $\blacksquare$

(c) We apply the equation $\beta = -\dfrac{\Delta\rho}{\rho\Delta T}$ for the specific case of water:

$$\beta = -\frac{\left(1.000\,0\ \text{g/cm}^3 - 0.999\,7\ \text{g/cm}^3\right)}{\left(1.000\,0\ \text{g/cm}^3\right)(4.00°\text{C} - 10.0°\text{C})}$$

Calculating, we find that $\beta = 5.0 \times 10^{-5}\,°\text{C}^{-1}.$ $\blacksquare$

(d) Between 0 and 4°C, water contracts with increasing temperature, so we find a negative coefficient of volume expansion by the same method:

$$\beta = -\frac{\Delta\rho}{\rho\Delta T} = -\frac{1.000\,0\ \text{g/cm}^3 - 0.999\,9\ \text{g/cm}^3}{\left(0.999\,9\ \text{g/cm}^3\right)\left(4.0°\text{C} - 0.0°\text{C}\right)} = -2.5 \times 10^{-5}\,°\text{C}^{-1} \quad\blacksquare$$

Finalize: The volume expansion coefficients of water in this range are less than many other liquids.

---

67. For a Maxwellian gas, use a computer or programmable calculator to find the numerical value of the ratio $N_v(v)/N_v(v_{mp})$ for the following values of v: (a) $v = (v_{mp}/50.0)$, (b) $(v_{mp}/10.0)$, (c) $(v_{mp}/2.00)$, (d) v_{mp}, (e) $2.00v_{mp}$, (f) $10.0v_{mp}$, and (g) $50.0v_{mp}$. Give your results to three significant figures.

Solution

Conceptualize: We are finding coordinates of points on the graph of the Maxwell–Boltzmann speed distribution function. The values will be small for speed values well below and well above the most probable speed, with, by definition, the maximum value at v_{mp}.

Categorize: In each case we substitute and evaluate. Getting the numerical value in each case emphasizes that the graph has a definite shape.

Analyze: We first substitute the most probable speed

$$v_{\text{mp}} = \left(\frac{2k_B T}{m_o} \right)^{1/2}$$

into the distribution function

$$N_v(v) = 4\pi N \left(\frac{m_o}{2\pi k_B T} \right)^{3/2} v^2 e^{-m_o v^2 / 2k_B T}$$

and evaluate $N_v(v_{\text{mp}}) = 4\pi N \left(\dfrac{m_o}{2\pi k_B T} \right)^{3/2} \dfrac{2k_B T}{m_o} e^{-1}.$

Simplifying, $N_v\left(v_{\text{mp}}\right) = 4\pi^{-1/2} e^{-1} N \left(\dfrac{m_o}{2k_B T} \right)^{1/2} = 4\pi^{-1/2} e^{-1} N v_{\text{mp}}^{-1}.$

Then the fraction we are assigned to think about has the simpler expression

$$\frac{N_v(v)}{N_v\left(v_{\text{mp}}\right)} = \frac{4\pi^{-1/2} N v_{\text{mp}}^{-3} v^2 e^{-v^2/v_{\text{mp}}^2}}{4\pi^{-1/2} N v_{\text{mp}}^{-1} e^{-1}} = \left(\frac{v^2}{v_{\text{mp}}^2} \right) e^{1-v^2/v_{\text{mp}}^2}$$

(a) For $v = \dfrac{v_{\text{mp}}}{50}$: $\dfrac{N_v(v)}{N_v\left(v_{\text{mp}}\right)} = \left(\dfrac{1}{50} \right)^2 e^{1-(1/50)^2} = 1.09 \times 10^{-3}$ ■

(b) For $v = \dfrac{v_{\text{mp}}}{10}$: $\dfrac{N_v(v)}{N_v\left(v_{\text{mp}}\right)} = \left(\dfrac{1}{10} \right)^2 e^{1-(1/10)^2} = 2.69 \times 10^{-2}$ ■

(c) For $v = \dfrac{v_{\text{mp}}}{2}$: $\dfrac{N_v(v)}{N_v\left(v_{\text{mp}}\right)} = \left(\dfrac{1}{2} \right)^2 e^{1-(1/2)^2} = 0.529$ ■

(d) For $v = v_{\text{mp}}$: $\dfrac{N_v(v)}{N_v\left(v_{\text{mp}}\right)} = (1)^2 e^{1-1^2} = 1.00$ ■

Similarly we fill in the rest of the table:

v/v_{mp}	$N_v(v)/N_v(v_{mp})$
1/50	1.09×10^{-3}
1/10	2.69×10^{-2}
1/2	0.529
1	1.00
(e) 2	0.199
(f) 10	1.01×10^{-41}
(g) 50	$1.25 \times 10^{-1\,082}$

To find the last value in the table, we calculate

$$\left(50^2\right)e^{1-2\,500} = \left(50^2\right)e^{-2\,499} = 10^{\log 2\,500}\,e^{(\ln\,10)(-2\,499/\ln\,10)} = 10^{(\log 2\,500-2\,499/\ln\,10)}$$
$$= 10^{-1\,081.904}$$

Thus, $(50^2)e^{1-2\,500} = 10^{-1\,081.904} = 10^{0.096\,0}\,10^{-1\,082} = 1.25 \times 10^{-1\,082}$. ■

Finalize: The distribution function gets very small as it goes to zero for speeds approaching zero. And the distribution function gets most remarkably small for speeds several times larger than the most probable speed. Think about the distribution function for wealth among the human population. If it were like that for speeds among gas molecules that exchange energy in random collisions, no one in the world would have more than a few times the average amount of money.

Chapter 17
Energy in Thermal Processes: The First Law of Thermodynamics

NOTES FROM SELECTED CHAPTER SECTIONS

Section 17.1   Heat and Internal Energy

When two systems at different temperatures are in contact with each other, energy will transfer between them until they reach the same temperature (that is, until they are in thermal equilibrium with each other). *The term heat refers to energy transfer as a consequence of a temperature difference.*

One unit of heat is the **calorie** (cal), defined as the amount of heat necessary to increase the temperature of 1 g of water from 14.5°C to 15.5°C. The mechanical equivalent of heat, first measured by Joule, is given by 1 cal = 4.186 J.

Section 17.2   Specific Heat

The **specific heat**, c, of any substance is defined as the amount of energy required to increase the temperature of 1 kg of that substance by one Celsius degree. The units of specific heat are $J/kg \cdot °C$.

Section 17.3   Latent Heat

The **heat of fusion** is the amount of energy required to melt (or freeze) 1 kg of a specific substance, with no temperature change occurring in the substance. The **heat of vaporization** characterizes the liquid-to-gas change in a similar manner. Both of these parameters have units of J/kg.

Section 17.4   Work in Thermodynamic Processes

State variables (e.g., pressure, volume, internal energy and temperature) are used to describe the *state* of a thermodynamic system that is in equilibrium.

Transfer variables (e.g., heat and work) are associated with a *change in the state* of a thermodynamic system by transfer of energy across the system boundary.

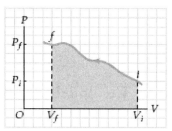

Work on gas = –(Area under curve)

A *PV* diagram showing compression of a gas.

A **PV diagram** is a graphical representation that shows the path followed by a gas as it progresses from an initial to a final state. The work done on a gas in a process that takes the gas between initial and final states is the **negative of the area** under the curve on a *PV* diagram. The graph shown above illustrates the case for compression of a gas.

- If the gas is compressed, $V_f < V_i$, the area $\left(\int_i^f P \, dV \right)$ will be negative and the work on the gas will be positive. *Remember that the work equals the negative of the area under the curve.*

- If the gas expands, $V_f > V_i$, the area will be positive, and the work on the gas will be negative.

- If the gas expands at *constant pressure*, called an isobaric process, then the work done on the gas will be $W = -P(V_f - V_i)$.

The work done on a system depends on the process by which the system goes from the initial to the final state; the work done depends on the initial, final, and intermediate states of the system.

Section 17.5   The First Law of Thermodynamics

In the first law of thermodynamics, $\Delta E_{int} = Q + W$:

ΔE_{int} = change in internal energy of the system
Q = energy *added* to the system by heat
W = work done *on* the system

The initial and final states must be equilibrium states; however, the intermediate states often are not equilibrium states since the thermodynamic coordinates may be impossible to determine during the thermodynamic process. For an infinitesimal change of the system, the first law of thermodynamics becomes $dE_{int} = dQ + dW$. Note that dQ and dW are not exact differentials, since both Q and W are not functions of the system coordinates. *Both Q and W depend on the path taken between the initial and final equilibrium states, during which time the system interacts with its environment.* On the other hand, dE_{int} is an exact differential and the internal energy E_{int} is a state function.

Section 17.6   Some Applications of the First Law of Thermodynamics

An **isolated system** is one that does not interact with its surroundings. In such a system, $Q = W = 0$, so it follows from the first law that $\Delta E_{int} = 0$. *The internal energy of an isolated system cannot change.*

A **cyclic process** is one that originates and ends at the same state. In this situation, $\Delta E_{int} = 0$, so from the first law we see that $Q = -W$. *That is, the work done per cycle equals the energy added by heat to the system per cycle.* This is important to remember when dealing with heat engines in the next chapter.

An **adiabatic process** is a process in which no energy enters or leaves the system by heat; that is, $Q = 0$. The first law applied to this process gives $\Delta E_{int} = W$. A system may undergo an adiabatic process if it is thermally insulated from its surroundings, or if the process is so rapid that negligible energy has time to flow by heat.

An **isobaric process** is a process that occurs at constant pressure.

An **isovolumetric process** is one that occurs at constant volume. By definition, $W = 0$ for such a process (since $dV = 0$), so from the first law it follows that $\Delta E_{int} = Q$. *That is, all of the energy added by heat to a system* *kept at constant volume goes into increasing the internal energy of the system.*

An **isothermal process** is one that occurs at constant temperature. A plot of P versus V at constant temperature for an ideal gas yields a hyperbolic curve called an **isotherm.** *The internal energy of an ideal gas is a function of temperature only.* Hence, in an isothermal process of an ideal gas, $\Delta E_{\text{int}} = 0$.

Section 17.7   Molar Specific Heats of Ideal Gases
Section 17.8   Adiabatic Processes for an Ideal Gas

A quantity of gas can be taken from an initial state at temperature T_i to a final state at temperature T_f by different processes. The energy transferred by heat to accomplish the change of state depends on the path taken. For this reason it is necessary to define two related quantities: molar specific at constant volume (C_V) and molar specific at constant pressure (C_P).

In a constant-volume process no work is done; all the energy added to the gas goes to increasing the internal energy.

In a constant-pressure process part of the energy transferred in by heat is transferred out by work.

During an adiabatic process no energy is transferred by heat into or out of the system. The state variables (pressure, volume, and temperature) change in value; however *some combination of those values remain constant.* The temperature of a gas decreases during an adiabatic expansion and increases during an adiabatic compression.

Section 17.10   Energy Transfer Mechanisms in Thermal Processes

There are three basic processes of thermal energy transfer: conduction, convection, and radiation.

Conduction is an energy transfer process that occurs in a substance when there is a temperature gradient across the substance. That is, conduction of energy occurs only when the temperature of the substance is not uniform. For example, if you position a metal rod with one end in a flame, energy will flow from the hot end to the colder end. The rate of flow of heat along the rod is proportional to the cross-sectional area of the rod, the temperature gradient, and k, the thermal conductivity of the material of which the rod is made.

Convection is a process of energy transfer by the motion of material, such as the mixing of hot and cold fluids. Convection can be due to a difference in density (natural convection) or due to movement of a substance by a fan or pump (forced convection). Convection heating is used in conventional hot-air and hot-water heating systems.

Radiation is energy transfer by the emission of electromagnetic radiation.

EQUATIONS AND CONCEPTS

The **mechanical equivalent of heat** was first measured by Joule. This is the definition of the calorie as an energy unit.

$$1 \text{ cal} \equiv 4.186 \text{ J} \tag{17.1}$$

Specific heat is a characteristic of a substance and is a measure of the quantity of energy required to change the temperature of 1 kg of the substance by 1°C. *Q represents the quantity of energy transferred by heat between a system and its environment.*

$$c \equiv \frac{Q}{m\Delta T} \tag{17.2}$$

Calorimetry is a technique in which an isolated system of two or more substances is specified. Energy is transferred by heat from the warmer substance(s) to the cooler substance(s) until thermal equilibrium of the system is achieved.

$$Q_{\text{cold}} = -Q_{\text{hot}} \tag{17.4}$$

Q_{cold} = energy gained by cooler substance(s)

Q_{hot} = energy transferred from warmer substance(s)

The **energy Q required to change the phase** of a mass m depends on a thermal property of the material called the **latent heat**, L. *The phase change process occurs at constant temperature. Choose the positive sign when energy is transferred into the system during melting or evaporation. Choose the negative sign when energy leaves the system during condensation or freezing.*

$$Q = L\Delta m \tag{17.6}$$

For specific phase change use:

L_f = heat of fusion

L_v = heat of vaporization

The **work done on a gas** that undergoes an expansion or compression from volume V_i to volume V_f depends on the path between the initial and final states. The pressure is generally not constant, so you must exercise care in evaluating W using this equation.

$$W = -\int_{V_i}^{V_f} P\, dV \qquad (17.7)$$

The work done is the negative of the area under the PV curve bounded by V_i and V_f.

The **first law of thermodynamics** is a special case of the more general law of conservation of energy. Equation 17.8 can be used in situations when the only energy considerations are heat, work, and internal energy.
The quantity $(Q + W)$ is independent of the path taken between the initial and final states.

$$\Delta E_{int} = Q + W \qquad (17.8)$$

Q is positive when energy is transferred to a system.

W is positive when work is done on a System

The thermodynamic processes described below will be important in applications of the laws of thermodynamics.

In an **isolated system** $Q = W = 0$, so that $\Delta E_{int} = 0$. *The internal energy of an isolated system remains constant.*

Isolated system
A system that does not interact with its surroundings.

In a **cyclic process**, $\Delta E_{int} = 0$ and $Q = -W$. *The net work done per cycle by the gas equals the area enclosed by the path representing the process on a PV diagram.*

Cyclic process
A process for which the initial and final states are the same.

At **constant volume**, $\Delta V = 0$ and $W = 0$.

The net energy added by heat at constant volume goes into increasing the internal energy.

Isovolumetric process
A process which occurs at constant volume.

$$\Delta E_{int} = Q \qquad (17.11)$$

$$Q = nC_V \Delta T \qquad (17.13)$$

$$C_V = \frac{3}{2}R \text{ (monatomic gas) } (17.17)$$

In an **isothermal process**, $\Delta E_{int} = 0$. *The internal energy is constant. Any energy transferred to the gas by work is transferred by heat from the gas to its surroundings.*

Isothermal process

A process that occurs at constant temperature.

$$W = -nRT \ln\left(\frac{V_f}{V_i}\right) \qquad (17.12)$$

In an **isobaric process**, the work done and the energy transferred by heat are both nonzero.

Isobaric process

A process that occurs at constant pressure.

$$W = P(V_f - V_i)$$
$$Q = nC_p\Delta T \qquad (17.14)$$

In an **adiabatic process**, $Q = 0$ and, therefore, the change in internal energy equals the work done on the system. *A system can undergo an adiabatic process if it is thermally insulated from its surroundings. An expansion or compression that occurs rapidly is approximately adiabatic.*

Adiabatic process

A process in which no energy is transferred by heat.

$$\Delta E_{int} = W \qquad (17.10)$$

$$P_i V_i^{\gamma} = P_f V_f^{\gamma} \qquad (17.25)$$

$$T_i V_i^{\gamma-1} = T_f V_f^{\gamma-1}$$

$$W = \frac{1}{\gamma - 1}\left(P_f V_f - P_i V_i\right) \qquad (17.27)$$

The **internal energy** E_{int} of N molecules (or n moles) of a monatomic ideal gas is proportional to the absolute temperature.

$$E_{int} = \tfrac{3}{2}nRT \qquad (17.15)$$

$$E_{int} = \tfrac{3}{2}Nk_B T$$

The **molar specific heat at constant pressure** is greater than the **molar specific heat at constant volume** by an amount R.

$$C_P - C_V = R \text{ (any ideal gas)} (17.21)$$

The **ratio of the specific heats** C_P and C_V, (γ), is a dimensionless quantity that is equal to 1.67 for a monatomic ideal gas.

$$\gamma = \frac{C_P}{C_V} \quad \text{(always)} \qquad (17.22)$$

For monatomic gas: $\gamma = 1.67$

This is the basic **law of thermal conduction**. The constant k is called the *thermal conductivity* and is characteristic of a particular material. Conduction of energy occurs only when there is a temperature gradient between two points in the conducting material.

$$P = kA\left(\frac{T_h - T_c}{L}\right) \qquad (17.35)$$

Stefan's law expresses the rate of emission of heat energy by **radiation** (the power radiated). *The radiated power is proportional to the fourth power of the absolute temperature.*

$$P = \sigma A e T^4 \qquad (17.36)$$

$$\sigma = 5.696 \times 10^{-8} \text{ W/m}^2 \cdot \text{K}^4$$

A = surface area in m^2
e = emissivity (values from 0 to 1)
T = temperature in kelvins.

Net radiated $(T > T_0)$ **or absorbed** $(T < T_0)$ **power** occurs when an object at temperature T is in surroundings at temperature T_0. *At thermal equilibrium* $(T = T_0)$, *an object radiates and absorbs energy at the same rate and the temperature of the object remains constant.*

$$P_{\text{net}} = \sigma A e\left(T^4 - T_0^{\,4}\right) \qquad (17.37)$$

SUGGESTIONS, SKILLS, AND STRATEGIES

Problem-Solving Strategy: Calorimetry Problems

If you are having difficulty with calorimetry problems, consider the following factors:

- Be sure your units are consistent throughout. For instance, if you are using specific heats in cal/g·°C, be sure that masses are in grams and temperatures are Celsius throughout.

- Energy losses and gains are found by using $Q = mc\Delta T$ only for those intervals in which no phase changes are occurring. The equations $Q = \pm mL_f$ and $Q = \pm mL_v$ are to be used only when phase changes **are** taking place.

- Often sign errors occur in calorimetry equations. Remember the negative sign in the equation $Q_{cold} = -Q_{hot}$.

Calculating Specific Heats

It is important to remember that for gases, two different values of molar heat capacity can be calculated: C_P, molar heat capacity at constant pressure and C_V, molar heat capacity at constant volume. For any ideal gas $C_P - C_V = R$, the universal gas constant and the ratio of $\dfrac{C_P}{C_V} = \gamma$ has a value that depends on the nature of the gas (see Table 17.3 of the text). The two equations above can be combined to find the following expressions for C_V and C_P:

$$C_V = \frac{R}{\gamma - 1} \quad \text{and} \quad C_P = \frac{\gamma R}{\gamma - 1}$$

These may be used for ideal gases whether or not they are monatomic.

REVIEW CHECKLIST

- Understand the concepts of heat, internal energy, and thermodynamic processes.

- Define and discuss the calorie, specific heat, and latent heat.

- Understand how work is defined when a system undergoes a change in state, and the fact that the work required to take a system from an initial to a final state depends on the path taken by the system.

- State the first law of thermodynamics ($\Delta E_{int} = Q + W$), and explain the meaning of the three energy terms related through this statement.

- Discuss the implications of the first law of thermodynamics as applied to (a) an isolated system, (b) a cyclic process, (c) an adiabatic process, and (d) an isothermal process.

- Recognize that the internal energy of an ideal gas is proportional to the absolute temperature, and be able to derive the specific heat of an ideal gas at constant volume from the first law of thermodynamics.

- Define an adiabatic process, and be able to derive the expression $PV^{\gamma} = \text{constant}$, which applies to an adiabatic process for an ideal gas.

- Use Equation 17.4 ($Q_{cold} = -Q_{hot}$) to make calculations involving calorimetry.

- Recognize thermodynamic processes as displayed on a *PV* diagram.

ANSWER TO AN OBJECTIVE QUESTION

12. If a gas is compressed isothermally, which of the following statements is true? (a) Energy is transferred into the gas by heat. (b) No work is done on the gas. (c) The temperature of the gas increases. (d) The internal energy of the gas remains constant. (e) None of those statements is true.

Answer (d) is true for a gas that is far from liquefaction. By definition, the temperature is constant in an isothermal process. For an ideal gas the internal energy is a function of temperature only and will also stay constant in this process. Choice (c) is untrue, because "isothermal" means constant-temperature. Choice (b) is untrue, because in any compression process the environment does positive work on the gas. To keep constant internal energy, the gas must then transfer energy out by heat, so choice (a) is untrue.

ANSWERS TO SELECTED CONCEPTUAL QUESTIONS

1. What is wrong with the following statement: "Given any two bodies, the one with the higher temperature contains more heat."

Answer The statement shows a misunderstanding of the concept of heat. Heat is a process by which energy is transferred, not a form of energy that is held or contained. If you wish to speak of energy that is "contained," you speak of **internal energy**, not **heat**.

Further, even if the statement used the term "internal energy," it would still be incorrect, since the effects of specific heat and mass are both ignored. A 1-kg mass of water at 20°C has more internal energy than a 1-kg mass of air at 30°C. Similarly, the Earth has far more internal energy than a drop of molten titanium metal.

Correct statements would be: (1) "Given any two bodies in thermal contact, the one with the higher temperature will transfer energy to the other by heat." (2) "Given any two bodies of equal mass, the one with the higher product of absolute temperature and specific heat contains more internal energy."

□   □   □   □

5. Using the first law of thermodynamics, explain why the *total* energy of an isolated system is always constant.

Answer The first law of thermodynamics says that the net change in internal energy of a system is equal to the energy added by heat, plus the work done on the system.

$$\Delta E_{int} = Q + W$$

It applies to a system for which the internal energy is the only energy that is changing, and work and heat are the only energy-transfer processes that are occurring. On the other hand, an isolated system is defined as an object or set of objects for which there is no exchange of energy with its surroundings. In the case of the first law of thermodynamics, this means that $Q = W = 0$, so the change in internal energy of the system at all times must be zero.

As it stays constant in amount, an isolated system's energy may change from one form to another or move from one object to another within the system. For example, a "bomb calorimeter" is a closed system that consists of a sturdy steel container, a water bath, an item of food, and oxygen, all inside a jacket of thermal insulation. The food is burned in the presence of an excess of oxygen, and chemical energy is converted to energy of more rapid random molecular motion. In the process of oxidation, energy is transferred by heat to the water bath, raising its temperature. The change in temperature of the water bath is used to determine the caloric content of the food. The process works specifically **because** the total energy remains unchanged in a closed system.

☐ ☐ ☐ ☐

SOLUTIONS TO SELECTED END-OF-CHAPTER PROBLEMS

3. A 55.0-kg woman cheats on her diet and eats a 540-Calorie (540 kcal) jelly doughnut for breakfast. (a) How many joules of energy are the equivalent of one jelly doughnut? (b) How many steps must the woman climb on a very tall stairway to change the gravitational potential energy of the woman–Earth system by a value equivalent to the food energy in one jelly doughnut? Assume the height of a single stair is 15.0 cm. (c) If the human body is only 25.0% efficient in converting chemical potential energy to mechanical energy, how many steps must the woman climb to work off her breakfast?

Solution

Conceptualize: 540 Calories is about one fourth of the average daily Calorie intake for adults. At about 4 000 joules per calorie, we expect an answer of about

2 million joules for part (a), which would require climbing a very large number of stairs, probably in the thousands or tens of thousands, and even more if the human body operates at 25% efficiency.

Categorize: We use Equation 17.1 for the mechanical equivalent of heat, and Equation 6.19 for gravitational potential energy.

Analyze:

(a) The energy equivalent of 540 Calories is found from

$$Q = 540 \text{ Cal} \left(\frac{10^3 \text{ cal}}{1 \text{ Cal}} \right) \left(\frac{4.186 \text{ J}}{1 \text{ cal}} \right) = 2.26 \times 10^6 \text{ J}$$ ∎

(b) The work done lifting her weight mg up one stair of height h is $W_1 = mgh$. Thus, the total work done in climbing N stairs is $W = Nmgh$, and we have $Q = Nmgh$, or

$$N = \frac{Q}{mgh} = \frac{2.26 \times 10^6 \text{ J}}{(55.0 \text{ kg})(9.80 \text{ m/s}^2)(0.150 \text{ m})} \simeq 2.80 \times 10^4 \text{ stairs}$$ ∎

(c) If only 25% of the energy from the donut goes into mechanical energy, the woman needs to climb one-fourth the number of stairs found in part (b), or

$$N = \frac{0.25Q}{mgh} = 0.25 \left(\frac{Q}{mgh} \right) = 0.25 (2.80 \times 10^4 \text{ stairs}) = 6.99 \times 10^3 \text{ stairs}$$ ∎

Finalize: It takes a lot of effort to burn off the calories from a single donut. At typical StairMaster speeds of 26 to 174 steps per minute, this effort would require (at 25% efficiency) 40 to 270 minutes of exercise!

---

4. The temperature of a silver bar rises by 10.0°C when it absorbs 1.23 kJ of energy by heat. The mass of the bar is 525 g. Determine the specific heat of silver from these data.

Solution

Conceptualize: Silver is far down in the periodic table, so we expect its specific heat to be on the order of one-tenth of that of water, or some hundreds of J/kg·°C.

Categorize: We find its specific heat from the definition, which is...

Analyze: ...contained in the equation $Q = mc_{silver}\Delta T$ for energy input by heat to produce a temperature change. Solving, we have

$$c_{silver} = \frac{Q}{m\Delta T}$$

$$c_{silver} = \frac{1.23 \times 10^3 \text{J}}{(0.525 \text{ kg})(10.0°\text{C})} = 234 \text{ J/kg} \cdot °\text{C}$$ ∎

Finalize: We were right about the order of magnitude.

—————

11. A 1.50-kg iron horseshoe initially at 600°C is dropped into a bucket containing 20.0 kg of water at 25.0°C. What is the final temperature of the water-horseshoe system? Ignore the heat capacity of the container and assume a negligible amount of water boils away.

Solution

Conceptualize: Even though the horseshoe is much hotter than the water, the mass of the water is significantly greater, so we might expect the water temperature to rise less than 10°C.

Categorize: The energy lost by the iron will be gained by the water. From complementary expressions for this single quantity of heat, the change in water temperature can be found.

Analyze: We assume that the water-horseshoe system is thermally isolated (insulated) from the environment for the short time required for the horseshoe to cool off and the water to warm up. Then the total energy input from the surroundings is zero, as expressed by $Q_{iron} + Q_{water} = 0$,

or $(mc\Delta T)_{iron} + (mc\Delta T)_{water} = 0$,

or $m_{Fe}c_{Fe}(T - 600°\text{C}) + m_w c_w(T - 25.0°\text{C}) = 0$.

Note that the first term in this equation is a negative number of joules, representing energy lost by the originally hot subsystem, and the second term is a positive number with the same absolute value, representing energy gained by heat by the cold stuff. Solving for the final temperature gives

$$T = \frac{m_w c_w(25.0°\text{C}) + m_{Fe}c_{Fe}(600°\text{C})}{m_{Fe}c_{Fe} + m_w c_w}$$

$$T = \frac{(20.0 \text{ kg})(4\ 186 \text{ J/kg} \cdot {}^\circ\text{C})(25.0^\circ\text{C}) + (1.50 \text{ kg})(448 \text{ J/kg} \cdot {}^\circ\text{C})(600^\circ\text{C})}{(1.50 \text{ kg})(448 \text{ J/kg} \cdot {}^\circ\text{C}) + (20.0 \text{ kg})(4\ 186 \text{ J/kg} \cdot {}^\circ\text{C})}$$

$$= 29.6^\circ\text{C}$$ ∎

Finalize: The temperature only rose about 5°C, so our answer seems reasonable. The specific heat of the water is about 10 times greater than the iron, so this effect also reduces the change in water temperature. In this problem, we assumed that a negligible amount of water boiled away. In reality, the final temperature of the water would be somewhat less than we calculated, since some of the energy would be used to vaporize a bit of water.

───────────────

15. In an insulated vessel, 250 g of ice at 0°C is added to 600 g of water at 18.0°C. (a) What is the final temperature of the system? (b) How much ice remains when the system reaches equilibrium?

Solution

Conceptualize: It takes a lot of energy to melt ice at 0°C without changing its temperature. If any bit of the ice remains solid when the originally warmer stuff reaches 0°C, the final temperature will be 0°C.

Categorize: The liquid water is losing energy by heat according to $mc\Delta T$. The ice is gaining energy described by $L\Delta m$. We must compare quantities to decide whether all the ice melts.

Analyze:

(a) If all 250 g of ice is melted it must absorb energy

$$Q_f = L_f \Delta m = (0.250 \text{ kg})(3.33 \times 10^5 \text{ J/kg}) = 83.3 \text{ kJ}$$

The energy released when 600 g of water cools from 18.0°C to 0°C is in absolute value,

$$|Q| = |mc\Delta T| = (0.600 \text{ kg})(4\ 186 \text{ J/kg} \cdot {}^\circ\text{C})(18.0^\circ\text{C}) = 45.2 \text{ kJ}$$

Since the energy required to melt 250 g of ice at 0°C **exceeds** the energy released by cooling 600 g of water from 18.0°C to 0°C, not all the ice melts and the final temperature of the system (water + ice) must be 0°C.   ∎

(b) The originally warmer water will cool all the way to 0°C, putting 45.2 kJ into the ice. This energy lost by the water will melt a mass of ice Δm, where $Q = L_f \Delta m$.

Solving for the mass,

$$\Delta m = \frac{Q}{L_f} = \frac{45.2 \times 10^3 \text{ J}}{3.33 \times 10^5 \text{ J/kg}} = 0.136 \text{ kg}$$

Therefore, the ice remaining is

$$m' = 0.250 \text{ kg} - 0.136 \text{ kg} = 0.114 \text{ kg}$$ ■

Finalize: Step-by-step reasoning is essential for solving a problem like this. It would be easy enough to write the equation $Lm_{ice} + m_{ice} c(T - 0) + m_{water} c(T - 10) = 0$. With it we would be assuming that all the ice melts. We would get a nonzero answer for final temperature T with no direct indication that it is wrong, and we would have made no progress toward solving the problem. The answer to part (a) was not read from a calculator, but identified by logic.

═══════════════

17. A 3.00-g lead bullet at 30.0°C is fired at a speed of 240 m/s into a large block of ice at 0°C, in which it becomes embedded. What quantity of ice melts?

Solution

Conceptualize: The amount of ice that melts is probably small, maybe only a few grams based on the size, speed, and initial temperature of the bullet.

Categorize: We will assume that all of the initial kinetic and excess internal energy of the bullet goes into internal energy to melt the ice, the mass of which can be found from the latent heat of fusion. Because the ice does not all melt, in the final state everything is at 0°C.

Analyze: At thermal equilibrium, the energy lost by the bullet equals the energy gained by the ice: $|\Delta K_b| + |Q_b| = Q_{ice}$ gives

$$\frac{1}{2} m_b v_b^2 + m_b c_{pb} |\Delta T| = L_f \Delta m_{melted} \quad \text{or} \quad m_{melted} = m_b \left(\frac{\frac{1}{2} v_b^2 + c_{Pb} |\Delta T|}{L_f} \right)$$

$$\Delta m_{melted} = \left(3.00 \times 10^{-3} \text{ kg} \right) \left[\frac{\frac{1}{2} (240 \text{ m/s})^2 + (128 \text{ J/kg} \cdot °\text{C})(30.0°\text{C})}{3.33 \times 10^5 \text{ J/kg}} \right]$$

$$\Delta m_{melted} = \frac{86.4 \text{ J} + 11.5 \text{ J}}{3.33 \times 10^5 \text{ J/kg}} = 2.94 \times 10^{-4} \text{ kg} = 0.294 \text{ g}$$ ■

Finalize: The amount of ice that melted is less than a gram, which agrees with our prediction. It appears that most of the energy used to melt the ice comes from the kinetic energy of the bullet (88%), while the excess internal energy of the bullet only contributes 12% to melt the ice. Small chips of ice probably fly off when the bullet makes impact. Some of the energy is transferred to their kinetic energy, so in reality, the amount of ice that would melt should be less than what we calculated. If the block of ice were colder than 0°C (as is often the case), then the melted ice would refreeze.

---

21. An ideal gas is taken through a quasi-static process described by $P = \alpha V^2$, with $\alpha = 5.00$ atm/m^6, as shown in Figure P17.21. The gas is expanded to twice its original volume of 1.00 m^3. How much work is done on the expanding gas in this process?

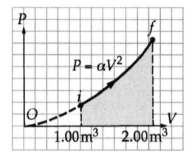

Figure P17.21

Solution

Conceptualize: The final pressure is $5 \cdot 4 = 20$ atm ≈ 2 MPa. The average pressure during expansion we could estimate as 1 MPa, and then the work −(1 MPa)(1 m^3) = −1 MJ.

Categorize: The work done on the gas is the negative of the area under the curve $P = \alpha V^2$, from V_i to V_f. The work *on* the gas is negative, to mean that the expanding gas *does* positive work. We will find its amount by doing the integral...

Analyze: ... $W = -\int_{V_i}^{V_f} P \, dV$ with $V_f = 2V_i = 2(1.00 \text{ m}^3) = 2.00 \text{ m}^3$:

$$W = -\int_{V_i}^{V_f} \alpha V^2 \, dV = -\tfrac{1}{3}\alpha\left(V_f^3 - V_i^3\right)$$

$$W = -\tfrac{1}{3}\left(5.00 \text{ atm/m}^6\right)\left(1.013 \times 10^5 \text{ Pa/atm}\right)\left[\left(2.00 \text{ m}^3\right)^3 - \left(1.00 \text{ m}^3\right)^3\right]$$

$$= -1.18 \times 10^6 \text{ J} \qquad \blacksquare$$

Finalize: Our estimate was good. To do the integral of $P dV$, we start by identifying how P depends on V, just as to do the integral of $y dx$ in mathematics you start by identifying how y depends on x.

---

23. An ideal gas is enclosed in a cylinder with a movable piston on top of it. The piston has a mass of 8 000 g and an area of 5.00 cm^2 and is free to slide up and down, keeping the pressure of the gas constant. How much work is done on the gas as the temperature of 0.200 mol of the gas is raised from 20.0°C to 300°C?

Solution

Conceptualize: The gas is expanding to put out a positive quantity of work in lifting the piston. The work *on* the gas is this number of joules but negative. The original internal energy of the gas is some multiple of nRT_i = (0.2 mol) (8.314 J/mol · K)(293 K) ~ 10^3 J. The work the gas puts out does not come just from this fund of internal energy, but from some of the energy that must be put into the sample by heat to raise its temperature. But still we can estimate the work as on the order of –1 kJ.

Categorize: The integral of PdV is easy because the pressure is constant. Then the equation of state will let us evaluate what we need about the pressure and the change in volume.

Analyze: For constant pressure, $W = -\int_{V_i}^{V_f} PdV = -P\Delta V = -P(V_f - V_i)$. Rather than evaluating the pressure numerically from atmospheric pressure plus the pressure due to the weight of the piston, we can just use the ideal gas law to write in the volumes, obtaining

$$W = -P\left(\frac{nRT_h}{P} - \frac{nRT_c}{P}\right) = -nR\left(T_h - T_c\right)$$

Therefore, $W = -nR\Delta T = -(0.200 \text{ mol})(8.314 \text{ J/mol·K})(280 \text{ K}) = -466 \text{ J}.$ ■

Finalize: We did not need to use the mass of the piston or its face area. The answer would be the same if the gas had to lift a heavier load over a smaller volume change or a lighter load through a larger volume change. The calculation turned out to be simple, but only because the pressure stayed constant during the expansion. Do not try to use $-P\Delta V$ for work in any case other than constant pressure, and never try to use $-V\Delta P$.

---

27. A thermodynamic system undergoes a process in which its internal energy decreases by 500 J. Over the same time interval, 220 J of work is done on the system. Find the energy transferred from it by heat.

Solution

Conceptualize: Think of a corporation's net worth decreasing by $500 on a day when it also pays its employees a combined salary of $220. It is clear that the corporation's net worth decreases by $720, analogous to this system transferring out 720 J by heat.

Categorize: It is the first law that keeps account of energy inputs and outputs by heat and work.

Analyze: We have $\Delta E_{int} = Q + W$, where W is positive, because work is done **on** the system. Then

$$Q = \Delta E_{int} - W = -500 \text{ J} - 220 \text{ J} = -720 \text{ J}$$

Thus, + 720 J of energy is transferred **from** the system by heat. ∎

Finalize: The system can be visualized as a gas in a cylinder with the piston being pushed in to compress it while it is put into a cold environment and drops in temperature.

---

31. An ideal gas initially at 300 K undergoes an isobaric expansion at 2.50 kPa. If the volume increases from 1.00 m³ to 3.00 m³ and 12.5 kJ is transferred to the gas by heat, what are (a) the change in its internal energy and (b) its final temperature?

Solution

Conceptualize: The gas pressure is much less than one atmosphere. This could be managed by having the gas in a cylinder with a piston on the bottom end, supporting a constant load that hangs from the piston. The (large) cylinder is put into a warmer environment to make the gas expand as its temperature rises to 900 K and its internal energy rises by thousands of joules.

Categorize: The first law will tell us the change in internal energy. The ideal gas law will tell us the final temperature.

Analyze: We use the energy version of the nonisolated system model.

(a) $\Delta E_{int} = Q + W$

   where $W = -P\Delta V$ for a constant-pressure process, so that

   $$\Delta E_{int} = Q - P\Delta V$$

$$\Delta E_{int} = 1.25 \times 10^4 \text{ J} - (2.50 \times 10^3 \text{ N/m}^2)(3.00 \text{ m}^3 - 1.00 \text{ m}^3) = 7\,500 \text{ J} \quad \blacksquare$$

(b) Since pressure and quantity of gas are constant, we have from the equation of state:

$$\frac{V_1}{T_1} = \frac{V_2}{T_2} \quad \text{and} \quad T_2 = \frac{V_2}{V_1}T_1 = \frac{3.00 \text{ m}^3}{1.00 \text{ m}^3}(300 \text{ K}) = 900 \text{ K} \quad \blacksquare$$

Finalize: We could have found the quantity of gas n from the initial pressure, volume, and temperature, but our solution was shorter.

---

34. (a) How much work is done on the steam when 1.00 mol of water at 100°C boils and becomes 1.00 mol of steam at 100°C at 1.00 atm pressure? Assume the steam to behave as an ideal gas. (b) Determine the change in internal energy of the system of the water and steam as the water vaporizes.

Solution

Conceptualize: Imagine the liquid water vaporizing inside a cylinder with a piston. The material expands greatly to lift the piston and do positive work. Many hundreds of joules of negative work is done on the material. Even more positive energy is put into it by heat, to produce a net increase in its internal energy.

Categorize: We use the idea of the negative integral of PdV to find the work, identifying the volumes of the liquid and the gas as we do so. Then $L\Delta m$ will tell us the heat input and the first law of thermodynamics will tell us the change in internal energy.

Analyze:

(a) We choose as a system the H_2O molecules that all participate in the phase change. For a constant-pressure process, $W = -P\Delta V = -P(V_s - V_w)$,

   where V_s is the volume of the steam and V_w is the volume of the liquid water.

   We can find them respectively from

$$PV_s = nRT \quad \text{and} \quad V_w = m/p = nM/p$$

   Calculating each work term,

$$PV_s = (1.00 \text{ mol})\left(8.314\frac{\text{J}}{\text{K}\cdot\text{mol}}\right)(373 \text{ K}) = 3101 \text{ J}$$

$$PV_w = (1.00 \text{ mol})(18.0 \text{ g/mol})\left(\frac{1.013\times10^5 \text{ N/m}^2}{1.00\times10^6 \text{ g/m}^3}\right) = 1.82 \text{ J}$$

Thus the work done is $W = -3\ 101 \text{ J} + 1.82 \text{ J} = -3.10 \text{ kJ}.$ ■

(b) The energy input by heat is $Q = L_V\Delta m = (18.0 \text{ g})(2.26 \times 10^6 \text{ J/kg}) = 40.7 \text{ kJ},$

so the change in internal energy is $\Delta E_{int} = Q + W = 37.6 \text{ kJ}.$ ■

Finalize: Steam at 100°C is on the point of liquefaction, so it probably behaves differently from the ideal gas we were told to assume. Still, the volume increase by more than a thousand times means that a significant amount of the "heat of vaporization" is coming out of the sample as it boils, as work done on the environment in its expansion. For many students the hardest thing to remember is that we start with 18 cubic centimeters of liquid water. Visualize it!

---

37. A 2.00-mol sample of helium gas initially at 300 K and 0.400 atm is compressed isothermally to 1.20 atm. Noting that the helium behaves as an ideal gas, find (a) the final volume of the gas, (b) the work done on the gas, and (c) the energy transferred by heat.

Solution

Conceptualize: The final volume will be one-third of the original, several liters. The positive work will be hundreds of joules. For the temperature to stay constant, the sample must put out as many joules by heat as it takes in by work.

Categorize: The ideal gas law can tell us the original and final volumes. The negative integral of PdV will tell us the work input, and the first law will tell us the energy output by heat.

Analyze:

(a) Rearranging $PV = nRT$ we get $V_i = \dfrac{nRT}{P_i}$

The initial volume is

$$V_i = \frac{(2.00 \text{ mol})(8.314 \text{ J/mol}\cdot\text{K})(300 \text{ K})}{(0.400 \text{ atm})(1.013 \times 10^5 \text{ Pa/atm})}\left(\frac{1 \text{ Pa}}{\text{N/m}^2}\right) = 0.123 \text{ m}^3$$

For isothermal compression, PV is constant, so $P_iV_i = P_fV_f$ and the final volume is

$$V_f = V_i\left(\frac{P_i}{P_f}\right) = (0.123 \text{ m}^3)\left(\frac{0.400 \text{ atm}}{1.20 \text{ atm}}\right) = 0.041\ 0 \text{ m}^3 \qquad \blacksquare$$

(b) $W = -\int P\,dV = -\int \dfrac{nRT}{V}\,dV = -nRT\ln\left(\dfrac{V_f}{V_i}\right) = -(4\ 988 \text{ J})\ln\left(\dfrac{1}{3}\right) = +5.48 \text{ kJ} \qquad \blacksquare$

(c)   The ideal gas keeps constant temperature so $\Delta E_{int} = 0 = Q + W$ and the heat is $Q = -5.48$ kJ $\qquad \blacksquare$

Finalize: Visualize squashing a piston down on the sample in a cylinder surrounded by a constant-temperature bath. With all that compression, the gas would tend to rise in temperature, so energy must leave it by heat for the process to be isothermal.

---

39. A 1.00-mol sample of hydrogen gas is warmed at constant pressure from 300 K to 420 K. Calculate (a) the energy transferred to the gas by heat, (b) the increase in its internal energy, and (c) the work done on the gas.

Solution

Conceptualize: The gas expands in this process, so it puts out positive work. The work done *on* it is negative. In absolute value, we expect some hundreds of joules for all three answers. The heat input will be the largest of the three, enough to supply the other two together.

Categorize: The specific heats of hydrogen at constant pressure and at constant volume will give us the first two answers. Their difference will be the energy escaping as work.

Analyze:

(a)   Since this is a constant-pressure process, $Q = nC_p\Delta T$.

The temperature rises by $\Delta T = 420 \text{ K} - 300 \text{ K} = 120 \text{ K}$:

$$Q = (1.00 \text{ mol})\,(28.8 \text{ J/mol} \cdot \text{K})(120 \text{ K}) = 3.46 \text{ kJ} \qquad \blacksquare$$

(b)   For any gas $\Delta E_{int} = nC_v\Delta T$,

so $\Delta E_{int} = (1.00 \text{ mol})(20.4 \text{ J/mol} \cdot \text{K})(120 \text{ K}) = 2.45 \text{ kJ}.$ $\qquad \blacksquare$

(c) The first law says $\Delta E_{int} = Q + W$,

so $W = \Delta E_{int} - Q = 2.45 \text{ kJ} - 3.46 \text{ kJ} = -1.01 \text{ kJ}.$ ∎

Finalize: The gas puts out work by expanding. Another way to find the work in this special constant-pressure case is to calculate

$$W = -\int_i^f P\,dV = -P\int_i^f dV = -P\left(V_f - V_i\right) = -nRT_f + nRT_i = -nR\Delta T$$
$$= (1 \text{ mol})(8.341 \text{ J/mol}\cdot\text{K})(120 \text{ K}) = 0.998 \text{ kJ}$$

in good agreement with our first-law result.

────────────

46. A 2.00-mol sample of a diatomic ideal gas expands slowly and adiabatically from a pressure of 5.00 atm and a volume of 12.0 L to a final volume of 30.0 L. (a) What is the final pressure of the gas? (b) What are the initial and final temperatures? Find (c) Q, (d) E_{int}, and (e) W for the gas during this process.

Solution

Conceptualize: The volume gets 2.5 times larger in the expansion. If the gas were at constant temperature, the pressure would become 2.5 times smaller (and ΔE_{int} would be zero, and Q would be positive). In the adiabatic process Q is zero. No heat comes in to supply the joules of work output, so the temperature will drop and the final pressure will be significantly smaller than 2 atm.

Categorize: We use the equation $PV^{\gamma} = $ constant for an adiabatic path. Then we use the ideal gas law to find the temperatures. From the temperatures we can find ΔE_{int}. With the known $Q = 0$, the first law will give us W.

Analyze:

(a) In an adiabatic process $P_i V_i^{\gamma} = P_f V_f^{\gamma}$:

$$P_f = P_i\left(\frac{V_i}{V_f}\right) = (5.00 \text{ atm})\left(\frac{12.0 \text{ L}}{30.0 \text{ L}}\right)^{1.40} = 1.39 \text{ atm}$$ ∎

(b) The initial temperature is

$$T_i = \frac{P_i V_i}{nR} = \frac{(5.00 \text{ atm})(1.03 \times 10^5 \text{ Pa/atm})(12.0 \times 10^{-3} \text{ m}^3)}{(2.00 \text{ mol})(8.314 \text{ N}\cdot\text{m/mol K})} = 366 \text{ K}$$ ∎

and similarly the final temperature is $T_f = \dfrac{P_f V_f}{nR} = 253 \text{ K}.$ ∎

(c) This is an adiabatic process, so by the definition $Q = 0$.   ■

(d) For any process, $\Delta E_{int} = nC_V \Delta T$, and for this diatomic ideal gas,

$$C_V = \frac{R}{\gamma - 1} = \frac{5}{2}R$$

Thus, $\Delta E_{int} = \dfrac{5}{2}(2.00 \text{ mol})(8.314 \text{ J/mol} \cdot \text{K})(253 \text{ K} - 366 \text{ K}) = -4\ 660 \text{ J}.$ ■

(e) Now, $W = \Delta E_{int} - Q = -4\ 660 \text{ J} - 0 = -4\ 660 \text{ J}.$ ■

Finalize: The work done *on* the gas is negative, so positive work is done *by* the gas on its environment as the gas expands.

---

51. Air in a thundercloud expands as it rises. If its initial temperature is 300 K and no energy is lost by thermal conduction on expansion, what is its temperature when the initial volume has doubled?

Solution

Conceptualize: The air should cool as it expands, so we should expect $T_f < 300$ K.

Categorize: The air expands adiabatically, losing no heat but dropping in temperature as it does work on the air around it, so we have PV^γ = constant, where $\gamma = 1.40$ for an ideal diatomic gas. Even as the triatomic water molecules condense in a thundercloud, the overwhelming majority of molecules are oxygen and nitrogen far from liquefaction, so a diatomic ideal gas is a good model for the air.

Analyze: Combining PV^γ = constant with the ideal gas law gives one of the textbook equations describing adiabatic processes, $T_1 V_1^{\gamma - 1} = T_2 V_2^{\gamma - 1}$:

$$T_2 = T_1 \left(\frac{V_1}{V_2} \right)^{\gamma - 1} = 300 \text{ K} \left(\frac{1}{2} \right)^{(1.40 - 1)} = 227 \text{ K}$$ ■

Finalize: The air does cool, but the temperature is not inversely proportional to the volume. The temperature drops 24% while the volume doubles. The pressure drops also in the expansion.

---

57. The relationship between the heat capacity of a sample and the specific heat of the sample material is discussed in Section 17.2. Consider a sample containing 2.00 mol of an ideal diatomic gas. Assuming the molecules rotate but do not vibrate, find (a) the total heat capacity of the sample at constant volume and (b) the total heat capacity at constant pressure. (c) **What If?** Repeat parts (a) and (b), assuming the molecules both rotate and vibrate.

Solution

Conceptualize: The molar heat capacity of an ideal gas at constant volume with single-atom molecules is (3/2)R. With two moles the total heat capacity will be twice as large. Having to supply energy to drive the diatomic molecules' rotation will raise the heat capacity farther. The heat capacity will be still larger if the molecules vibrate. And the heat capacity at constant pressure is always larger than that for a constant-volume process, because some of the energy put in as heat leaks out of the expanding gas as work while the gas is warming up in a constant-pressure process.

Categorize: We use the kinetic theory structural model of an ideal gas.

Analyze:

(a) Count degrees of freedom. A diatomic molecule oriented along the y axis can possess energy by moving in x, y, and z directions and by rotating around x and z axes. Rotation around the y axis does not represent an energy contribution because the moment of inertia of the molecule about this axis is essentially zero. The molecule will have an average energy $\frac{1}{2}k_BT$ for each of these five degrees of freedom.

The gas with N molecules will have internal energy

$$E_{int} = N\left(\frac{5}{2}\right)k_BT = nN_A\left(\frac{5}{2}\right)\left(\frac{R}{N_A}\right)T = \left(\frac{5}{2}\right)nRT$$

so the constant-volume heat capacity of the whole sample is

$$\frac{\Delta E_{int}}{\Delta T} = \frac{5}{2}nR = \frac{5}{2}(2.00 \text{ mol})(8.314 \text{ J/mol} \cdot \text{K}) = 41.6 \text{ J/K} \qquad \blacksquare$$

(b) For one mole, $C_P = C_V + R$. For the sample, $nC_P = nC_V + nR$.

With P constant,

$$nC_p = 41.6 \text{ J/K} + (2.00 \text{ mol})(8.314 \text{ J/mol} \cdot \text{K}) = 58.2 \text{ J/K} \qquad \blacksquare$$

(c) Vibration adds a degree of freedom for kinetic energy and a degree of freedom for elastic energy.

Now the molecule's average energy is $\dfrac{7}{2} k_B T$,

the sample's internal energy is $N \dfrac{7}{2} k_B T = \dfrac{7}{2} nRT$,

and the sample's constant-volume heat capacity is

$$\frac{7}{2} nR = \frac{7}{2} (2.00 \text{ mol})(8.314 \text{ J/mol} \cdot \text{K}) = 58.2 \text{ J/K} \qquad \blacksquare$$

At constant pressure, its heat capacity is

$$nC_V + nR = 58.2 \text{ J/K} + (2.00 \text{ mol})(8.314 \text{ J/mol} \cdot \text{K}) = 74.8 \text{ J/K} \qquad \blacksquare$$

Finalize: It is the randomness of the collisions among molecules that makes the average energy equal for each different kind of motion, for each degree of freedom. One can say that the diatomic molecule does not possess any energy of rotation about an axis along the line joining the atoms because any outside force has no lever arm to set the massive atomic nuclei into rotation about this axis. When we study energy quantization at the end of the book we will be able to give a better account of the reason.

─────────

59. A bar of gold (Au) is in thermal contact with a bar of silver (Ag) of the same length and area (Fig. P17.59). One end of the compound bar is maintained at 80.0°C, and the opposite end is at 30.0°C. When the energy transfer reaches steady state, what is the temperature at the junction?

Solution

Conceptualize: Silver is a better conductor than gold, so the junction temperature will be a bit less than the 55°C that is numerically halfway between the temperatures of the ends.

Categorize: We must think of the temperature difference across each bar as the driving force for an energy current of heat going through the bar by conduction.

Analyze: Call the gold bar Object 1 and the silver bar Object 2. Each is a nonisolated system in steady state. When energy transfer by heat reaches a steady state, the flow rate through each will be the same, so that the junction can stay at constant temperature thereafter, with as much heat coming in through the gold as goes out through the silver.

$$P_1 = P_2 \quad \text{or} \quad \frac{k_1 A_1 \Delta T_1}{L_1} = \frac{k_2 A_2 \Delta T_2}{L_2}$$

In this case, $L_1 = L_2$ and $A_1 = A_2$,

so $k_1 \Delta T_1 = k_2 \Delta T_2$.

Let T_3 be the temperature at the junction; then

$$k_1(80.0°C - T_3) = k_2(T_3 - 30.0°C)$$

Rearranging, we find

$$T_3 = \frac{(80.0°C)\, k_1 + (30.0°C)k_2}{k_1 + k_2}$$

$$T_3 = \frac{(80.0°C)(314\ \text{W/m} \cdot °C) + (30.0°C)(427\ \text{W/m} \cdot °C)}{(314\ \text{W/m} \cdot °C) + (427\ \text{W/m} \cdot °C)} = 51.2°C \qquad \blacksquare$$

Finalize: We did not need to know the value of the length or area of either bar, or the value of the heat current, to identify the current as equal in the two materials in steady state. And that equality was enough to predict the junction temperature. Solving a problem like this about the voltage at the junction between electric resistors in a series circuit will be routine later in the course.

---

69. An aluminum rod 0.500 m in length and with a cross-sectional area of 2.50 cm^2 is inserted into a thermally insulated vessel containing liquid helium at 4.20 K. The rod is initially at 300 K. (a) If one half of the rod is inserted into the helium, how many liters of helium boil off by the time the inserted half cools to 4.20 K? Assume the upper half does not yet cool. (b) If the circular surface of the upper end of the rod is maintained at 300 K, what is the approximate boil-off rate of liquid helium after the lower half has reached 4.20 K? (Aluminum has thermal conductivity of 3 100 W/m K at 4.20 K; ignore its temperature variation. The density of liquid helium is 125 kg/m^3.)

Solution

Conceptualize: Demonstrations with liquid nitrogen give us some indication of the phenomenon described. Since the rod is much hotter than the liquid helium and of significant size (almost 2 cm in diameter), a substantial volume (maybe as much as a liter) of helium will boil off before thermal equilibrium is reached.

Likewise, since aluminum conducts rather well, a significant amount of helium will continue to boil off as long as the upper end of the rod is maintained at 300 K.

Categorize: Until thermal equilibrium is reached, the excess internal energy of the rod will go into vaporizing liquid helium, which is already at its boiling point (so there is no change in the temperature of the helium).

Analyze: As you solve this problem, be careful not to confuse L (the **conduction length** of the rod) with L_v (the **heat of vaporization** of the helium), or with L, the symbol for the unit liter.

(a) Before heat conduction has time to become important, we suppose the energy lost by heat by half the rod equals the energy gained by the helium. Therefore,

$$\left(L_v \Delta m\right)_{He} = \left|mc\Delta T\right|_{Al} \quad \text{or} \quad \left(\rho V L_v\right)_{He} = \left|\rho V c\Delta T\right|_{Al}$$

So $\quad V_{He} = \dfrac{\left|\rho V c\Delta T\right|_{Al}}{\left(\rho L_v\right)_{He}} = \dfrac{\left(2.70 \text{ g/cm}^3\right)\left(62.5 \text{ cm}^3\right)\left(900 \text{ J/kg}\cdot{}^\circ\text{C}\right)(295.8\ {}^\circ\text{C})}{\left(0.125 \text{ g/cm}^3\right)(2.09\times10^4 \text{ J/kg})}$,

$$V_{He} = 1.72\times10^4 \text{ cm}^3 = 17.2 \text{ liters} \qquad\blacksquare$$

(b) Heat will be conducted along the rod at the rate

$$\frac{dQ}{dt} = P = \frac{kA\Delta T}{L} \qquad\qquad\qquad\text{[1]}$$

and will boil off helium according to

$$Q = L_v\Delta m \quad\text{so}\quad \frac{dQ}{dt} = \left(\frac{dm}{dt}\right)L_v. \qquad\qquad\text{[2]}$$

Combining [1] and [2] gives us the "boil-off" rate: $\dfrac{dm}{dt} = \dfrac{kA\Delta T}{L\cdot L_v}$

Set the conduction length $L = 25$ cm, and use $k = 3\,100$ W/m · K

to find $\quad \dfrac{dm}{dt} = \dfrac{(3\,100 \text{ W/100 cm}\cdot\text{K})\left(2.50 \text{ cm}^2\right)(295.8 \text{ K})}{(25.0 \text{ cm})(20.9 \text{ J/g})} = 43.9\,\text{g/s}$

or $\quad \dfrac{dV}{dt} = \dfrac{43.9 \text{ g/s}}{0.125 \text{ g/cm}^3} = 351 \text{ cm}^3/\text{s} = 0.351 \text{ liters/s.} \qquad\blacksquare$

Finalize: The volume of helium boiled off initially is much more than expected. If our calculations are correct, that sure is a lot of liquid helium that is wasted!

Since liquid helium is much more expensive than liquid nitrogen, most low-temperature equipment is designed to avoid unnecessary loss of liquid helium by surrounding the liquid with a Dewar inside a container of liquid nitrogen.

══════════════

84. In a cylinder, a sample of an ideal gas with number of moles n undergoes an adiabatic process. (a) Starting with the expression $W = -\int P\,dV$ and using the condition $PV^{\gamma} = $ constant, show that the work done on the gas is

$$W = \left(\frac{1}{\gamma - 1}\right)\left(P_f V_f - P_i V_i\right)$$

(b) Starting with the first law of thermodynamics, show that the work done on the gas is equal to $nC_V(T_f - T_i)$. (c) Are these two results consistent with each other? Explain.

Solution

Conceptualize: The gas does positive work if it expands. Its environment does positive work on it if the gas contracts.

Categorize: We will start as instructed with the integral expression for work in part (a) and with $\Delta E_{int} = Q + W$ in part (b). In each case we specialize to an adiabatic process and evaluate.

Analyze:

(a) For the adiabatic process $PV^{\gamma} = k$, a constant.

The work is

$$W = -\int_i^f P\,dV = -k\int_{V_i}^{V_f} \frac{dV}{V^{\gamma}} = \left.\frac{-kV^{1-\gamma}}{1-\gamma}\right|_{V_i}^{V_f}$$

For k we can substitute $P_i V_i^{\gamma}$ and also $P_f V_f^{\gamma}$ to have

$$W = -\frac{P_f V_f^{\gamma} V_f^{1-\gamma} - P_i V_i^{\gamma} V_i^{1-\gamma}}{1-\gamma} = \frac{P_f V_f - P_i V_i}{\gamma - 1} \qquad ∎$$

(b) For an adiabatic process $\Delta E_{int} = Q + W$ and $Q = 0$.

Therefore, $W = \Delta E_{int} = nC_V\Delta T = nC_V(T_f - T_i)$. $\qquad ∎$

(c) They are consistent. To prove consistency between these two equations, consider that $\gamma = \dfrac{C_p}{C_V}$ and $C_p - C_V = R$,

so that $\dfrac{1}{\gamma - 1} = \dfrac{C_V}{R}$.

Thus part (a) becomes $\quad W = \left(P_f V_f - P_i V_i\right)\dfrac{C_V}{R}$.

Then, $\quad \dfrac{PV}{R} = nT \quad$ so that $\quad W = nC_V\left(T_f - T_i\right)$. ∎

Finalize: In an adiabatic compression the temperature rises, and the final value of PV is greater than the initial value. Thus both expressions give positive results as required.

———————

89. Water in an electric teakettle is boiling. The power absorbed by the water is 1.00 kW. Assuming the pressure of vapor in the kettle equals atmospheric pressure, determine the speed of effusion of vapor from the kettle's spout if the spout has a cross-sectional area of 2.00 cm^2. Model the steam as an ideal gas.

Solution

Conceptualize: A story says that James Watt observed that the jet from his mother's teakettle could do work, and went on to invent the steam engine that started the Industrial Revolution. Not many grams will come out of the spout each second, but the density of the steam is low enough for us to suppose that the speed is a few meters per second.

Categorize: The temperature of the teakettle is staying constant at 100°C. We can call the water and steam inside a nonisolated system, in energy terms, in steady state, gaining energy from the electric resistor by heat and losing energy by matter transfer. The equality of the rates of energy input and output will give us the answer. We use *Power* as a symbol for the rate of energy transfer, and P_0 to represent atmospheric pressure.

Analyze: From $Q = L_V\,\Delta m$, the rate of boiling is described by $Power = \dfrac{Q}{\Delta t} = \dfrac{L_V \Delta m}{\Delta t}$, so that the mass flow rate of steam from the kettle is

$$\dfrac{\Delta m}{\Delta t} = \dfrac{Power}{L_V}$$

The symbols Δm for mass vaporized and m for mass leaving the kettle have the same meaning, but recall that M represents the molar mass. Even though it is on the point of liquefaction, we model the water vapor as an ideal gas. The volume flow rate $V/\Delta t$ of the fluid is the cross-sectional area of the spout multiplied by the speed of flow, forming the product Av.

$$P_0 V = nRT = \left(\frac{m}{M}\right)RT$$

$$\frac{P_0 V}{\Delta t} = \frac{m}{\Delta t}\left(\frac{RT}{M}\right)$$

$$P_0 Av = \frac{Power}{L_V}\left(\frac{RT}{M}\right)$$

$$v = \frac{Power\, RT}{M L_V P_0 A}$$

$$= \frac{(1\,000\text{ W})(8.314\text{ J/mol}\cdot\text{K})(373\text{ K})}{(0.018\,0\text{ kg/mol})(2.26\times10^6\text{ J/kg})(1.013\times10^5\text{ N/m}^2)(2.00\times10^{-4}\text{ m}^2)}$$

$$v = 3.76\,\text{m/s} \qquad\blacksquare$$

Finalize: The speed is on the order of 10 m/s, not just 1 m/s. In Chapter 20 we will treat in some detail the practical application of thermodynamics to heat engines like this one. Heat engines are devices that convert energy input by heat into mechanical energy output. They always have limited efficiency.

Chapter 18
Heat Engines, Entropy, and the Second Law of Thermodynamics

Section 18.1   Heat Engines and the Second Law of Thermodynamics

A heat engine is a device that takes in energy as heat and puts out energy in other useful forms, such as mechanical or electrical energy. A practical heat engine carries some working substance through a cyclic process during which (1) energy is absorbed by heat from a source at a high temperature, (2) work is done by the engine, and (3) energy is expelled by heat from the engine to a reservoir at a lower temperature.

The engine absorbs a quantity of energy, $|Q_h|$, from a hot reservoir, does work $|W| = W_{eng}$, and then gives up energy $|Q_c|$ to a cold reservoir. Because the working substance goes through a cycle, its initial and final internal energies are equal, so $\Delta E_{int} = 0$. From the first law the net work done by a heat engine equals the net energy absorbed by the engine: $W_{eng} = |Q_h| - |Q_c|$.

If the working substance is a gas, the net work done for a cyclic process is the area enclosed by the curve representing the process on a PV diagram.

The **thermal efficiency**, *e*, of a heat engine is the ratio of the net work done by the engine to the energy absorbed by heat at the higher temperature during one cycle.

The **second law of thermodynamics** can be stated in several ways:

- **Clausius Statement:** Energy will not flow spontaneously by heat from a cold object to a hot object. *No thermodynamic process can occur* *whose only result is to transfer energy from a colder to a hotter body by heat.* Such a process (refrigeration) is only possible if work is done on the system.

- **Kelvin-Planck Statement:** It is impossible for a cyclic thermodynamic process to occur whose only result is the complete conversion of energy extracted by heat from a hot reservoir into work. *That is, it is impossible to construct a heat engine that, operating in a cycle, produces no other effect than the absorption of energy by heat from a reservoir and the performance of an equal amount of work.* A heat engine must also release energy by heat into a cold reservoir. The efficiency of a heat engine must be less than 100%. The engine's energy output by heat is often called heat exhaust, wasted heat, rejected heat, expelled heat, or thermal pollution.

- **Entropy statement:** The second law of thermodynamics can be stated in terms of entropy as follows: *The total entropy of an isolated system always increases in time if the system undergoes an irreversible process.* If an isolated system undergoes a reversible process, the total entropy remains constant.

Section 18.2   Reversible and Irreversible Processes

A process is **reversible** if the system passes from the initial to the final state through a succession of equilibrium states. Then, the process can be made to run in the opposite direction at any point by an infinitesimal change in conditions. Otherwise the process is irreversible.

Section 18.3   The Carnot Engine

An ideal reversible cyclic process, called the **Carnot cycle**, is described in the *PV* diagram shown at right. The Carnot cycle consists of two adiabatic and two isothermal processes, all being reversible.

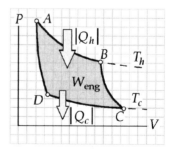

PV diagram for the Carnot cycle. The work done during one cycle equals the area enclosed by the path on the *PV* diagram.

- The process $A \to B$ is an **isotherm** (constant *T*), during which time the gas expands at constant temperature T_h, and absorbs energy $\left| Q_h \right|$ by heat from the hot reservoir.

- The process $B \to C$ is an **adiabatic expansion** ($Q = 0$), during which time the gas expands and cools to a temperature T_c.

- The process $C \to D$ is a second **isotherm**, during which time the gas is compressed at constant temperature T_c, and gives up energy $|Q_c|$ by heat to the cold reservoir.

- The final process $D \to A$ is an **adiabatic compression** in which the gas temperature increases to a final temperature of T_h.

No working engine is 100% efficient, even when losses such as friction are neglected. The theoretical limit on the efficiency of a real engine can determined by comparison with the ideal Carnot engine. A reversible engine is one that will operate with the same efficiency in the forward and reverse directions. The Carnot engine is one example of a reversible engine.

Carnot's theorems, which can be proved from the first and second laws of thermodynamics, can be stated as follows:

- **Theorem I:** No real (irreversible) engine can have an efficiency greater than that of a reversible engine operating between the same two temperatures.

- **Theorem II:** All reversible engines operating between T_h and T_c have the same efficiency, given by Equation 18.4, $e_C = 1 - \dfrac{T_c}{T_h}$.

A schematic diagram of a heat engine is shown on the left in the figure below, where $|Q_h|$ is the energy extracted from the hot reservoir by heat at temperature T_h, $|Q_c|$ is the energy rejected to the cold reservoir by heat at temperature T_c, and W_{eng} is the work done by the engine.

Section 18.4   Heat Pumps and Refrigerators

A schematic representation of a refrigerator is shown on the right in the figure below. The refrigerator is a heat engine operating in reverse. During one cycle of operation, the refrigerator absorbs energy $|Q_c|$ by heat from the cold reservoir, expels energy $|Q_h|$ by heat to the hot reservoir, and the work done on the system is $W = |Q_h| - |Q_c|$.

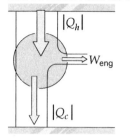

Hot Reservoir at T_h

W_{eng}

Cold Reservoir at T_c
Heat engine

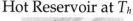
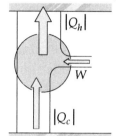

Hot Reservoir at T_h

W

Cold Reservoir at T_c
Refrigerator

Section 18.6   Entropy
Section 18.7   Entropy and the Second Law of Thermodynamics
Section 18.8   Entropy Changes in Irreversible Processes

Entropy is a quantity used to measure the degree of disorder in a system. For example, the molecules of a gas in a container at a high temperature are in a more disordered state (higher entropy) than the same molecules at a lower temperature.

When energy is added by heat to a system in an incremental reversible process, dQ is positive and the entropy increases. When energy is removed, dQ is negative and the entropy decreases. Note that *only changes in entropy* are defined by Equation 18.8; therefore, the concept of entropy is most useful when a system undergoes a change in its state.

When using Equation 18.12, $\Delta S = nR \ln\left(\dfrac{V_f}{V_i}\right)$, to calculate entropy changes, note that ΔS may be obtained even if the process is irreversible, since ΔS depends only on the initial and final equilibrium states, not on the path. In order to calculate ΔS for an irreversible process, you must devise a reversible process (or sequence of reversible processes) between the initial and final states, and compute $\dfrac{dQ_r}{T}$ for the reversible process.

The entropy change for the irreversible process is the same as that of the reversible process between the same initial and final equilibrium states.

EQUATIONS AND CONCEPTS

The **net work done by a heat engine during one cycle** equals the net energy absorbed by the engine. During each cycle energy is transferred from a reservoir at a high temperature, work is done by the engine, and energy is expelled by the engine to a reservoir at a low temperature.

$$W_{eng} = |Q_h| - |Q_c| \qquad (18.1)$$

Q_h = quantity of energy absorbed from the high temperature reservoir

Q_c = quantity of energy expelled to the low temperature reservoir

The **thermal efficiency**, e, of a heat engine is defined as the ratio of the net work done to the energy absorbed as heat during one cycle of the process.

$$e = \frac{W_{eng}}{|Q_h|} = 1 - \frac{|Q_c|}{|Q_h|} \qquad (18.2)$$

The **Carnot efficiency** is the limiting efficiency for an engine operating reversibly in a Carnot cycle between two given temperatures.

$$e_C = 1 - \frac{T_c}{T_h} \qquad (18.4)$$

The **coefficient of performance**, COP, of a heat pump, operating in the heating mode, is the ratio of the energy transferred as heat at the high temperature to the work done on the pump.

$$\text{COP (heat pump)} = \frac{|Q_h|}{W} \qquad (18.5)$$

The **coefficient of performance** of a refrigerator (or a heat pump operating in the cooling mode) is defined by the ratio of the energy absorbed by heat Q_c, to the work done. A good refrigerator has a high coefficient of performance.

$$\text{COP (refrigerator)} = \frac{|Q_c|}{W} \qquad (18.6)$$

The **change in entropy** as a system changes from one equilibrium state to another, under a reversible, quasi-static process, is defined by Equation 18.8.

$$dS = \frac{dQ_r}{T} \qquad (18.8)$$

dQ_r = quantity of energy added (or removed)

T = absolute temperature

Entropy, S, is a thermodynamic variable that characterizes the degree of disorder in a system. *All physical processes tend toward a state of increasing entropy; the entropy of the Universe increases in all natural processes.*

$$S \equiv k_B \ln W \qquad (18.9)$$

$W =$ the number of possible microstates of the system

The **change in entropy** of a system that undergoes a reversible process between the states i and f depends only on the properties of the initial and final equilibrium states.

$$\Delta S = \int_i^f \frac{dQ_r}{T} \quad \text{(reversible path)} \qquad (18.10)$$

For any **arbitrary reversible cycle** the change in entropy of a system is identically zero.

$$\oint \frac{dQ_r}{T} = 0 \qquad (18.11)$$

During a **free expansion**, the entropy of a gas increases.

$$\Delta S = nR \ln \left(\frac{V_f}{V_i} \right) \qquad (18.12)$$

REVIEW CHECKLIST

- Understand the basic principle of the operation of a heat engine, and be able to define and discuss the thermal efficiency of a heat engine.

- State the second law of thermodynamics, and discuss the difference between reversible and irreversible processes. Discuss the importance of the first and second laws of thermodynamics as they apply to various forms of energy conversion and thermal pollution.

- Describe the processes that take place in an ideal heat engine taken through a Carnot cycle.

- Calculate the efficiency of a Carnot engine, and note that the efficiency of real heat engines is always less than the Carnot efficiency.

- Calculate entropy changes for reversible processes (such as one involving an ideal gas).

- Calculate entropy changes for irreversible processes, recognizing that the entropy change for an irreversible process is equivalent to that of a reversible process between the same two equilibrium states.

ANSWER TO AN OBJECTIVE QUESTION

5.   Consider cyclic processes completely characterized by each of the following net energy inputs and outputs. In each case, the energy transfers listed are the *only* ones occurring. Classify each process as (a) possible, (b) impossible according to the first law of thermodynamics, (c) impossible according to the second law of thermodynamics, or (d) impossible according to both the first and second laws. **(i)** Input is 5 J of work, and output is 4 J of work. **(ii)** Input is 5 J of work, and output is 5 J of energy transferred by heat. **(iii)** Input is 5 J of energy transferred by electrical transmission, and output is 6 J of work. **(iv)** Input is 5 J of energy transferred by heat, and output is 5 J of energy transferred by heat. **(v)** Input is 5 J of energy transferred by heat, and output is 5 J of work. **(vi)** Input is 5 J of energy transferred by heat, and output is 3 J of work plus 2 J of energy transferred by heat.

Answer   Each process is cyclic, returning the sample to its starting point. The internal energy of the working substance undergoes no net change. The first law then says that the total energy input must be equal in amount to the total energy output. The first law does not refer to forms of energy. The second law, we can say, implies that if the input energy is transferred by heat, at least some of the output energy must be transferred by heat. Process **(i)** is (b) impossible according to the first law because 5 J is different from 4 J. Process **(ii)** is (a) possible: for example, give a box a push to set it sliding across the floor, wait for it to stop, and wait some more for the rubbing surfaces to cool off to reach the environmental temperature again. Process **(iii)** is (b) an example of imagined energy creation, with 6 J different from 5 J. Process **(iv)** is (a) possible: warm up some soup on the stove and let it cool off again in the bowl of a child who won't eat it. Process **(v)** is (c) an impossible heat engine with 100% efficiency. Process **(vi)** is (a) possible, representing a 60%-efficient heat engine. We could imagine a process violating both laws: let a sample take in 5 J by heat and put out 6 J carried by mechanical waves as it returns to its initial state.

ANSWERS TO SELECTED CONCEPTUAL QUESTIONS

1.   The device shown in Figure CQ18.1, called a thermoelectric converter, uses a series of semiconductor cells to transform internal energy to electric potential energy, which we will study in Chapter 20. In the picture on the left, both legs of the device are at the same temperature and no electric potential energy is produced. When one leg is at a higher temperature than the other as shown in the

Figure CQ18.1

picture on the right, however, electric potential energy is produced as the device extracts energy from the hot reservoir and drives a small electric motor. (a) Why is the difference in temperature necessary to produce electric potential energy in this demonstration? (b) In what sense does this intriguing experiment demonstrate the second law of thermodynamics?

Answer

(a) The semiconductor converter operates essentially like a thermocouple, which is a pair of wires of different metals, with a junction at each end. When the junctions are at different temperatures, a small voltage appears around the loop, so that the device can be used to measure temperature or (here) to drive a small motor.

(b) The second law states that an engine operating in a cycle cannot absorb energy by heat from one reservoir and expel it entirely by work. This exactly describes the first situation, where both legs are in contact with a single reservoir, and the thermocouple fails to produce electrical work. To expel energy by work, the device must transfer energy by heat from a hot reservoir to a cold reservoir, as in the second situation. The device must have heat exhaust.

☐ ☐ ☐ ☐

6. A steam-driven turbine is one major component of an electric power plant. Why is it advantageous to have the temperature of the steam as high as possible?

Answer The most optimistic limit on efficiency in a steam engine is the ideal (Carnot) efficiency. This can be calculated from the high and low temperatures of the steam, as it expands:

$$e_C = \frac{T_h - T_c}{T_h} = 1 - \frac{T_c}{T_h}$$

The engine will be most efficient when the low temperature is extremely low, and the high temperature is extremely high. However, since the electric power plant is typically placed on the surface of the Earth, the temperature and pressure of the surroundings limit how much the steam can expand, and how low the lower temperature can be. The only way to further increase the efficiency, then, is to raise the temperature of the hot steam, as high as possible.

☐ ☐ ☐ ☐

7. What are some factors that affect the efficiency of automobile engines?

Answer A gasoline engine does not exactly fit the definition of a heat engine. It takes in energy by mass transfer. Nevertheless, it converts this chemical energy into internal energy, so it can be modeled as taking in energy by heat. Therefore, Carnot's limit on the efficiency of a heat engine applies to a gasoline engine. The fundamental limit on its efficiency is $1 - T_c/T_h$, set by the maximum temperature the engine block can stand and the temperature of the surroundings into which exhaust heat must be dumped.

The Carnot efficiency can only be attained by a reversible engine. To run with any speed, a gasoline engine must carry out irreversible processes and have an efficiency below the Carnot limit. In order for the compression and power strokes to be adiabatic, we would like to minimize irreversibly losing heat to the engine block. We would like to have the processes happen very quickly, so that there is negligible time for energy to be conducted away. But in a standard piston engine the compression and power strokes must take one-half of the total cycle time, and the angular speed of the crankshaft is limited by the time intervals required to open and close the valves to get the fuel into each cylinder and the exhaust out.

Other limits on efficiency are imposed by irreversible processes like friction, both inside and outside the engine block. If the ignition timing is off, then the effective compressed volume will not be as small as it could be. If the burning of the gasoline is not complete, then some of the chemical energy will never enter the process.

The assumption used in obtaining $1 - T_c/T_h$, that the intake and exhaust gases have the same value of γ, hides other possibilities. Combustion breaks up the gasoline and oxygen molecules into a larger number of simpler water and CO_2 molecules, raising γ. By increasing γ and N, combustion slightly improves the efficiency. For this reason a gasoline engine can be more efficient than a methane engine, and also more efficient after the gasoline is refined to remove double carbon bonds (which would raise γ for the fuel).

□   □   □   □

9. Discuss the change in entropy of a gas that expands (a) at constant temperature and (b) adiabatically.

Answer

(a) The expanding gas is doing work. If it is ideal, its constant temperature implies constant internal energy, and it must be taking in energy by heat equal in amount to the work it is doing. As energy enters the gas by heat its entropy increases. The change in entropy is in fact $\Delta S = nR \ln(V_f/V_i)$.

(b) In a reversible adiabatic expansion there is no entropy change. We can say this is because the heat input is zero, or we can say it is because the temperature drops to compensate for the volume increase. In an irreversible adiabatic expansion the entropy increases. In a free expansion the change in entropy is again $\Delta S = nR \ln(v_f/v_i)$.

SOLUTIONS TO SELECTED END-OF-CHAPTER PROBLEMS

5. A particular heat engine has a mechanical power output of 5.00 kW and an efficiency of 25.0%. The engine expels 8.00×10^3 J of exhaust energy in each cycle. Find (a) the energy taken in during each cycle and (b) the time interval for each cycle.

Solution

Conceptualize: Visualize the input energy as dividing up into work output and wasted output. The exhaust by heat must be the other three-quarters of the energy input, so the input will be a bit more than 10 000 J in each cycle. If each cycle took one second, the mechanical output would be around 3 000 J each second. A cycle must take less than a second for the useful output to be 5 000 J each second.

Categorize: We use the first law and the definition of efficiency to solve part (a). We use the definition of power for part (b).

Analyze: In $Q_h = W_{eng} + |Q_c|$ we are given that $|Q_c| = 8\ 000$ J.

(a) We have $e = \dfrac{W_{eng}}{|Q_h|} = \dfrac{|Q_h| - |Q_c|}{|Q_h|} = 1 - \dfrac{|Q_c|}{|Q_h|} = 0.250$.

Isolating $|Q_h|$, we have $|Q_h| = \dfrac{|Q_c|}{1-e} = \dfrac{8\ 000 \text{ J}}{1-0.250} = 10.7$ KJ. ∎

(b) The work per cycle is $W_{eng} = |Q_h| - |Q_c| = 2\ 667$ J.

From the definition of output power, $P = \dfrac{W_{eng}}{\Delta t}$,

we have the time for one cycle $\Delta t = \dfrac{W_{eng}}{P} = \dfrac{2\ 667 \text{ J}}{5\ 000 \text{ J/s}} = 0.533$ s. ∎

Finalize: It would be correct to write efficiency as useful output power divided by total input power. This would amount to thinking about energy transfer per second instead of energy transfer per cycle.

═══════════════

7.   One of the most efficient heat engines ever built is a coal-fired steam turbine in the Ohio River valley, operating between 1 870°C and 430°C. (a) What is its maximum theoretical efficiency? (b) The actual efficiency of the engine is 42.0%. How much mechanical power does the engine deliver if it absorbs 1.40×10^5 J of energy each second from its hot reservoir?

Solution

Conceptualize: The brakes on your car can have 100% efficiency in converting kinetic energy into internal energy. We do not get so much as 50% efficiency from this heat engine. Its useful output power will be far less than 10^5 joules each second.

Categorize: We use the Carnot expression for maximum possible efficiency, and the definition of efficiency to find the useful output.

Analyze: The engine is a steam turbine in an electric generating station with

$$T_c = 430°C = 703 \text{ K} \qquad \text{and} \qquad T_h = 1\,870°C = 2\,143 \text{ K}$$

(a) $e_C = \dfrac{\Delta T}{T_h} = \dfrac{1\,440 \text{ K}}{2\,143 \text{ K}} = 0.672$ or   67.2%   ∎

(b) $e = W_{eng}/|Q_h| = 0.420$ and $|Q_h| = 1.40 \times 10^5$ J

for one second of operation, so $W_{eng} = 0.420|Q_h| = 5.88 \times 10^4$ J.

and the power is

$$P = \frac{W_{eng}}{\Delta t} = \frac{5.88 \times 10^4 \text{ J}}{1 \text{ s}} = 58.8 \text{ kW} \qquad ∎$$

Finalize: Your most likely mistake is forgetting to convert the temperatures to kelvin. Consumers buying electricity in the Midwest pay for the fuel and the operation of the turbine and the generator it runs. The wasted heat output makes a very small contribution to global warming, and the carbon dioxide output makes a larger contribution. Special technology may be required to get coal to burn so hot. The plant may put out sulfur and nitrogen oxides that contribute to acid rain in the northeastern states and the Maritime Provinces.

========

11. An ideal gas is taken through a Carnot cycle. The isothermal expansion occurs at 250°C, and the isothermal compression takes place at 50.0°C. The gas takes in 1.20×10^3 J of energy from the hot reservoir during the isothermal expansion. Find (a) the energy expelled to the cold reservoir in each cycle and (b) the net work done by the gas in each cycle.

Solution

Conceptualize: The two answers will add up to 1 200 J, and may be approximately equal to each other.

Categorize: We will use just the Carnot expression for the maximum efficiency. The temperatures quoted are the maximum and minimum temperatures.

Analyze:

(a) For a Carnot cycle, $\quad e_c = 1 - \dfrac{T_c}{T_h}$.

For any engine, $\quad e = \dfrac{W_{eng}}{|Q_h|} = 1 - \dfrac{|Q_c|}{|Q_h|}$.

Therefore, for a Carnot engine, $\quad 1 - \dfrac{T_c}{T_h} = 1 - \dfrac{|Q_c|}{|Q_h|}$.

Then we have $\quad |Q_c| = |Q_h|\left(\dfrac{T_c}{T_h}\right) = (1\ 200\ \text{J})\left(\dfrac{323\ \text{K}}{523\ \text{K}}\right) = 741\ \text{J}.$ ∎

(b) The work we can calculate as

$$W_{eng} = |Q_h| - |Q_c| = 1\ 200\ \text{J} - 741\ \text{J} = 459\ \text{J}$$ ∎

Finalize: We could equally well compute the efficiency numerically as our first step. We could equally well use $W_{eng} = eQ_h$ in the last step. Seeing that 459 J is considerably less than 741 J, we see that the efficiency of this Carnot cycle is considerably less than 50%.

========

22. An ideal refrigerator or ideal heat pump is equivalent to a Carnot engine running in reverse. That is, energy $|Q_c|$ is taken in from a cold reservoir and energy $|Q_h|$ is rejected to a hot reservoir. (a) Show that the work that must be supplied to run the refrigerator or heat pump is

$$W = \frac{T_h - T_c}{T_c}|Q_c|$$

(b) Show that the coefficient of performance (COP) of the ideal refrigerator is

$$COP = \frac{T_c}{T_h - T_c}$$

Solution

Conceptualize: The Carnot cycle makes the highest-performance refrigerator as well as the highest-efficiency heat engine.

Categorize: We solve the problem by using $|Q_h|/|Q_c| = T_h/T_c$, the first-law equation accounting for total energy input equal to energy output, and the definition of refrigerator COP.

Analyze:

(a) For a complete cycle $\Delta E_{int} = 0$, and

$$W = |Q_h| - |Q_c| = |Q_c|\left(\frac{|Q_h|}{|Q_c|} - 1\right)$$

For a Carnot cycle (and only for a Carnot cycle), $|Q_h|/|Q_c| = T_h/T_c$.

Then, $W = |Q_c|(T_h - T_c)/T_c$.   ∎

(b) The coefficient of performance of a refrigerator is defined as $COP = \dfrac{|Q_c|}{W}$,

so its best possible value is $COP = \dfrac{T_c}{T_h - T_c}$.   ∎

Finalize: The result of part (b) says that it can be easy to pump heat across a small temperature difference, but the refrigerator's effectiveness goes to zero as the low temperature it produces approaches absolute zero. Does thermodynamics seem to pull a rabbit out of a hat? From Joule and Kelvin meeting by chance at a Swiss waterfall, through Planck and Einstein at the beginning of the twentieth century, to you, physicists have been impressed with how an axiom-based chain of logic can have such general things to say about order and disorder, energy, and practical devices.

---

26. How much work does an ideal Carnot refrigerator require to remove 1.00 J of energy from liquid helium at 4.00 K and expel this energy to a room-temperature (293 K) environment?

Solution

Conceptualize: The refrigerator must lift the joule of heat across a large temperature difference, so many joules of work must be put in to make it run.

Categorize: We use the definition of a refrigerator's coefficient of performance and the identification of its maximum possible value.

Analyze: As the chapter text notes and as problem 20 proves,

$$(\text{COP})_{\text{Carnot refrig}} = \frac{T_c}{\Delta T} = \frac{4.00 \text{ K}}{293 \text{ K} - 4.00 \text{ K}} = 0.013\ 8 = \frac{|Q_c|}{W}$$

so

$$W = \frac{|Q_c|}{\text{COP}} = \frac{1.00 \text{ J}}{0.013\ 8} = 72.2 \text{ J}$$ ∎

Finalize: Liquid nitrogen at 77 K sells for about the same price as beer, but a cryogenic refrigerator at liquid-helium temperatures is an expensive thing to run.

---

29. Calculate the change in entropy of 250 g of water warmed slowly from 20.0°C to 80.0°C. (*Suggestion:* Note that $dQ = mc\, dT$.)

Conceptualize: Following the suggestion, we note that a 60°C change in temperature is also a 60 K change in temperature. With $c \sim 4\ 000$ J/kg · K and ¼ kg of water, we estimate an entropy change of ~ 200 J/K.

Categorize: We use the energy version of the nonisolated system model. To do the heating reversibly, put the water pot successively into contact with reservoirs at temperatures $20.0°C + \delta$, $20.0°C + 2\delta$, ... 80.0°C, where δ is some small increment.

Analyze: The change in entropy is given by

$$\Delta S = \int_i^f \frac{dQ}{T} = \int_{T_i}^{T_f} mc \frac{dT}{T}$$

Here T means the absolute temperature. We would ordinarily think of dT as the change in the Celsius temperature, but one Celsius degree of temperature change is the same size as one kelvin of change, so dT is also the change in absolute T.

$$\Delta S = mc \ln T \big|_{T_i}^{T_f} = mc \ln\left(\frac{T_f}{T_i}\right)$$

$$\Delta S = (0.250 \text{ kg})(4\,186 \text{ J/kg} \cdot \text{K}) \ln\left(\frac{353 \text{ K}}{293 \text{ K}}\right) = 195 \text{ J/K}$$

Finalize: Our estimate was good.

---

31. Prepare a table like Table 28.1 by using the same procedure (a) for the case in which you draw three marbles from your bag rather than four and (b) for the case in which you draw five marbles rather than four.

Solution

Conceptualize: At one roulette wheel in Monte Carlo in the 1920s, the red won twenty-something times in succession. Identifying possibilities in this problem will show how that sort of spontaneous order becomes less and less probable when the number of particles goes up.

Categorize: Recall that the table describes the results of an experiment of drawing a marble from the bag, recording its color, and then returning the marble to the bag before the next marble is drawn. Therefore the probability of the result of one draw is constant. There is only one way for the marbles to be all red, but we will list several ways for more even breaks to show up.

Analyze:

(a)

Result	Possible Combinations	Total
All red	RRR	1
2R, 1G	RRG, RGR, GRR	3
1R, 2G	RGG, GRG, GGR	3
All green	GGG	1

(b)

Result	Possible Combinations	Total
All red	RRRRR	1
4R, 1G	RRRRG, RRRGR, RRGRR, RGRRR, GRRRR	5
3R, 2G	RRRGG, RRGRG, RGRRG, GRRRG, RRGGR, RGRGR, GRRGR, RGGRR, GRGRR, GGRRR	10
2R, 3G	GGGRR, GGRGR, GRGGR, RGGGR, GGRRG, GRGRG, RGGRG, GRRGG, RGRGG, RRGGG	10
1R, 4G	RGGGG, GRGGG, GGRGG, GGGRG, GGGGR	5
All green	GGGGG	1

Finalize: The 'possible combinations' in the middle column are microstates and the 'results' in the first column are macrostates. The table in (b) shows that it is a good bet that drawing five marbles will yield three of one color and two of the other.

41. A 1.00-mol sample of H_2 gas is contained in the left side of the container shown in Figure P18.41, which has equal volumes left and right. The right side is evacuated. When the valve is opened, the gas streams into the right side. (a) What is the entropy change of the gas? (b) Does the temperature of the gas change? Assume the container is so large that the hydrogen behaves as an ideal gas.

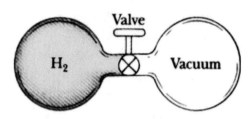

Figure P18.41

Solution

Conceptualize: One mole is only a couple of grams, so we expect an entropy increase of only a few joules per kelvin.

Categorize: The process is irreversible, so it must create entropy. We must think of a reversible process that carries the gas to the same final state. Then the definition of entropy change applied to that reversible process will give us the answer. We need not think about the environment outside the container, because it is unchanged.

Analyze:

(a)   This is an example of free expansion; from the chapter text we have

$$\Delta S = nR \ln\left(\frac{V_f}{V_i}\right)$$

$$\Delta S = (1.00 \text{ mole})(8.314 \text{ J/mole} \cdot \text{K}) \ln\left(\frac{2}{1}\right)$$

$$\Delta S = 5.76 \text{ J/K} \quad \text{or} \quad \Delta S = 1.38 \text{ cal/K}$$

(b)   The gas is expanding into an evacuated region. Therefore, $W = 0$. It expands so fast that energy has no time to flow by heat: $Q = 0$. But $\Delta E_{int} = Q + W$, so in this case $\Delta E_{int} = 0$. For an ideal gas, the internal energy is a function of the temperature and no other variables, so with $\Delta E_{int} = 0$, the temperature remains constant.

Finalize: When the gas was confined in one half of the container, we could have let it blow through a turbine as it expanded to double in volume. The amount of its entropy increase in the process considered here can be thought of as a measure of the loss of opportunity to extract work from the original system.

45. Energy transfers by heat through the exterior walls and roof of a house at a rate of 5.00×10^3 J/s = 5.00 kW when the interior temperature is 22.0°C and the outside temperature is 5.00°C. (a) Calculate the electric power required to maintain the interior temperature at 22.0°C if the power is used in electric resistance heaters that convert all the energy transferred in by electrical transmission into internal energy. (b) **What If?** Calculate the electric power required to maintain the interior temperature at 22.0°C if the power is used to drive an electric motor that operates the compressor of a heat pump that has a coefficient of performance equal to 60.0% of the Carnot-cycle value.

Solution

Conceptualize: To stay at constant temperature, the house must take in 5 000 J by heat every second, to replace the 5 kW that the house is losing to the cold environment. The electric heater should be 100% efficient, so P = 5 kW in part (a). It sounds as if the heat pump is only 60% efficient, so we might expect P = 9 kW in (b).

Categorize: Power is the rate of energy transfer per unit of time, so we can find the power in each case by examining the energy input as heat required for the house as a nonisolated system in steady state.

Analyze:

(a) We know that $P_{\text{electric}} = T_{ET}/\Delta t$, where T_{ET} stands for energy transmitted electrically. All of the energy transferred into the heater by electrical transmission becomes internal energy, so

$$P_{\text{electric}} = \frac{T_{ET}}{\Delta t} = 5.00 \text{ kW} \qquad \blacksquare$$

(b) Now let T stand for absolute temperature.

For a heat pump, $(\text{COP})_{\text{Carnot}} = \dfrac{T_h}{\Delta T} = \dfrac{295 \text{ K}}{27.0 \text{ K}} = 10.93.$

$$\text{Actual COP} = (0.600)(10.93) = 6.56 = \frac{|Q_h|}{W} = \frac{|Q_h|/\Delta t}{W/\Delta t}$$

Therefore, to bring 5 000 W of heat into the house only requires input power

$$P_{\text{heat pump}} = \frac{W}{\Delta t} = \frac{|Q_h|/\Delta t}{\text{COP}} = \frac{5\,000 \text{ W}}{6.56} = 763 \text{ W} \qquad \blacksquare$$

Finalize: The result for the electric heater's power is consistent with our prediction, but the heat pump actually requires **less** power than we expected. Since both types of heaters use electricity to operate, we can now see why it is more cost effective to use a heat pump even though it is less than 100% efficient!

═══════════

47. In 1816, Robert Stirling, a Scottish clergyman, patented the *Stirling engine,* which has found a wide variety of applications ever since. Fuel is burned externally to warm one of the engine's two cylinders. A fixed quantity of inert gas moves cyclically between the cylinders, expanding in the hot one and contracting in the cold one. Figure P18.47 represents a model for its thermodynamic cycle. Consider n moles of an ideal monatomic gas being taken once through the cycle, consisting of two isothermal processes at temperatures $3T_i$ and T_i and two constant-volume processes. Let us find the efficiency of this engine. (a) Find the energy transferred by heat into the gas during the

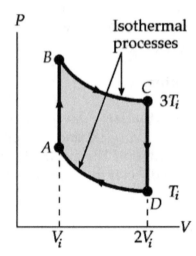

Figure P18.47

isovolumetric process *AB*. (b) Find the energy transferred by heat into the gas during the isothermal process *BC*. (c)Find the energy transferred by heat into the gas during the isovolumetric process *CD*. (d) Find the energy transferred by heat into the gas during the isothermal process *DA*. (e) Identify which of the results from parts (a) through (d) are positive and evaluate the energy input to the engine by heat. (f) From the first law of thermodynamics, find the work done by the engine. (g) From the results of parts (e) and (f), evaluate the efficiency of the engine. A Stirling engine is easier to manufacture than an internal combustion engine or a turbine. It can run on burning garbage. It can run on the energy transferred by sunlight and produce no material exhaust. Stirling engines are not presently used in automobiles due to long startup times and poor acceleration response.

Solution

Conceptualize: A Carnot engine operating between these temperatures would have efficiency 2/3 = 67%. The actual engine must have lower efficiency, maybe 30%.

Categorize: We must think about energy inputs and outputs by heat and work for each process, add up heat input and work output for the whole cycle, and use the definition of efficiency.

Analyze: The internal energy of a monatomic ideal gas is $E_{\text{int}} = \dfrac{3}{2}nRT$.

(a) In the constant-volume processes the work is zero, so

$$Q_{AB} = \Delta E_{\text{int},AB} = nC_V\Delta T = \frac{3}{2}nR(3T_i - T_i) = 3nRT_i \qquad \blacksquare$$

(b) For an isothermal process $\Delta E_{\text{int}} = 0$ and

$$Q = -W = nRT\ln\left(\frac{V_f}{V_i}\right)$$

Therefore, $Q_{BC} = nR\,(3T_i)\ln 2$. $\qquad \blacksquare$

(c) Similarly, $Q_{CD} = \Delta E_{\text{int},CD} = \dfrac{3}{2}nR(T_i - 3T_i) = -3nRT_i$, $\qquad \blacksquare$

(d) and $Q_{DA} = nRT_i\ln\left(\dfrac{1}{2}\right) = -nRT_i\ln 2$. $\qquad \blacksquare$

(e) Q_h is the sum of the positive contributions to Q_{net}. In processes AB and BC energy is taken in from the hot source, amounting to

$$Q_h = Q_{AB} + Q_{BC} = 3nRT_i + 3nRT_i\ln 2 = 3nRT_i(1 + \ln 2) \qquad \blacksquare$$

(f) Since the change in temperature for the complete cycle is zero,

$$\Delta E_{\text{int}} = 0 \quad \text{and} \quad W_{\text{eng}} = Q_{\text{net}} = Q_{AB} + Q_{BC} + Q_{CD} + Q_{DA},$$

so $\quad W_{\text{eng}} = 2nRT_i\ln 2$. $\qquad \blacksquare$

(g) Therefore the efficiency is $e = \dfrac{W_{\text{eng}}}{Q_h} = \dfrac{2\ln 2}{3(1 + \ln 2)} = 0.273 = 27.3\%$. $\qquad \blacksquare$

Finalize: Our estimate was good. The Sterling engine need have no material exhaust, but it has energy exhaust.

48. An idealized diesel engine operates in a cycle known as the *air-standard diesel* cycle shown in Figure P18.48. Fuel is sprayed into the cylinder at the point of maximum compression, *B*. Combustion occurs during the expansion *B→C*, which is modeled as an isobaric process. Show that the efficiency of an engine operating in this idealized diesel cycle is

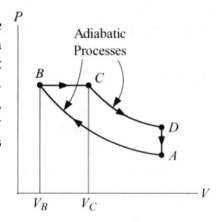

Figure P18.48

$$e = 1 - \frac{1}{\gamma}\left(\frac{T_D - T_A}{T_C - T_B}\right)$$

Solution

Conceptualize: One reasonable feature of the result to be derived is that it is less than 1.

Categorize: We will do an analysis of heat input and output for each process, then add up heat input and work output for the whole cycle, and use the definition of efficiency.

Analyze: The energy transfers by heat over the paths *CD* and *BA* are zero since they are adiabats.

Over path *BC* : $Q_{BC} = nC_p(T_c - T_B) > 0$

Over path *DA* : $Q_{DA} = nC_V(T_A - T_D) < 0$

Therefore, $|Q_c| = |Q_{DA}|$ and $|Q_h| = Q_{BC}$.

Hence, the efficiency is

$$e = 1 - \frac{|Q_c|}{Q_h} = 1 - \left(\frac{T_D - T_A}{T_c - T_B}\right)\frac{C_V}{C_P}$$

$$e = 1 - \frac{1}{\gamma}\left(\frac{T_D - T_A}{T_C - T_B}\right)$$

Finalize: Easier than you expected? Shorter, at least? The high temperature is T_C. Since it is in the bottom of the fraction in the negative term, raising that temperature will raise the efficiency, just as in a Carnot cycle.

58. A 1.00-mol sample of a monatomic ideal gas is taken through the cycle shown in Figure P18.58. At point A, the pressure, volume, and temperature are P_i, V_i, and T_i, respectively. In terms of R and T_i, find (a) the total energy entering the system by heat per cycle, (b) the total energy leaving the system by heat per cycle, and (c) the efficiency of an engine operating in this cycle. (d) Explain how the efficiency compares with that of an engine operating in a Carnot cycle between the same temperature extremes.

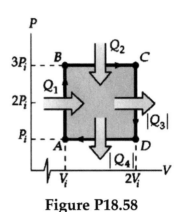

Figure P18.58

Solution

Conceptualize: This real engine will have lower efficiency than a comparable Carnot engine.

Categorize: The *PV* diagram shows the pressures and volumes at the beginning and end of each process. We will find the temperature at each corner. Then we will find the heat, input or output, for each process. That will tell us enough to identify the work output and heat input for the cycle, so that we can find the efficiency from its definition.

Analyze: At point A, $P_i V_i = nRT_i$ with $n = 1.00$ mol.

At point B, $3P_i V_i = nRT_B$ so $T_B = 3T_i$.

At point C, $(3P_i)(2V_i) = nRT_C$ and $T_C = 6T_i$.

At point D, $P_i(2V_i) = nRT_D$ and $T_D = 2T_i$.

We find the energy transfer by heat for each step in the cycle using

$C_V = \dfrac{3}{2}R$ and $C_P = \dfrac{5}{2}R$:

$$Q_1 = Q_{AB} = nC_V(3T_i - T_i) = 3nRT_i$$

$$Q_2 = Q_{BC} = nC_P(6T_i - 3T_i) = 7.5nRT_i$$

$$Q_3 = Q_{CD} = nC_V(2T_i - 6T_i) = -6nRT_i$$

$$Q_4 = Q_{DA} = nC_P(T_i - 2T_i) = -2.5nRT_i$$

(a) Therefore, $Q_{in} = |Q_h| = Q_{AB} + Q_{BC} = 10.5nRT_i$. ∎

(b) $Q_{out} = |Q_c| = |Q_{CD} + Q_{DA}| = 8.5nRT_i$ ■

(c) $e = \dfrac{|Q_h| - |Q_c|}{|Q_h|} = 0.190 = 19.0\%$ ■

(d) The Carnot efficiency is $e_C = 1 - \dfrac{T_c}{T_h} = 1 - \dfrac{T_i}{6T_i} = 0.833 = 83.3\%.$ ■

Finalize: The net work output is only $10.5nRT_i - 8.5nRT_i = 2nRT_i$. The actual efficiency is a lot less than the Carnot efficiency. Irreversible processes go on as the gas absorbs and gives up heat to hot and cold reservoirs across large temperature differences.

---

62. A system consisting of n moles of an ideal gas with molar specific heat at constant pressure C_P undergoes two reversible processes. It starts with pressure P_i and volume V_i, expands isothermally, and then contracts adiabatically to reach a final state with pressure P_i and volume $3V_i$. (a) Find its change in entropy in the isothermal process. (The entropy does not change in the adiabatic process.) (b) **What If?** Explain why the answer to part (a) must be the same as the answer to problem 60. (You do not need to solve Problem 60 to answer this question.)

Solution

Conceptualize: The final state has larger volume than the initial state. The molecules' motion has more randomness, so the entropy will be higher.

Categorize: We will use the equations describing the shape of the isothermal and adiabatic curves to find the volume after the isothermal expansion. Then $nR \ln(V_f/V_i)$ will tell us the change of entropy in the isothermal process.

Analyze: The diagram shows the isobaric process considered in problem 60 as AB. The processes considered in this problem are AC and CB.

(a) For the isotherm (AC), $P_A V_A = P_C V_C.$

For the adiabat (CB), $P_C V_C^{\gamma} = P_B V_B^{\gamma}.$

Combining these equations by substitution gives

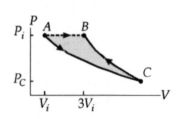

$$V_C = \left(\frac{P_B V_B^{\gamma}}{P_A V_A}\right)^{1/(\gamma-1)} = \left[\left(\frac{P_i}{P_i}\right)\frac{(3V_i)^{\gamma}}{V_i}\right]^{1/(\gamma-1)} = \left(3^{\gamma/(\gamma-1)}\right)V_i$$

Therefore,

$$\Delta S_{AC} = nR \ln\left(\frac{V_C}{V_A}\right) = nR \ln\left[3^{\gamma/(\gamma-1)}\right] = \frac{nR\gamma \ln 3}{\gamma - 1}$$ ∎

(b) Since the change in entropy is path independent, $\Delta S_{AB} = \Delta S_{AC} + \Delta S_{CB}$.

But because (CB) is adiabatic, $\Delta S_{CB} = 0$.

Then $\Delta S_{AB} = \Delta S_{AC}$.

The answer to problem 60 was stated as $\Delta S_{AB} = nC_P \ln 3$.

Because $\gamma = C_P/C_V$,   we have $C_V = C_P/\gamma$,
along with $C_P - C_V = R$,   this gives $C_P - C_P/\gamma = R$.
So $\gamma C_P - C_P = \gamma R$ and $C_P = \gamma R / (\gamma - 1)$.

Thus, the answers to problems 60 and 62 are in fact equal. ∎

Finalize: For one case this has been an explicit demonstration of the general principle that entropy change depends only on the initial and final states, not on the path taken between them. Entropy is a *function of state*. Corollaries: The total change in entropy for a system undergoing a cycle must be zero. The change in a system's entropy for a process $A \rightarrow B$ is the negative of its change in entropy for the process $B \rightarrow A$.

Chapter 19
Electric Forces and Electric Fields

NOTES FROM SELECTED CHAPTER SECTIONS

Section 19.2   Properties of Electric Charges

Electric charge has the following important properties:

- There are two kinds of charges in nature, positive and negative, with the property that unlike charges attract one another and like charges repel one another.

- Charge is conserved.

- Charge is quantized.

Section 19.3   Insulators and Conductors

Conductors are materials in which electric charges move freely under the influence of an electric field; **insulators** are materials that do not readily transport charge.

Section 19.4   Coulomb's Law

Experiments show that an electric force between a pair of point charges is:

- of equal magnitude and opposite direction on the two charges;

- attractive if the charges are of opposite sign and repulsive if the charges have the same sign;

- inversely proportional to the square of the separation, r, between the two charges and is along the line joining them;

- proportional to the product of the magnitudes of the charges.

Section 19.5   Electric Fields

An electric field exists at some point if a test charge placed at that point experiences an electrical force.

The electric field vector $\vec{\mathbf{E}}$ at some point in space is defined as the electric force $\vec{\mathbf{F}}_e$ acting on a positive test charge placed at that point divided by the magnitude of the test charge q_0.

At any point the total electric field created by a group of discrete point charges equals the vector sum of the electric fields due to each of the individual charges.

Section 19.6   Electric Field Lines

Electric field lines are a convenient graphical representation of the electric field in the vicinity of a group of charges or a charged object. At any point along a field line, the tangent to the line is in the direction of the electric field vector at that point.

Lines are drawn so that:

- lines begin on a positive charge (or at infinity) and terminate on a negative charge (or at infinity);

- the number of lines leaving a positive charge (or approaching a negative charge) is proportional to the magnitude of the charge;

- no two field lines can cross;

- the electric field vector is tangent to an electric field line at each point;

- The number of lines per unit area through a surface perpendicular to the field lines is proportional to the magnitude of the electric field in that region; $\vec{\mathbf{E}}$ is large when the field lines are close together and small when they are far apart.

Section 19.9   Gauss's Law

Gauss's law states that the net electric flux through a closed gaussian surface is equal to the net charge inside the surface divided by ϵ_0.

When using Gauss's law to calculate an electric field, choose the gaussian surface so that it has the same symmetry as the charge distribution.

Section 19.11   Conductors in Electrostatic Equilibrium

A conductor in electrostatic equilibrium has the following properties:

- The electric field is zero everywhere inside the conductor.

- Any excess charge on an isolated conductor resides entirely on its surface.

- The electric field just outside a charged conductor is perpendicular to the conductor's surface and has a magnitude σ/ϵ_0, where σ is the charge per unit area at that point.

- On an irregularly shaped conductor, charge tends to accumulate at locations where the radius of curvature of the surface is the smallest, that is, at sharp points.

EQUATIONS AND CONCEPTS

The **magnitude of the electrostatic force** between two stationary point charges, q_1 and q_2, separated by a distance r is given by **Coulomb's law**.

$$F_e = k_e \frac{|q_1||q_2|}{r^2} \qquad (19.1)$$

$$k_e = \frac{1}{4\pi\epsilon_0} = 8.99 \times 10^9 \ \text{N} \cdot \text{m}^2/\text{C}^2$$

$$\epsilon_0 = 8.85 \times 10^{-12} \ \text{C}^2/\text{N} \cdot \text{m}^2$$

The **direction of the electrostatic force** on each charge is determined from the experimental observation that like sign charges experience forces of mutual repulsion and unlike sign charges attract each other. *By virtue of Newton's third law, the magnitude of the force on each of the two charges is the same regardless of the relative magnitude of the charges.*

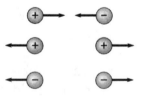

Like sign charges repel; unlike sign charges attract.

The **vector form of Coulomb's law** includes a unit vector. The unit vector $\hat{\mathbf{r}}_{12}$ is directed from q_1 to q_2 and $\vec{\mathbf{F}}_{12}$ is the force on q_2 due to q_1. *Regardless of the relative magnitudes of the two charges, $\vec{\mathbf{F}}_{21} = -\vec{\mathbf{F}}_{12}$; this follows from Newton's third law.*

$$\vec{\mathbf{F}}_{12} = k_e \frac{q_1 q_2}{r^2} \hat{\mathbf{r}}_{12} \qquad (19.2)$$

The **resultant force** on any given charge within a group of charges is the vector sum of the forces exerted on that charge by the remaining individual charges present. This is an example of the *principle of superposition.*

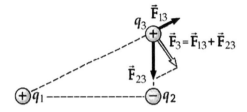

The open arrow shows the resultant force on q_3 due to q_1 and q_2.

The **electric field at any point** in space is defined as the ratio of electric force per unit charge exerted on a small positive test charge placed at the point where the field is to be determined. In the figure $\vec{\mathbf{E}}$ is the electric field produced by the *source charge +Q* and not the field due to the *test charge q_0.*

$$\vec{\mathbf{E}} \equiv \frac{\vec{\mathbf{F}}_e}{q_0} \qquad (19.3)$$

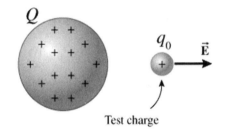

Test charge

The direction of the electric field at any point is the direction of force on a positive test charge placed at the point.

The **electric field a distance r from a point charge** is given by Equation 19.5. The unit vector $\hat{\mathbf{r}}$ is directed away from q and toward the point where the field is to be calculated.

$$\vec{\mathbf{E}} = k_e \frac{q}{r^2} \hat{\mathbf{r}} \qquad (19.5)$$

The direction of the electric field is radially outward from a positive point charge and radially inward toward a negative charge.

The **superposition principle** holds when the electric field at a point is due to a number of point charges.

$$\vec{\mathbf{E}} = k_e \sum_i \frac{q_i}{r_i^2} \hat{\mathbf{r}}_i \qquad \text{(vector sum)} (19.6)$$

The **electric field** of a **continuous charge distribution** is found by integrating over the entire region that contains the charge. This is a vector operation and can usually be carried out easily when the charge is distributed uniformly along a line, over a surface, or throughout a volume.

$$\vec{\mathbf{E}} = k_e \int \frac{dq}{r^2} \hat{\mathbf{r}} \qquad (19.7)$$

The concept of **charge density** is utilized to perform the integration described above. It is convenient to represent a charge increment dq as the product of an element of length, area, or volume and the charge density over that region.

For an element of:

length (dx): $dq = \lambda\, dx$
area (dA): $dq = \sigma\, dA$
volume (dV): $dq = \rho\, dV$

For **uniform charge distributions** the charge densities can be calculated from the total charge and the total volume, area, or length.

$$\lambda \equiv \frac{Q}{\ell} \qquad \sigma \equiv \frac{Q}{A} \qquad \rho \equiv \frac{Q}{V}$$

For non-uniform distributions:
λ, σ, and ρ must be stated
as functions of position.

Electric flux equals the product of the magnitude of the field and the projection of the area onto a plane perpendicular to the direction of the field. Equation 19.13 gives the value of the flux through a plane area when a uniform field makes an angle θ with the normal to the surface. *Electric flux is a scalar quantity and has SI units of* N $\cdot$ m²/C.

$$\Phi_E = EA\cos\theta \qquad (19.13)$$

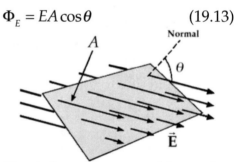

Electric flux is a measure of the number of electric field lines that penetrate a surface.

The general definition of electric flux is stated in terms of an integral of a dot product which must be evaluated over the surface penetrated by the electric field.

$$\Phi_E = \int_{\text{surface}} \vec{\mathbf{E}} \cdot d\vec{\mathbf{A}} \qquad (19.14)$$

The **net flux through a closed surface** is proportional to the *net number of lines leaving the surface.*

$$\Phi_E = \oint E_n\, dA \qquad (19.15)$$

Gauss's law states that the net flux through any closed surface equals the net charge enclosed by the surface divided by the constant ϵ_0. The net flux is independent of the shape and size of the surface. *The surface is called a "gaussian" surface.*

$$\Phi_E = \oint \vec{E} \cdot d\vec{A} = \frac{q_{in}}{\epsilon_0} \qquad (19.17)$$

The **magnitude of the electric field due to symmetric and uniform charge distributions** can be determined by evaluating Equation 19.17 over an appropriate Gaussian surface.
Some examples are given below.

For a point P located:

The electric field is given by:

a **distance** r **from a point charge** q

$$E = k_e \frac{q}{r^2}$$

exterior to a uniformly charged insulating sphere of radius a and charge Q (for $r > a$)

$$E = k_e \frac{Q}{r^2}$$

interior to a uniformly charged insulating sphere with total charge Q (for $r < a$)

$$E = k_e \left(\frac{Q}{a^3} \right) r$$

outside a thin uniformly charged spherical shell of radius a and charge Q (for $r > a$)

$$E = k_e \frac{Q}{r^2}$$

a **distance** r **from an infinitely long uniform line of charge** with linear charge density λ

$$E = 2k_e \frac{\lambda}{r}$$

any distance from an infinite plane of charge with surface charge density σ

$$E = \frac{\sigma}{2\epsilon_0}$$

just outside the surface of a charged conductor in equilibrium with surface charge density σ

$$E_n = \frac{\sigma}{\epsilon_0}$$

SUGGESTIONS, SKILLS, AND STRATEGIES

Problem-Solving Hints for Electric Forces and Fields

- **Units:** When performing calculations that involve the use of the Coulomb constant k_e that appears in Coulomb's law, charges must be in coulombs and distances in meters. If they are given in other units, you must convert them to SI units.

- **Applying Coulomb's law to point charges:** It is important to use the superposition principle properly when dealing with a collection of interacting point charges. When several charges are present, the resultant force on any one of them is found by finding the individual force that every other charge exerts on it and then finding the vector sum of all these forces. The magnitude of the force that any charged object exerts on another is given by Coulomb's law, and the direction of the force is found by noting that the forces are repulsive between like charges and attractive between unlike charges.

- **Calculating the electric field due to a group of point charges:** Remember that the superposition principle can be applied to electric fields, which are also vector quantities. To find the total electric field at a given point, first calculate the electric field at the point due to each individual charge. The resultant field at the point is the vector sum of the fields due to the individual charges.

- **Calculating the electric field due to a continuous charge distribution:** To evaluate the electric field of a continuous charge distribution, it is convenient to employ the concept of charge density. Charge density can be written in different ways: charge per unit volume, ρ; charge per unit area, σ; or charge per unit length, λ. The total charge distribution is then subdivided into a small element of volume dV, area dA, or length dx. Each element contains an increment of charge dq (equal to ρdV, σdA, or λdx). If the charge is nonuniformly distributed over the region, then the charge densities must be written as functions of position.

  For example, if the charge density along a line or long bar is proportional to the distance from one end of the bar, then the linear charge density could be written as $\lambda = bx$ (where b is the constant of proportionality) and the charge increment dq becomes $dq = bxdx$.

- **Symmetry:** Whenever dealing with either a distribution of point charges or a continuous charge distribution, take advantage of any symmetry in the system to simplify your calculations.

Problem-Solving Hints for Gauss's Law

Gauss's law is a very powerful theorem, which relates any charge distribution to the resulting electric field at any point in the vicinity of the charge. In this chapter you should learn how to apply Gauss's law to those *cases in which the charge distribution has a sufficiently high degree of symmetry*. As you review each of the examples of Gauss's law presented in **Section 19.10** of the text, observe carefully how the gaussian surface was chosen to have the characteristics listed below. The Gaussian surface:

- **must be closed;**

- should have the **same symmetry as the charge distribution** (e.g. line, plane, sphere or point charge);

- must include the **point where the electric field is to be calculated**.

In each of the examples notice the manner in which the surface was divided into regions of area *such that over each region the electric field has a constant value*. Each of the separate regions should satisfy one or more of the following conditions:

- $\vec{E} \perp d\vec{A}$ so that $EdA \cos \theta = 0$.

- $\vec{E} \parallel d\vec{A}$ so that $EdA \cos \theta = EdA$.

- $\vec{E}$ and $d\vec{A}$ are oppositely directed so that $EdA \cos \theta = -EdA$.

- $\vec{E} = 0$ (as would be the case over a surface inside a conductor).

In calculating the total charge enclosed by the Gaussian surface it is often convenient to represent an element of charge (dq) in terms of the charge density (λ, σ, or ρ):

$dq = \lambda dx$ for a line of charge

$dq = \sigma dA$ for a chaged surface

$dq = \rho dV$ for volume of charge

The left and right sides of Gauss's law ($\Phi_E = \oint \vec{E} \cdot d\vec{A} = \dfrac{q_{in}}{\epsilon_0}$) can now be evaluated.

You can then calculate the electric field on the gaussian surface, assuming the charge distribution is given in the problem. Conversely, if the electric field is known, you can calculate the charge distribution that produces the field.

REVIEW CHECKLIST

- Describe the fundamental properties of electric charge and the nature of electrostatic forces between charged bodies.

  Use Coulomb's law to determine the net electrostatic force on a point electric charge due to a known distribution of a finite number of point charges.

- Calculate the electric field $\vec{E}$ (magnitude and direction) at a specified location in the vicinity of a group of point charges.

- Calculate the electric field due to a continuous charge distribution. The charge may be distributed uniformly or nonuniformly along a line, over a surface, or throughout a volume.

- Calculate the electric flux through a surface; in particular, find the net electric flux through a closed surface.

- Understand that a gaussian surface is a closed mathematical surface; the surface may be on or within a conductor, an insulator, or in space. Also remember that the net electric flux through a closed gaussian surface is equal to the net charge enclosed by the surface divided by the constant ϵ_0.

  Use Gauss's law to evaluate the electric field at points in the vicinity of charge distributions which exhibit spherical, cylindrical, or planar symmetry.

  Review carefully Examples 19.10, 19.11, and 19.12 in Section 19.10 of the text.

ANSWER TO AN OBJECTIVE QUESTION

4. A particle with charge q is located inside a cubical gaussian surface. No other charges are nearby. **(i)** If the particle is at the center of the cube, what is the flux through each one of the faces of the cube? (a) 0 (b) $q/2\epsilon_0$ (c) $q/6\epsilon_0$ (d) $q/8\epsilon_0$ (e) depends on the size of the cube **(ii)** If the particle can be moved to any point within the cube, what maximum value can the flux through one face approach?

Choose from the same possibilities as in part (i).

Answer **(i)** The cube has six faces (front, back, left, right, top, and bottom). The particle at the center of the cube sends equal amounts of electric flux through each face. Gauss's law says that the total flux is q/ϵ_0. Then, the flux through each separate face is $q/6\epsilon_0$, answer (c).

(ii) To maximize the flux through the top face of the cube, move the charge to a position just below the center of the top face, still barely inside the cube. Just half of the flux created by the charge is associated with an electric field with an upward vertical component, and the other half is flux of a field that has a downward component. As the distance from the charge up to the top of the cube approaches zero, the top of the cube will intercept essentially all of the flux that goes through a hemisphere above the charge. The amount of this flux is one half of q/ϵ_0, so the answer is (b).

ANSWERS TO SELECTED CONCEPTUAL QUESTIONS

5. (a) Would life be different if the electron were positively charged and the proton were negatively charged? (b) Does the choice of signs have any bearing on physical and chemical interactions? Explain your answers.

Answer (a) No, life would not be different.

(b) The character and effect of electric forces is defined by (1) the fact that there are only two types of electric charge—positive and negative, and (2) the fact that opposite charges attract, while like charges repel. The choice of which sign is which is completely arbitrary.

As a related exercise, you might consider how the world would be very different if there were three types of electric charge, or if opposite charges repelled and like charges attracted.

9. A person is placed in a large hollow metallic sphere that is insulated from ground. (a) If a large charge is placed on the sphere, will the person be harmed upon touching the inside of the sphere? (b) Explain what will happen if the person also has an initial charge whose sign is opposite that of the charge on the sphere.

Answer (a) The metallic sphere is a good conductor, so any excess charge on the sphere will reside on the outside of the sphere. From Gauss's law, we know that the field inside the sphere will then be zero. As a result, when the uncharged person touches the inside of the sphere, no charge will be exchanged between the person and the sphere, and the person will not be harmed.

(b) What happens, then, if the person has an initial charge? Regardless of the sign of the person's initial charge, the charges in the conducting surface will redistribute themselves to maintain a net zero charge within the **conducting metal.** Thus, if the person has a 5.00-μC charge on his skin, exactly –5.00 μC will gather on the inner surface of the sphere, so that the electric field inside the metal will be zero. When the person touches the metallic sphere, then he will receive a shock due only to the charge on his own skin.

□   □   □   □

12. If the total charge inside a closed surface is known but the distribution of the charge is unspecified, can you use Gauss's law to find the electric field? Explain.

Answer   No. If we wish to use Gauss's law to find the electric field, we must be able to bring the electric field $\vec{E}$ out of the integral. This can be done in some cases—when the field is constant, for example. However, since we do not know the charge distribution, we cannot claim that the field is constant, and thus cannot find the electric field.

To illustrate this point, consider a sphere that contains a net charge of 100 μC. The charges could be located near the center, or they could all be grouped at the northernmost point within the sphere. Between the two cases, the net electric flux would be the same, but the electric field would vary greatly.

□   □   □   □

15. A common demonstration involves charging a rubber balloon, which is an insulator, by rubbing it on your hair, and touching the balloon to a ceiling or wall, which is also an insulator. Because of the electrical attraction between the charged balloon and the neutral wall, the balloon sticks to the wall. Imagine now that we have two infinitely large flat sheets of insulating material. One is charged and the other is neutral. If these sheets are brought into contact, does an attractive force exist between them, as there was for the balloon and the wall?

Answer   There will not be an attractive force. There are two factors to consider in the attractive force between a balloon and a wall, or between any pair of charged and neutral objects. The first factor is that the molecules in the wall will orient themselves with their positive ends toward the balloon, and their negative ends pointing away from the balloon. The second factor to consider is that the balloon is of finite size, and thus the molecules in the wall are in a nonuniform electric field. Therefore the nearby "positive ends" of the molecules in the wall will experience an attractive electrostatic force that will be greater in magnitude than the repulsive force exerted on the more distant "negative ends" of the molecules. The net result is an overall force of attraction.

Now consider the infinite sheets brought into contact. The polarization of the molecules in the neutral sheet will indeed occur, as in the wall. But the electric field from the charged sheet is **uniform**, and therefore is independent of the distance from the sheet. Thus, both the negative and positive charges in the neutral sheet will experience the same electric field and the same magnitude of electric force. The attractive force on the positive charges will cancel with the repulsive force on the negative charges, and there will be no net force.

□   □   □   □

SOLUTIONS TO SELECTED END-OF-CHAPTER PROBLEMS

3. Nobel laureate Richard Feynman (1918–1988) once said that if two persons stood at arm's length from each other and each person had 1% more electrons than protons, the force of repulsion between them would be enough to lift a "weight" equal to that of the entire Earth. Carry out an order-of-magnitude calculation to substantiate this assertion.

Solution

Conceptualize: We can think of this problem as showing that in a kind of controlled-experiment fair comparison, the electrical force exerted by one ordinary object on another is much stronger than the gravitational force. Stated differently, the problem will demonstrate that the imbalances between electron and proton numbers in macroscopic charged objects are much less than one-percent imbalances.

Categorize: We will estimate the charges on the two people in coulombs and then use Coulomb's law to calculate the magnitude of the force each exerts on the other.

Analyze: Suppose each person has mass 70 kg. In terms of elementary charges, each person consists of precisely equal numbers of protons and electrons and a nearly equal number of neutrons. The electrons comprise very little of the mass, so for each person we find the total number of protons and neutrons, taken together:

$$(70 \text{ kg}) \left(\frac{1 \text{ u}}{1.67 \times 10^{-27} \text{ kg}} \right) = 4 \times 10^{28} \text{ u}$$

Of these, nearly one half, 2×10^{28}, are protons, and 1% of this is 2×10^{26}, constituting a charge of $(2 \times 10^{26})(1.60 \times 10^{-19} \text{ C}) = 3 \times 10^7 \text{ C}$.

Thus, Feynman's force has magnitude

$$F = \frac{k_e q_1 q_1}{r^2} = \frac{(8.99 \times 10^9 \text{ N} \cdot \text{m}^2/\text{C}^2)(3 \times 10^7 \text{ C})^2}{(0.5 \text{ m})^2} \sim 10^{26} \text{ N}$$

where we have used a half-meter arm's length. According to the particle in a gravitational field model, if the Earth were in an externally-produced uniform gravitational field of magnitude 9.80 m/s², it would weigh $F_g = mg = (6 \times 10^{24} \text{ kg})(10 \text{ m/s}^2) \sim 10^{26}$ N.

Thus, the forces are of the same order of magnitude.

Finalize: In the interactions of electrons and protons, the electrical force is vastly larger than the gravitational force. For the gravitational force that the Earth exerts on you to be larger than an electrical force, the net charges on you and the Earth must be very close to zero.

─────────────

9. Three charged particles are located at the corners of an equilateral triangle as shown in Figure P19.9. Calculate the total electric force on the 7.00-μC charge.

Solution

Conceptualize: The 7.00-μC charge experiences a repulsive force $\vec{F}_1$ due to the 2.00-μC charge, and an attractive force $\vec{F}_2$ due to the –4.00-μC charge, where $F_2 = 2F_1$. If we sketch vectors representing $\vec{F}_1$ and $\vec{F}_2$ and their sum, $\vec{F}$ (see the second diagram), we find that the resultant appears to be about the same magnitude as F_2 and is directed to the right about 30.0° below the horizontal.

Categorize: We can find the net electric force by adding the two separate forces acting on the 7.00-μC charge. These individual forces can be found by applying Coulomb's law to each pair of charges.

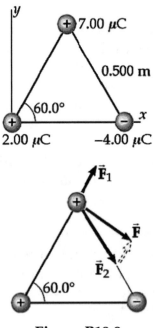

Figure P19.9

Analyze: The force exerted on the 7.00-μC charge by the 2.00-μC charge is

$$\vec{\mathbf{F}}_1 = k_e \frac{q_1 q_2}{r^2}\hat{\mathbf{r}}$$

$$= \frac{(8.99 \times 10^9 \text{ N} \cdot \text{m}^2/\text{C}^2)(7.00 \times 10^{-6} \text{ C})(2.00 \times 10^{-6} \text{ C})}{(0.500 \text{ m})^2}$$

$$\times (\cos 60°\hat{\mathbf{i}} + \sin 60°\hat{\mathbf{j}})$$

$$\vec{\mathbf{F}}_1 = (0.252\,\hat{\mathbf{i}} + 0.436\,\hat{\mathbf{j}}) \text{ N}$$

Similarly, the force on the 7.00-μC charge by the –4.00-μC charge is

$$\vec{\mathbf{F}}_2 = k_e \frac{q_1 q_3}{r^2}\hat{\mathbf{r}}$$

$$= -\frac{(8.99 \times 10^9 \text{ N} \cdot \text{m}^2/\text{C}^2)(7.00 \times 10^{-6} \text{ C})(-4.00 \times 10^{-6} \text{ C})}{(0.500 \text{ m})^2}$$

$$\times (\cos 60°\hat{\mathbf{i}} - \sin 60°\hat{\mathbf{j}})$$

$$\vec{\mathbf{F}}_2 = (0.503\hat{\mathbf{i}} - 0.872\,\hat{\mathbf{j}}) \text{ N}$$

Thus, the total force on the 7.00-μC charge, expressed as a set of components, is

$$\vec{\mathbf{F}} = \vec{\mathbf{F}}_1 + \vec{\mathbf{F}}_2 = (0.755\,\hat{\mathbf{i}} - 0.436\,\hat{\mathbf{j}}) \text{ N}$$

We can also write the total force as

$$\vec{\mathbf{F}} = \sqrt{(0.755 \text{ N})^2 + (0.436 \text{ N})^2} \text{ at } \tan^{-1}\left(\frac{0.436 \text{ N}}{0.755 \text{ N}}\right) \text{ below the } +x \text{ axis}$$

$$\vec{\mathbf{F}} = 0.872 \text{ N at } 30.0° \text{ below the } +x \text{ axis} \qquad\blacksquare$$

Finalize: Our calculated answer agrees with our initial estimate. An equivalent approach to this problem would be to find the net electric field at the location of the upper charge due to the two lower charges, and then apply $\vec{\mathbf{F}} = q\vec{\mathbf{E}}$ to find the force on the upper charge in this electric field.

---

15. A uniformly charged ring of radius 10.0 cm has a total charge of 75.0 μC. Find the electric field on the axis of the ring at (a) 1.00 cm, (b) 5.00 cm, (c) 30.0 cm, and (d) 100 cm from the center of the ring.

Solution

Conceptualize: At the very center ($x = 0$) the field is zero because it has no direction to point in. At an infinite distance the field is zero again. We should see evidence for a maximum field somewhere between answers (b) and (c). The field points away from the ring along its axis, because the ring has uniformly distributed positive charge. Different elements of charge in the ring contribute to the total field components that sum to zero parallel to the face of the ring. The components perpendicular to the face of the ring add together to form the total field along the axis.

Categorize: We repeatedly evaluate the result of Example 19.6.

Analyze: We may particularize the result of Example 19.6 to

$$E = \frac{k_e x Q}{\left(x^2 + a^2\right)^{3/2}}$$

$$= \frac{(8.99 \times 10^9 \text{ N} \cdot \text{m}^2/\text{C}^2)(75.0 \times 10^{-6} \text{ C})x}{\left(x^2 + (0.100 \text{ m})^2\right)^{3/2}}$$

$$= \frac{\left(6.74 \times 10^5 \text{ N} \cdot \text{m}^2/\text{C}\right)x}{\left(x^2 + (0.100 \text{ m})^2\right)^{3/2}}$$

Now just use your calculator to substitute different values for x.

(a) At $x = 0.010\,0$ m, $\vec{\mathbf{E}} = 6.64 \times 10^6 \hat{\mathbf{i}}$ N/C. ∎

(b) At $x = 0.050\,0$ m, $\vec{\mathbf{E}} = 2.41 \times 10^7 \hat{\mathbf{i}}$ N/C. ∎

(c) At $x = 0.300$ m, $\vec{\mathbf{E}} = 6.40 \times 10^6 \hat{\mathbf{i}}$ N/C. ∎

(d) At $x = 1.00$ m, $\vec{\mathbf{E}} = 6.64 \times 10^5 \hat{\mathbf{i}}$ N/C. ∎

Finalize: We have specified the field as being in the positive x direction at points on the positive x axis. The field points away from the ring on both sides. The 24 meganewtons per coulomb at 5 cm is indeed larger than the 6 or 0.6 MN/C at other points. Make sure you have gotten enough practice in evaluating a quantity and raising it to the 3/2 power.

18. A thin rod of length ℓ and uniform charge per unit length λ lies along the x axis, as shown in Figure P19.18. (a) Show that the electric field at P, a distance d from the rod along its perpendicular bisector, has no x component and is given by $E = 2k_e\lambda \sin\theta_0/d$. (b) **What if?** Using your result to part (a), show that the field of a rod of infinite length is $E = 2k_e\lambda/d$.

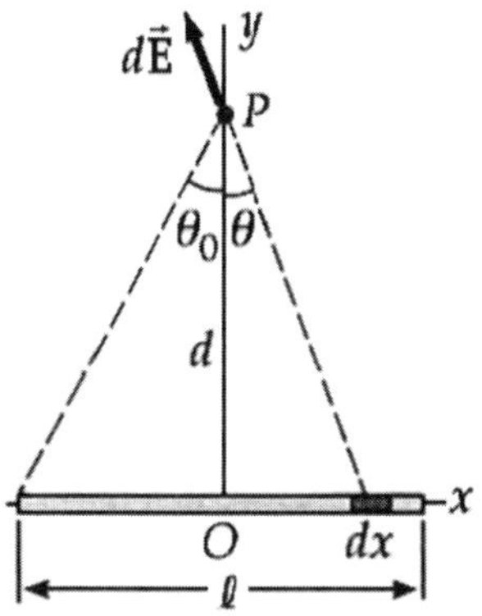

Figure P19.18

Solution

Conceptualize: The proportionalities of the field to the charge and to the Coulomb's-law constant k_e are reasonable. The proportionality to $1/d$ instead of $1/d^2$ is a bit remarkable.

Categorize: We calculate the field at P due to an element of the rod of length dx, which has a charge λdx. Then we do an integral to include the contributions of all bits of the charge to the field. We could say it is obvious that the x component of the field above the rod's midpoint is zero, but we will also prove it explicitly.

Analyze:

(a) The segment of rod from x to $(x + dx)$ has a charge of λdx, and creates an electric field upward along the line from dq to P:

$$d\vec{\mathbf{E}} = \frac{k_e dq}{r^2}\hat{\mathbf{r}} = \frac{k_e\lambda dx}{d^2 + x^2}\hat{\mathbf{r}}$$

This bit of field has an x component

$$dE_x = \frac{k_e\lambda dx}{d^2 + x^2}(-\sin\theta) = \frac{-k_e\lambda x\, dx}{\left(d^2 + x^2\right)^{3/2}}$$

and a y component $\quad dE_y = \frac{k_e\lambda\, dx}{d^2 + x^2}(\cos\theta) = \frac{k_e\lambda d\, dx}{\left(d^2 + x^2\right)^{3/2}}.$

The total field has an x component $E_x = \int\limits_{\text{All } q} dE_x = \int\limits_{-\ell/2}^{\ell/2} -\frac{k_e\lambda x\, dx}{\left(d^2 + x^2\right)^{3/2}}.$

To integrate, we make the change of variables to θ, such that $x = d\tan\theta$.

When $x = -\ell/2$, $\theta = -\theta_0$, and when $x = -\ell/2$, $\theta = \theta_0$.

Further, $(d^2 + x^2)^{3/2} = (d^2 + d^2 \tan^2 \theta)^{3/2} = d^3 \sec^3 \theta$ and $dx = d \sec^2 \theta d\theta$.

Thus, $E_x = \int_{-\theta_0}^{\theta_0} \dfrac{-k_e \lambda d (\tan \theta) d \left(\sec^2 \theta \right) d\theta}{d^3 \sec^3 \theta} = -\dfrac{k_e \lambda}{d} \int_{-\theta_0}^{\theta_0} \sin \theta \; d\theta$:

$$E_x = +\frac{k_e \lambda \cos \theta}{d} \bigg|_{-\theta_0}^{+\theta_0} = \frac{k_e \lambda}{d} \Big[\cos \theta_0 - \cos(-\theta_0) \Big] = \frac{k_e \lambda}{d} \left(\cos \theta_0 - \cos \theta_0 \right) = 0$$

This answer has to be zero because each segment of rod on the left produces a field whose contribution cancels out that of the corresponding segment of rod on the right. But every incremental bit of charge produces at P a contribution to the field with upward y component:

$$E_y = \int_{\text{All } q} dE_y = \int_{-\ell/2}^{\ell/2} \frac{k_e \lambda d \; dx}{\left(d^2 + x^2 \right)^{3/2}}$$

Think of k_e, λ, and d as known constants. Now E_y is the unknown and x is the variable of integration, which we again change to θ, with $x = d \tan \theta$:

$$E_y = \int_{-\theta_0}^{\theta_0} \frac{k_e \lambda d \left(d \sec^2 \theta \right) d\theta}{d^3 \sec^3 \theta} = \frac{k_e \lambda}{d} \int_{-\theta_0}^{\theta_0} \cos \theta \; d\theta = +\frac{k_e \lambda \sin \theta}{d} \bigg|_{-\theta_0}^{+\theta_0}$$

$$E_y = \frac{k_e \lambda}{d} \Big[\sin \theta_0 - \sin(-\theta_0) \Big] = \frac{k_e \lambda}{d} \left(\sin \theta_0 + \sin \theta_0 \right) = \frac{2 k_e \lambda \sin \theta_0}{d} \qquad \blacksquare$$

(b) As ℓ goes to infinity, θ_0 goes to 90° and $\sin \theta_0$ becomes 1. Then the infinite amount of charge produces a finite field at P:

$$\vec{E} = 0\hat{i} + \frac{2 k_e \lambda}{d} \hat{j} \qquad \blacksquare$$

Finalize: The equations we were asked to prove are in fact true. In the next chapter we will see a different derivation of the $2 k_e \lambda / d = \lambda / 2\pi d \epsilon_0$ result.

21. A uniformly charged insulating rod of length 14.0 cm is bent into the shape of a semicircle, as shown in Figure P19.21. The rod has a total charge of –7.50 μC. Find (a) the magnitude and (b) the direction of the electric field at O, the center of the semicircle.

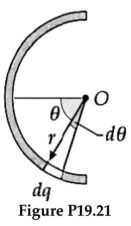

Figure P19.21

Solution

Conceptualize: The bottom and top halves of the shape will create downward and upward fields at O that will add to zero. We expect a net field of some millions of newtons per coulomb straight to the left, toward the midpoint of the negative charge.

Categorize: We will think of an incremental bit of charge as a charged particle, identify its contribution to the field at O, and then integrate to find the total field.

Analyze: Let λ be the charge per unit length. The diagram shows the bit of charge $dq = \lambda ds = \lambda r d\theta$.

The field it creates has magnitude $dE = \dfrac{k_e dq}{r^2} = \dfrac{k_e \lambda r d\theta}{r^2}$.

In component form, $E_y = 0$ (from symmetry) and $dE_x = dE \cos\theta$.

Integrating, $E_x = \displaystyle\int_{\text{all charge}} dE_x = \int \dfrac{k_e \lambda r \cos\theta}{r^2} d\theta$:

$$E_x = \frac{k_e \lambda}{r} \int_{-\pi/2}^{\pi/2} \cos\theta \, d\theta = \frac{k_e \lambda}{r} \sin\theta \Big|_{-\pi/2}^{\pi/2} = \frac{k_e \lambda}{r}[1-(-1)] = \frac{2k_e \lambda}{r}$$

But $Q_{\text{total}} = \lambda \ell$, where $\ell = 0.140$ m and $r = \ell/\pi$.

Thus, $E_x = \dfrac{2\pi k_e Q}{\ell^2} = \dfrac{2\pi\left(8.99\times10^9 \text{ N}\cdot\text{m}^2/\text{C}^2\right)\left(-7.50\times10^{-6} \text{ C}\right)}{(0.140 \text{ m})^2}$.

$\vec{\mathbf{E}} = \left(-2.16\times10^7 \text{ N/C}\right)\hat{\mathbf{i}}$ or (a) 21.6 MN/C (b) to the left in the diagram ∎

Finalize: We were correct about several million newtons per coulomb straight to the left. It is important to think of the field as a vector—that is where the cosine factor in the integrand comes from.

28. Three equal positive charges q are at the corners of an equilateral triangle of side a as shown in Figure P19.28. Assume that the three charges together create an electric field. (a) Sketch the field lines in the plane of the charges. (b) Find the location of one point (other than ∞) where the electric field is zero. What are (c) the magnitude and (d) the direction of the electric field at P due to the two charges at the base?

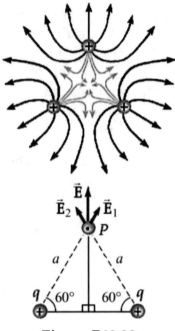

Figure P19.28

Solution

Conceptualize:

(a) The electric field has the general appearance shown by the curving black arrows in the figure. (b) This drawing indicates that $\vec{E} = 0$ at the center of the triangle, since a small positive charge placed at the center of this triangle will be pushed away from each corner equally strongly. This fact could be verified by vector addition as in parts (c) and (d) below.

∎

(c) and (d) The electric field at point P should be directed upwards and about twice the magnitude of the electric field due to just one of the lower charges as shown in the diagram at the right. For these parts of the problem, we must ignore the effect of the charge at point P, because a charge cannot exert a force on itself.

Categorize: The electric field at point P can be found by adding the electric field vectors due to each of the two lower charged particles: $\vec{E} = \vec{E}_1 + \vec{E}_2$.

Analyze:

(c) and (d) The electric field from a charged particle is $\vec{E} = k_e \dfrac{q}{r^2}\hat{\mathbf{r}}$.

As shown in the figure, $\vec{E}_1$ and $\vec{E}_2$ both point 60° above the horizontal,

$\vec{E}_1 = k_e \dfrac{q}{a^2}$ to the right and upward, $\vec{E}_2 = k_e \dfrac{q}{a^2}$ to the left and upward.

$$\vec{E} = \vec{E}_1 + \vec{E}_2 = k_e \frac{q}{a^2}\left[\left(\cos 60°\,\hat{\mathbf{i}} + \sin 60°\,\hat{\mathbf{j}}\right) + \left(-\cos 60°\,\hat{\mathbf{i}} + \sin 60°\,\hat{\mathbf{j}}\right)\right]$$

$$\vec{E} = k_e \frac{q}{a^2}\Big[2\big(\sin 60°\hat{j}\big)\Big] = 1.73 k_e \frac{q}{a^2}\hat{j}$$

■

Finalize: The net electric field at point *P* is indeed nearly twice the magnitude due to a single charge and is entirely vertical as expected from the symmetry of the configuration. In addition to the center of the triangle, the curving gray electric field lines in the figure above indicate three other points near the middle of each leg of the triangle where $\vec{E} = 0$, but they are more difficult to find mathematically.

---

35. A proton moves at 4.50×10^5 m/s in the horizontal direction. It enters a uniform vertical electric field with a magnitude of 9.60×10^3 N/C. Ignoring any gravitational effects, find (a) the time interval required for the proton to travel 5.00 cm horizontally, (b) its vertical displacement during the time interval in which it travels 5.00 cm horizontally, and (c) the horizontal and vertical components of its velocity after it has traveled 5.00 cm horizontally.

Solution

Conceptualize: The proton will maintain a constant horizontal velocity component. In less than a microsecond it could traverse a vacuum tube while moving with a huge vertical acceleration to get some measurable deflection.

Categorize: This is a projectile-motion problem. One component of the motion has zero acceleration and the other constant nonzero acceleration.

Analyze: $\vec{E}$ is directed along the *y* direction; therefore, $a_x = 0$ and $x = v_{xi}t$.

(a) $t = \dfrac{x}{v_{xi}} = \dfrac{0.050\ 0\ \text{m}}{4.50 \times 10^5\ \text{m/s}} = 1.11 \times 10^{-7}\ \text{s}$

■

(b) $a_y = \dfrac{qE_y}{m} = \dfrac{\big(1.60 \times 10^{-19}\ \text{C}\big)\big(9.60 \times 10^3\ \text{N/C}\big)}{1.67 \times 10^{-27}\ \text{kg}} = 9.20 \times 10^{11}\ \text{m/s}^2$

$y = v_{yi}t + \dfrac{1}{2}a_y t^2 = \Big(\dfrac{1}{2}\Big)\big(9.20 \times 10^{11}\ \text{m/s}^2\big)\big(1.11 \times 10^{-7}\ \text{s}\big)^2 = 5.68\ \text{mm}$

■

(c) $v_{xf} = v_{xi} = 4.50 \times 10^5\ \text{m/s}$

■

$v_{yf} = v_{yi} + a_y t = 0 + (9.20 \times 10^{11}\ \text{m/s}^2)(1.11 \times 10^{-7}\ \text{s}) = 1.02 \times 10^5\ \text{m/s}$

■

Finalize: Beyond the problem statement, we have direct evidence that gravity is negligible, from the huge size of the vertical acceleration caused by the electrical force. In a laboratory it is much easier to work with electron beams than with proton beams, but the equations are the same.

═══════════

37. A 40.0-cm-diameter loop is rotated in a uniform electric field until the position of maximum electric flux is found. The flux in this position is measured to be $5.20 \times 10^5 \, \text{N} \cdot \text{m}^2/\text{C}$. What is the magnitude of the electric field?

Solution

Conceptualize: Visualize orienting a flat-plate solar energy collector so that it intercepts maximum flux of sunlight.

Categorize: We use the definition of electric flux.

Analyze: For a uniform field the flux is $\Phi = \vec{E} \cdot \vec{A} = EA \cos \theta$.

The maximum value of the flux occurs when $\theta = 0$. This means, when the field is in the same direction as the area vector, which is defined as having the direction of the perpendicular to the area.

Therefore, we can calculate the field strength at this point as

$$E = \frac{\Phi_{max}}{A} = \frac{\Phi_{max}}{\pi r^2}$$

$$E = \frac{5.20 \times 10^5 \, \text{N} \cdot \text{m}^2/\text{C}}{\pi (0.200 \, \text{m})^2} = 4.14 \times 10^6 \, \text{N/C} \qquad \blacksquare$$

Finalize: This is a straightforward problem emphasizing the point that the flux depends on the strength of the field and the size of the area, but also on the orientation of the field to the area.

═══════════

38. A particle with charge Q is located a small distance δ immediately above the center of the flat face of a hemisphere of radius R as shown in Figure P19.38. What is the electric flux (a) through the curved surface and (b) through the flat face as $\delta \to 0$?

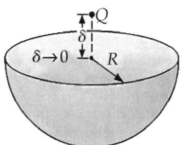

Solution

Conceptualize: From Gauss's law, the flux through a sphere with a charged particle in it should be Q/ϵ_0, so we should expect the electric

Figure P19.38

flux through a hemisphere to be half this value:

$$\Phi_{curved} = Q/2\epsilon_0$$

The flat section appears like an infinite plane to a point just above its surface. Then half of all the field lines from the charged particle are intercepted by the flat surface, so the flux through this section should also equal $Q/2\epsilon_0$.

Categorize: We can apply the definition of electric flux directly for part (a) and then use Gauss's law to find the flux for part (b).

Analyze:

(a) With δ very small, all points on the hemisphere are nearly at distance R from the charge, so the field everywhere on the curved surface is $k_e Q/R^2$ radially outward (normal to the surface). Therefore, the flux is this field strength times the area of half a sphere:

$$\Phi_{curved} = \int \vec{E} \cdot d\vec{A} = E_{local} A_{hemisphere}$$

$$\Phi_{curved} = \left(k_e \frac{Q}{R^2} \right)\left(\frac{1}{2} \right)\left(4\pi R^2 \right) = \frac{1}{4\pi \, \epsilon_0} Q(2\pi) = \frac{Q}{2 \, \epsilon_0} \qquad \blacksquare$$

(b) The closed surface encloses zero charge, so Gauss's law gives

$$\Phi_{curved} + \Phi_{flat} = 0 \qquad \text{or} \qquad \Phi_{flat} = -\Phi_{curved} = \frac{-Q}{2\epsilon_0} \qquad \blacksquare$$

Finalize: The direct calculations of the electric flux agree with our predictions, except for the negative sign in part (b), which comes from the fact that the area unit vector is defined as pointing outward from an enclosed surface, and in this case, the electric field has a component in the opposite direction (down).

48. Consider a long cylindrical charge distribution of radius R with a uniform charge density ρ. Find the electric field at distance r from the axis where $r < R$.

Solution

Conceptualize: At $r = 0$, the field should be zero because it could not point in one direction more than in any other direction. It would make sense for the field to be radially outward at all nonzero values of r, and to grow larger as r increases.

Categorize: We use our knowledge of the direction of the field to choose a gaussian surface. We evaluate both sides of Gauss's law to obtain an equation involving the unknown field magnitude, which we can then solve.

Analyze: If ρ is positive, the field must everywhere be radially outward. Choose as the gaussian surface a cylinder of length L and radius r, contained inside the charged rod. Its volume is $\pi r^2 L$ and it encloses charge $\rho \pi r^2 L$. The circular end caps have no electric flux through them; there $\vec{\mathbf{E}} \cdot d\vec{\mathbf{A}} = E\, dA \cos 90.0° = 0$. The curved surface has $\vec{\mathbf{E}} \cdot d\vec{\mathbf{A}} = E\, dA \cos 0°$, and E must be the same strength everywhere over the curved surface.

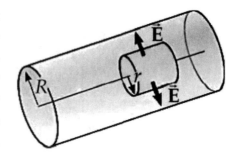

Then $\oint \vec{\mathbf{E}} \cdot d\vec{\mathbf{A}} = \dfrac{q_{inside}}{\epsilon_0}$ becomes $E \displaystyle\int_{\substack{\text{Curved} \\ \text{Surface}}} dA = \dfrac{\rho \pi r^2 L}{\epsilon_0}$.

Noting that $2\pi r^2 L$ is the lateral surface area of the cylinder, we have

$$E(2\pi r)L = \frac{\rho \pi r^2 L}{\epsilon_0}$$

Thus, $\vec{\mathbf{E}} = \dfrac{\rho r}{2\epsilon_0}$ radially away from the axis. ∎

Finalize: We do indeed have zero field on the axis of the rod. The field is strongest at the surface of the rod. Inside it grows proportionately to the radius. Outside it decreases in proportion to $1/r$.

---

53. Consider a thin spherical shell of radius 14.0 cm with a total charge of 32.0 μC distributed uniformly on its surface. Find the electric field (a) 10.0 cm and (b) 20.0 cm from the center of the charge distribution.

Solution

Conceptualize: The field inside the shell should be zero. The field outside the shell should be just like that created by a charged particle at its center.

Categorize: The field must be radially outward and uniform in size over any sphere centered at the center of the shell. Knowing this lets us use spherical gaussian surfaces through the field points to find the field magnitudes there.

Analyze:

(a) A gaussian sphere, radius 10.0 cm, encloses 0 charge. Then at the surface of this gaussian sphere, inside the charged shell, we have $\vec{\mathbf{E}} = 0$. ∎

(b) For a gaussian sphere of radius 20.0 cm, we apply $\oint \vec{\mathbf{E}} \cdot d\vec{\mathbf{A}} = \dfrac{q_{in}}{\epsilon_0}$.

The field is radially outward, and $4\pi r^2 E = q/\epsilon_0$:

$$E = \frac{k_e q}{r^2} = \frac{\left(8.99 \times 10^9 \ \text{N} \cdot \text{m}^2 / \text{C}^2\right)\left(32.0 \times 10^{-6} \ \text{C}\right)}{\left(0.200 \ \text{m}\right)^2} = 7.19 \times 10^6 \ \text{N/C}$$

so $\quad \vec{\mathbf{E}} = (7.19 \times 10^6 \ \text{N/C}) \, \hat{\mathbf{r}}$. ∎

Finalize: Our results agree with our predictions. The chapter-opener photograph in the textbook suggests the situation of the field lines starting on the surface of the sphere.

━━━━━━━━

59. A thin square conducting plate 50.0 cm on a side lies in the *xy* plane. A total charge of 4.00×10^{-8} C is placed on the plate. Find (a) the charge density on each face of the plate, (b) the electric field just above the plate, and (c) the electric field just below the plate. You may assume the charge density is uniform.

Solution

Conceptualize: The sheet of charge creates uniform away-pointing field, upward in the space above the sheet and downward in the space below.

Categorize: We compute the charge density from its definition. We compute the field from the equation that the chapter proves for a charged sheet.

Analyze: We ignore "edge" effects and assume that the total charge distributes itself uniformly over each side of the plate, with one half the total charge on each side.

(a) $\sigma = \dfrac{q}{A} = \dfrac{4.00 \times 10^{-8}\ \text{C}}{2(0.500\ \text{m})^2} = 8.00 \times 10^{-8}\ \text{C/m}^2$ ∎

(b)   Just above the plate,

$$E = \dfrac{\sigma}{\epsilon_0} = \dfrac{8.00 \times 10^{-8}\ \text{C/m}^2}{8.85 \times 10^{-12}\ \text{C}^2/\text{N} \cdot \text{m}^2} = 9.04 \times 10^3\ \text{N/C}\quad \text{upward}$$ ∎

(c)   Just below the plate, $E = \dfrac{\sigma}{\epsilon_0} = 9.04 \times 10^3\ \text{N/C}$ downward.   ∎

Finalize: This is not a trick question. The answers to (b) and (c) have the same numerical value and follow the same rule about direction (that positive charge creates away-pointing electric field throughout the surrounding space, like a crabby porcupine) but they are fair questions. Drawing a picture may help you, but the reasoning required is really three-dimensional reasoning. Think about the quills of a big group of porcupines who are crowded close together on a flat mesh hammock, and are taking care not to poke one another. Their quills point straight up and straight down.

────────────

60. A long, straight wire is surrounded by a hollow metal cylinder whose axis coincides with that of the wire. The wire has a charge per unit length of λ, and the cylinder has a net charge per unit length of 2λ. From this information, use Gauss's law to find (a) the charge per unit length on the inner surface of the cylinder, (b) the charge per unit length on the outer surface of the cylinder and (c) the electric field outside the cylinder, a distance r from the axis.

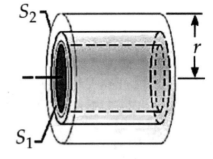

Solution

Conceptualize: Assume λ represents a positive number. The positively charged wire induces a negative charge on the inner surface of the metal shell. The shell has polarization contributing to the charge on its outer surface, as well as a net charge.

Categorize: By considering surface after nesting gaussian surface we will build up our answers.

Analyze:

(a) Use a cylindrical gaussian surface S_1 within the metal of the conducting cylinder. Here $E = 0$.

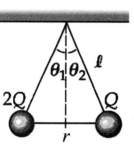

Thus, $\oint E_n \, dA = \left(\dfrac{1}{\epsilon_0} \right) q_{in} = 0$ and $\lambda_{inner} = -\lambda$.

(b) Next, the net charge on the metal pipe is described by

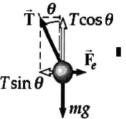

$$\lambda_{inner} + \lambda_{outer} = 2\lambda \quad \text{so} \quad \lambda_{outer} = 3\lambda.$$

(c) For a gaussian surface S_2 outside the conducting cylinder,

$$\oint E_n \, dA = \left(\frac{1}{\epsilon_0} \right) q_{in} \quad \text{or} \quad E(2\pi rL) = \frac{1}{\epsilon_0}(\lambda - \lambda + 3\lambda)L:$$

$$E = \frac{3\lambda}{2\pi \epsilon_0 r} \quad \text{radially away from the axis} \qquad \blacksquare$$

Finalize: In the space, empty of matter, between the wire and the inner wall of the cylinder, the field is $E = \dfrac{\lambda}{2\pi \epsilon_0 r}$ away from the axis. It is unaffected by the metal cylinder. The field must be zero within the metal material, but it can be nonzero in a hollow space enclosed by metal.

═══════════════

72. Two small spheres of mass m are suspended from strings of length ℓ that are connected at a common point. One sphere has charge Q and the other charge $2Q$. The strings make angles θ_1 and θ_2 with the vertical. (a) Explain how θ_1 and θ_2 are related. (b) Assume θ_1 and θ_2 are small. Show that the distance r between the spheres is approximately

$$r \approx \left(\frac{4k_e Q^2 \ell}{mg} \right)^{1/3}$$

Solution

Conceptualize: The formula we are to prove has these reasonable features: The distance between the particles will increase if their charges increase and decrease in a stronger gravitational field. We may be surprised about where the one-third power comes from.

Categorize: Part (a) is about how different the idea of force is from the idea of charge. In part (b) we use the particle in equilibrium model.

Analyze:

(a) The spheres have different charges, but each exerts an equal force on the other, given by $F_e = k_e(Q)(2Q)/r^2$, where r is the distance between them. Because their masses are also equal, $\theta_2 = \theta_1$.   ∎

(b) For equilibrium the equation $\Sigma F_y = 0$ becomes $T\cos\theta - mg = 0$, and similarly $\Sigma F_x = 0$ becomes $F_e - T\sin\theta = 0$.

To solve simultaneously we substitute $T = \dfrac{mg}{\cos\theta}$

to obtain $F_e = \dfrac{mg\sin\theta}{\cos\theta} = mg\tan\theta$.

For small angles, $\tan\theta \approx \sin\theta = \dfrac{r}{2\ell}$.

Therefore, $F_e \approx mg\dfrac{r}{2\ell}$.

The force F_e is $\dfrac{k_eQ(2Q)}{r^2} \approx mg\dfrac{r}{2\ell}$,

so that $4k_eQ^2\ell \approx mgr^3$ and $r \approx \left(\dfrac{4k_eQ^2\ell}{mg}\right)^{1/3}$.   ∎

Finalize: Let   us   check   for   dimensional   correctness: we   have $\left(\dfrac{N\cdot m^2 \cdot C^2 \cdot m}{C^2 \cdot N}\right)^{1/3} = m$ as required. The one-third power means physically that since the electrical force depends so strongly on the distance between the charges, this distance will not change by very much if we change the charge or the mass or the string length or the free-fall acceleration by some factor.

73. Two infinite, nonconducting sheets of charge are parallel to each other as shown in Figure P19.73. The sheet on the left has a uniform surface charge density σ, and the one on the right has a uniform charge density $-\sigma$. Calculate the electric field at points (a) to the left of, (b) in between, and (c) to the right of the two sheets. (d) **What if?** Find the electric fields in all three regions if both sheets have *positive* uniform surface charge densities of value σ.

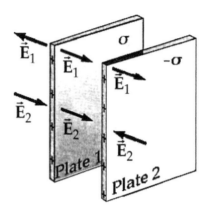

Figure P19.73

Solution

Conceptualize: When the sheets have opposite-sign charges with the same charge density, a positive test charge at a point midway between them will experience the same force in the same direction, as a force of attraction to one sheet and a force of repulsion from the other. Therefore, the electric field here will add up to $E = \sigma/\epsilon_0$, twice as large as a single sheet of charge would produce. Because each infinite sheet produces a field that is uniform in space, the test charge will experience an equally large force everywhere in the region between the plates. Outside the oppositely-charged sheets, the electric fields they separately produce will point in opposite directions, so they add to zero.

Categorize: The principle of superposition can be applied to add the electric field vectors due to each sheet of charge.

Analyze: Each sheet separately creates electric field at all points given by $\vec{E} = \sigma/2\epsilon_0$ directed away from the positive sheet and toward the negative sheet.

(a) At any point to the left of the two parallel sheets,

$$\vec{E} = E_1\left(-\hat{i}\right) + E_2\left(+\hat{i}\right) = -\frac{\sigma}{2\,\epsilon_0}\hat{i} + \frac{\sigma}{2\,\epsilon_0}\hat{i} = 0 \qquad \blacksquare$$

(b) At any point between the two sheets, $\vec{E} = E_1\hat{i} + E_2\hat{i} = 2E\hat{i} = \dfrac{\sigma}{\epsilon_0}\hat{i}.$ $\qquad \blacksquare$

(c) At any point to the right of the two sheets, $\vec{E} = E_1\hat{i} + E_2\left(-\hat{i}\right) = 0.$ $\qquad \blacksquare$

(d) If both sheets are positive, we must reverse all three arrows labeled $\vec{E}_2$ in the diagram. Now at a point to the left the addition is

$$\vec{\mathbf{E}} = E_1\left(-\hat{\mathbf{i}}\right) + E_2\left(-\hat{\mathbf{i}}\right) = -\frac{\sigma}{2\,\epsilon_0}\hat{\mathbf{i}} - \frac{\sigma}{2\,\epsilon_0}\hat{\mathbf{i}} = -\frac{\sigma}{\epsilon_0}\hat{\mathbf{i}}$$ ∎

In between, the total field is $\vec{\mathbf{E}} = E_1\hat{\mathbf{i}} + E_2\left(-\hat{\mathbf{i}}\right) = 0.$ ∎

At a point to the right of both sheets, $\vec{\mathbf{E}} = E_1\hat{\mathbf{i}} + E_2\hat{\mathbf{i}} = 2E\hat{\mathbf{i}} = \frac{\sigma}{\epsilon_0}\hat{\mathbf{i}}.$ ∎

Finalize: We essentially solved parts (a) through (c) in the Conceptualize step, so it is no surprise that the results are what we expected. In Chapter 20 we will identify a pair of oppositely-charged plates as a capacitor. The results of part (d) are complementary to the case where the plates are oppositely charged. Note that the field-creating influence of one sheet can be thought of as reaching right through the other sheet, its charge unimpeded.

═══════════

75. A solid, insulating sphere of radius a has a uniform charge density throughout its volume and a total charge Q. Concentric with this sphere is an uncharged, conducting hollow sphere whose inner and outer radii are b and c, as shown in Figure P19.75. We wish to understand completely the charges and electric fields at all locations. (a) Find the charge contained within a sphere of radius $r < a$. (b) From this value, find the magnitude of the electric field for $r < a$. (c) What charge is contained within a sphere of radius r where $a < r < b$? (d) From this value, find the magnitude of the electric field for r where $a < r < b$. (e) Now consider r where $b < r < c$. What is the magnitude of the electric field for this range of values of r? (f) From this value, what must be the charge on the inner surface of the hollow sphere? (g) From part (f), what must be the charge on the outer surface of the hollow sphere? (h) Consider the three spherical surfaces of radii a, b, and c. Which of these surfaces has the largest magnitude of surface charge density?

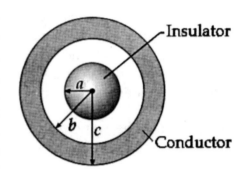

Figure P19.75

Solution

Conceptualize: Suppose Q is positive. The charge on the inner solid sphere will create outward field to polarize the metal shell, inducing negative charge on its inner surface. The net charge of the shell is zero, so it must have an equal amount (Q) of positive charge on its outer surface. The charge density will be predictably lower on the larger outer surface.

Categorize: We consider nesting gaussian surfaces from the inside out. We will use the identification of the field as zero between b and c. We will use the property that the metal has zero total charge.

Analyze: Choose as each gaussian surface a concentric sphere of radius r. The electric field will be perpendicular to its surface, and will be uniform in strength over its surface.

The density of charge in the insulating sphere is $\rho = Q / \left(\dfrac{4}{3} \pi a^3 \right)$.

(a)   The sphere of radius $r < a$ encloses charge

$$\rho = \left(\frac{4}{3} \pi r^3 \right) = Q \left(\frac{4}{3} \pi r^3 \right) / \left(\frac{4}{3} \pi a^3 \right) = Q(r/a)^3 \qquad \blacksquare$$

(b)   Applying Gauss's law to this sphere reveals the magnitude of the field at its surface.

Here $\Phi = q/\epsilon_0$ becomes $E \left(4\pi r^2 \right) = \dfrac{Q(r/a)^3}{\epsilon_0}$ so $E = \dfrac{Qr}{4\pi \epsilon_0 \, a^3} = \dfrac{k_e Q r}{a^3}$. $\qquad \blacksquare$

(c)   For a sphere of radius r with $a < r < b$, the whole insulating sphere is enclosed, so the charge within is Q. $\qquad \blacksquare$

(d)   Gauss's law for this sphere becomes

$$E \left(4\pi r^2 \right) = \frac{Q}{\epsilon_0} \text{ and } E = \frac{Q}{4\pi \epsilon_0 \, r^2} = k_e Q / r^2 \qquad \blacksquare$$

(e)   For $b < r < c$, we must have $E = 0$ because any nonzero field would be moving charges in the metal. $\qquad \blacksquare$

(f)   A spherical gaussian surface within the metal must enclose zero total charge. Part of it is charge Q on the insulating sphere, and the rest of it must be charge $-Q$ on the inner surface of the metal, at radius b. $\qquad \blacksquare$

This induced charge was deposited on the inner wall of the shell when charges were moving in the metal, when the shell was first set up to enclose the charged sphere.

(g)   The shell is uncharged as a whole. With charge $-Q$ on its inner surface and no charge in the bulk of the metal, charge $+Q$ is left on its outer surface, at radius c. $\qquad \blacksquare$

This charge was deposited here in the polarization process, when charges were moving in the metal to reduce the field within it to zero.

(h) The field is strongest at radius a, but the insulating sphere carries charge in the volume of its material. With zero thickness, the surface of radius a has zero volume and zero surface charge density. Then the surface of radius b, smaller in area than the surface of radius c, carries charge with the greatest surface density magnitude. ∎

Finalize: The answers are specified in terms of some of the symbols we take as given information, namely Q, r, a, b, c, and the universal constant $k_0 = 1/4\pi \epsilon_0$. When you pay careful attention to where the zeros are and to the reasons for them, you will be thinking carefully about charge and field, just as you should.

It would be natural to continue the logic and find the field outside the shell, for $r > c$. Here

$$\Phi = q/\epsilon_0 \text{ becomes } E(4\pi r^2) = \frac{Q+Q-Q}{\epsilon_0} \text{ and } E = \frac{Q}{4\pi \epsilon_0 r^2} = k_e Q/r^2.$$

---

76. Review. A negatively charged particle $-q$ is placed at the center of a uniformly charged ring, where the ring has a total positive charge Q as shown in Figure P19.76. The particle, confined to move along the x axis, is moved a small distance x along the axis (where $x \ll a$) and released. Show that the particle oscillates in simple harmonic motion with a frequency given by

$$f = \frac{1}{2\pi}\left(\frac{k_e q Q}{m a^3}\right)^{1/2}$$

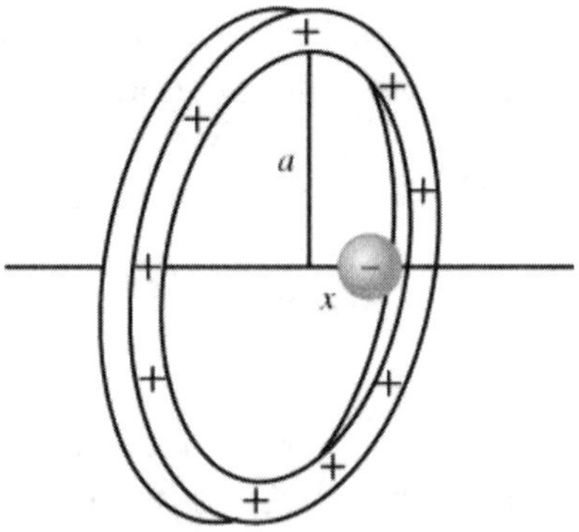

Figure P19.76

Solution

Conceptualize: The negative charge is attracted toward the center of the ring if it is located on the axis on either side, so the restoring force can be expected to cause vibration.

Categorize: To find the frequency we must examine the force law and compare it with $\dfrac{d^2\vec{x}}{dt^2} = -\omega^2\vec{x}$, the defining equation for simple harmonic motion.

Analyze: From Example 19.6, the electric field at points along the x axis is

$$\vec{E} = \frac{k_e x Q \hat{i}}{\left(x^2 + a^2\right)^{3/2}}$$

The field is zero at $x = 0$, so the negatively-charged particle is in equilibrium at this point. When it is displaced by an amount x that is small compared to a,

$$\sum \vec{F} = m\frac{d^2\vec{x}}{dt^2} \quad \text{becomes} \quad (-q)\frac{k_e x Q \hat{i}}{a^3} = m\frac{d^2\vec{x}}{dt^2}.$$

The particle's acceleration is proportional to its displacement (x) from the equilibrium position and is oppositely directed, so it moves in simple harmonic motion. ■

Since the angular frequency of any simple harmonic motion is set by $\frac{d^2\vec{x}}{dt^2} = -\omega^2\vec{x}$, we have by comparison $\omega = \left(\frac{k_e q Q}{ma^3}\right)^{1/2} = 2\pi f$ and

$$f = \frac{1}{2\pi}\left(\frac{k_e q Q}{ma^3}\right)^{1/2}.$$ ■

Finalize: A larger charge on either the particle or the ring would make the restoring force larger and increase the vibration frequency. Larger mass would make the frequency smaller. Larger ring size a would have a bigger effect in reducing the frequency.

Chapter 20
Electric Potential and Capacitance

Section 20.1   Electric Potential and Potential Difference

The electrostatic force is conservative; therefore, it is possible to define an electric potential energy function associated with this force. The change in potential between two points is proportional to the change in potential energy of a charge as it moves between the two points.

Electrical potential difference (a scalar quantity) is the work done to move a charge from point A to point B divided by the magnitude of the charge. Therefore, the SI units of potential are joules per coulomb, and are called volts (V).

A positive charge gains electrical potential energy when it is moved opposite the direction of the electric field. When a positive charge is released in an electric field, it moves in the direction of the field, from a point of high potential to a point of lower potential.

Section 20.2   Potential Differences in a Uniform Electric Field

Electric field lines always point in the direction of decreasing electric potential. When a positive charge moves along the direction of the electric field, the electric potential energy of the charge-field system decreases. An equipotential surface is any surface consisting of a continuous distribution of points having the same electric potential. Equipotential surfaces are perpendicular to electric field lines and no work is required to move any charge between two points on an equipotential surface.

Section 20.3 Electric Potential and Potential Energy Due to Point Charges

In electric circuits, a point of zero potential is often defined by grounding (connecting to Earth) some point in the circuit. In the case of a point charge, the point of zero potential is taken to be at an infinite distance from the charge. The electric potential at a given point in space due to a point charge q depends only on the value of the charge and the distance, r, from the charge to the specified point in space. An electric potential can exist at a point in space whether or not a test charge exists at that point.

Electric potential is a scalar quantity. *When using the superposition principle to determine the value of the electric potential at a point due to several point charges, the algebraic sum of the individual electric potentials must be used. Be careful not to confuse electric potential difference with electric potential energy.*

- Electric potential is a scalar property of the region surrounding an electric charge. It does not depend on the presence of a test charge in the field.

- Electric potential energy is a characteristic of a charge-field system. It is due to the interaction of the field and a charge located within the field.

- In the figure to the right an electric field, due to the presence of $+Q$ (the source charge), is present in the region surrounding the charge. A second charge $+q$ (the test charge) experiences a force due to the electric field produced by $+Q$.

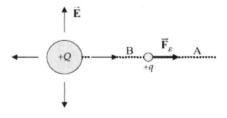

Electric field in vicinity of a point charge

Refer to the figure above as you consider the following statements:

- Electric field lines are directed along outward radials from the charge Q; points A and B are shown along one of the field lines.

- If a positive test charge q is placed at some point between A and B, the electric field will exert a force $F_E = qE$ on q directed along the field line and away from Q.

- If the test charge q is moved from point A to point B (where the potential is greater), the electric field will do work, $W = -q\Delta V = -q(V_B - V_A)$, and the electric potential energy of the two-charge system (Q and q) changes by an amount equal to $q\Delta V$. *The electric potential energy of a positive charge increases when the charge moves to a point closer to another isolated positive point charge (or along a direction opposite the direction of the electric field).*

- *Since the electrostatic force is a conservative force, the work done by the electric field as the test charge q moves from point **A** to point **B** will be the same regardless of the actual path taken between the two points.*

You should consider the consequences of changing either or both Q and q from a positive to a negative charge.

Section 20.4   Obtaining the Value of the Electric Field from the Electric Potential

If the electric potential (which is a scalar) is known as a function of coordinates (x, y, z), the components of the electric field (a vector quantity) can be obtained by taking the negative derivative of the potential with respect to the coordinates.

Section 20.5   Electric Potential Due to Continuous Charge Distributions
Section 20.6   Electric Potential due to a Charged Conductor

The electric potential is a constant everywhere on the surface of a charged conductor in electrostatic equilibrium.

The electric potential is constant everywhere inside a conductor and equal to its value at the surface. This is true because the electric field is zero inside a conductor and no work is required to move a charge between two points inside the conductor.

The electron volt (eV) is defined as the energy that an electron (or proton) gains or loses when accelerated through a potential difference of 1 V.

A surface on which every point has the same value of electric potential is an equipotential surface. The electric field at every point on an equipotential surface is perpendicular to the surface; otherwise work would be required to move a charge between two points on the surface.

Section 20.7 Capacitance

A capacitor is a device consisting of a pair of conductors (plates) separated by insulating material. A charged capacitor acts as a storehouse of charge and energy that can be reclaimed when needed for a specific application. The capacitance of a capacitor depends on the physical characteristics of the device (size, shape, plate separation, and the nature of the medium filling the region between the plates).

The capacitance, C, of a capacitor is defined as the ratio of the magnitude of the charge on either conductor to the magnitude of the potential difference between the conductors. Capacitance has SI units of coulombs per volt, called farads (F). The farad is a very large unit of capacitance. In practice, most capacitors have capacitances ranging from picofarads to microfarads.

Section 20.8 Combinations of Capacitors

Two or more capacitors can be connected in a circuit in several possible combinations. For example, three capacitors (each assumed to have the *same value of capacitance, C*) can be combined as illustrated in the figure at right to achieve four different values of equivalent capacitance.

(a) all in parallel

(b) all in series

(c) 2 in series, in parallel with a third

(d) 2 in parallel, in series with a third

$C_{eq} = 3C$

$C_{eq} = \frac{1}{3}C$

$C_{eq} = \frac{3}{2}C$

$C_{eq} = \frac{2}{3}C$

Four ways in which 3 capacitors can be connected. The equivalent values given in the figure are for the special case when all capacitors in the circuit have a value equal to C.

Can you determine how many different values of equivalent capacitance could be achieved by combining the three capacitors if they *each had a different value of capacitance, C_1, C_2, and C_3?*

Note in the circuit (a) above when a group of capacitors are connected in parallel, each capacitor in the group has two circuit points in common with each of the others.

When they are connected in series (b) they are arranged "end-to-end" and only adjacent capacitors have a single circuit point in common.

For capacitors connected in series:

- The reciprocal of the equivalent capacitance equals the sum of the reciprocals of the individual capacitances, as seen in Equation 20.28, $\dfrac{1}{C_{eq}} = \dfrac{1}{C_1} + \dfrac{1}{C_2} + \dfrac{1}{C_3} + \cdots$

- The potential difference across the series group equals the sum of the potential differences across the individual capacitors.

- Each capacitor of a group in series has the same charge (even if their capacitances are not the same) and this is also the value of the total charge on the series group.

For capacitors connected in parallel:

- The equivalent capacitance equals the sum of the individual capacitances, as seen in Equation 20.26, $C_{eq} = C_1 + C_2 + C_3 + \cdots$

- The potential difference across each capacitor has the same value.

- The total charge on the parallel group equals the sum of the charges on the individual capacitors.

Section 20.9   Energy Stored in a Charged Capacitor

The electrostatic potential energy stored in a charged capacitor equals the work done in the charging process—moving charges from one conductor at a lower potential to another conductor at a higher potential.

Each capacitor has a limiting voltage that depends on the capacitor's physical characteristics. When the potential difference between the plates of a capacitor exceeds that limiting voltage, a discharge will occur through the insulating material between the two plates. This is known as electrical breakdown. Therefore, the maximum energy that can be stored in a capacitor is limited by the breakdown voltage.

Section 20.10   Capacitors with Dielectrics

A dielectric is an insulating material, such as rubber, glass, or waxed paper. When a dielectric is inserted between the plates of a capacitor, the capacitance increases. If the dielectric completely fills the space between the plates, the capacitance is multiplied by a dimensionless factor called the dielectric constant. The dielectric constant (κ) is a property of the dielectric material and has a value of 1.000 for vacuum.

In general, the use of a dielectric has the following effects:

- It increases the capacitance.

- It increases the maximum operating voltage.

- It provides mechanical support for the two conductors.

EQUATIONS AND CONCEPTS

The **potential difference** between two points, $\Delta V = V_B - V_A$, is defined as the change in the potential energy of a charge-field system when a test charge, q_0, is moved from point A to point B divided by the test charge q_0. ΔV can be evaluated by integrating $\vec{\mathbf{E}} \cdot d\vec{\mathbf{s}}$ along any path from A to B. *Electric potential is a characteristic of an electric field, independent of any charges that may be placed in the field. The SI unit of electric potential is the volt (V).*

$$\Delta V \equiv \frac{\Delta U}{q_0} = -\int_A^B \vec{\mathbf{E}} \cdot d\vec{\mathbf{s}} \qquad (20.3)$$

$$1\ \text{V} = 1\ \text{J/C}$$

$$1\ \text{N/C} = 1\ \text{V/m}$$

The **electron volt** (eV) is a unit of energy that is defined as the energy that a charge-field system gains or loses when a charge of magnitude e is moved through a potential difference of 1 V.

$$1\ \text{eV} = 1.602 \times 10^{-19}\ \text{J} \qquad (20.5)$$

In a **uniform electric field**, the potential difference between two points depends on the displacement d along the direction *parallel* to $\vec{E}$ and on the magnitude of $\vec{E}$. *Electric field lines always point in the direction of decreasing electric potential.*

$$\Delta V = -E\int_A^B ds = -Ed \qquad (20.6)$$

(when $d \parallel \vec{E}$ in a uniform field)

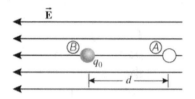

A uniform electric field directed to the left

The **change in potential energy**, ΔU, of a charge-field system as a charge moves from point A to point B in an electric field depends on the sign and magnitude of the charge as well as on the change in potential, ΔV. *The electric potential energy of a charge-field system decreases when a positive charge moves in the direction of the field.*

$$\Delta U = q_0 \Delta V = -q_0 Ed \qquad (20.7)$$

(when $d \parallel \vec{E}$ in a uniform field)

$$\Delta U = -q_0 \vec{E} \cdot \vec{s} \qquad (20.9)$$

(when $\vec{E}$ and $\vec{s}$ are not parallel)

The **electric potential due to a single point charge**, at a distance r from the charge, depends inversely on the distance from the charge. The potential is given by Equation 20.11 when the zero reference level for potential is taken to be at infinity.

$$V = k_e \frac{q}{r} \qquad (20.11)$$

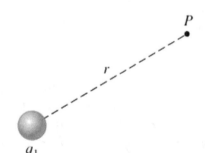

A potential exists at point P due to the presence of charge q_1

The **electric potential in the vicinity of several point charges** is the scalar sum of the potentials due to the individual charges.

$$V = k_e \sum_i \frac{q_i}{r_i} \qquad (20.12)$$

The **electric potential energy associated with a pair of point charges** separated by a distance r_{12}

represents the work required to assemble the charges from an infinite separation. The **negative of the potential energy** equals the minimum work required to separate them by an infinite distance with no final kinetic energy. *U is positive if the charges have the same sign; negative if the signs are opposite.*

$$U = k_e \frac{q_1 q_2}{r_{12}} \qquad (20.13)$$

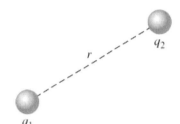

Two point charges separated by a distance r

The **total potential energy of a group of three charges** is found by calculating U for each pair of charges and summing the terms algebraically.

$$\Delta U = k_e \left(\frac{q_1 q_2}{r_{12}} + \frac{q_1 q_3}{r_{13}} + \frac{q_2 q_3}{r_{23}} \right)$$

The **components of the electric field** are the negative partial derivatives of the electric potential.

$$E_x = -\frac{\partial V}{\partial x}, \; E_y = -\frac{\partial V}{\partial y}, \; E_z = -\frac{\partial V}{\partial z}$$
$$(20.17)$$

In the case of a **spherical symmetric charge distribution** the electric field is radial.

$$E_r = -\frac{dV}{dr} \qquad (20.16)$$

For a continuous charge distribution, the electric potential (relative to zero at infinity) can be calculated by integrating the contribution due to a charge element dq over the line, surface, or volume that contains all the charge. It is convenient to represent dq in terms of the appropriate charge density.

$$V = k_e \int \frac{dq}{r} \qquad (20.19)$$

$dq = \lambda \, dx$ for a charged line
$dq = \sigma \, dA$ for a charged surface
$dq = \rho \, dV$ for a charged volume

The **capacitance** of a capacitor is defined as the ratio of the charge on *either conductor* (or plate) to the magnitude of the potential difference between the conductors.

$$C \equiv \frac{Q}{\Delta V} \qquad (20.20)$$

Capacitance has SI units of farads (F).

The **capacitance of an air-filled parallel-plate capacitor** is proportional to the area of the plates and inversely proportional to their separation. The charges on the two plates have equal magnitude and opposite sign.

$$C = \frac{\epsilon_0 A}{d} \qquad (20.22)$$

Parallel plate capacitor.

When a **dielectric material** completely fills the region between the plates of a capacitor, the capacitance increases by a factor κ. *Kappa, called the dielectric constant, is dimensionless and is characteristic of a particular material.*

$$C = \kappa \frac{\epsilon_0 A}{d} \qquad (20.33)$$

For capacitors connected in parallel:
- Each capacitor has two circuit points in common with each of the other capacitors in the group.

Capacitors in parallel

- The **equivalent capacitance** equals the algebraic sum of the individual capacitances.

$$C_{eq} = C_1 + C_2 + C_3 + \ldots \qquad (20.26)$$

- The **total charge** is the sum of charges on the individual capacitors.

$$Q_{total} = Q_1 + Q_2 + Q_3 + \ldots$$

- The **potential difference** across each capacitor is the same.

$$\Delta V_1 = \Delta V_2 = \Delta V_3 = \ldots$$

For capacitors connected in series:
- Adjacent capacitors have one circuit point in common.

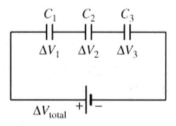

Capacitors in series

- The **inverse of the equivalent capacitance** equals the algebraic sum of the inverses of the capacitances of the individual capacitors.

$$\frac{1}{C_{eq}} = \frac{1}{C_1} + \frac{1}{C_2} + \frac{1}{C_3} + \ldots \quad (20.28)$$

- The **potential difference** across a series group of capacitors equals the sum of the potential differences across the individual capacitors.

$$\Delta V_{total} = \Delta V_1 + \Delta V_2 + \Delta V_3 \ldots$$

- The **total charge** on the group is equal to the charge on each capacitor.

$$Q_{total} = Q_1 = Q_2 = Q_3 \ldots$$

A **series-parallel combination** of capacitors can be arranged using three or more capacitors.

Using Equations 20.26 and 20.28 in sequence, each combination can be reduced to a single equivalent capacitor. If each capacitor in the figure at right has a value C:

on the left $C_{eq} = \dfrac{3C}{2}$,

and on the right $C_{eq} = \dfrac{2C}{3}$.

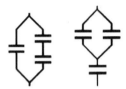

On the left 2 capacitors in series are combined in parallel with a third.

On the right 2 capacitors in parallel are combined in series with a third.

The **electrostatic energy** stored in the electrostatic field of a charged capacitor equals the work done (by a battery or other source) in charging the capacitor from $q = 0$ to $q = Q$.

$$U = \frac{Q^2}{2C} = \frac{1}{2}Q\Delta V = \frac{1}{2}C(\Delta V)^2 \quad (20.29)$$

The **energy density** (energy per unit volume) at any point in an electrostatic field is proportional to the square of the electric field intensity at that point.

$$u_E = \frac{1}{2}\epsilon_0 E^2 \quad (20.31)$$

SUGGESTIONS, SKILLS, AND STRATEGIES

The vector expression giving the components of the electric field over a region can be obtained from the scalar function that describes the electric potential, V, over the region by using the equations below:

$$E_x = -\frac{\partial V}{\partial x} \qquad\qquad E_y = -\frac{\partial V}{\partial y} \qquad\qquad E_z = -\frac{\partial V}{\partial z}$$

The derivatives in the above expressions are called partial derivatives. This means that when the derivative is taken with respect to any one coordinate, any other coordinates that appear in the expression for the potential function are treated as constants.

Since the electrostatic force is a conservative force, the work done by the electrostatic force in moving a charge q from an initial point A to a final point B depends only on the location of the two points and is independent of the path taken between A and B. When calculating potential difference, any path between A and B may be chosen to evaluate the integral in Equation 20.3. You should select a path for which the evaluation of the line integral will be as convenient as possible.

Problem-Solving Hints for Electric Potentials

- When working problems involving electric potential, remember that potential is a scalar quantity (rather than a vector quantity like the electric field), so there are no components to worry about. Therefore, when using the superposition principle to evaluate the electric potential at a point due to a system of point charges, you simply take the algebraic sum of the potentials due to each charge. However, you must keep track of signs.

- The potential created by each positive charge $\left(V = \frac{k_e q}{r} \right)$ is positive, while the potential for each negative charge is negative.

- Just as in mechanics, only changes in electric potential are significant, hence the point where you choose the potential to be zero is arbitrary. When dealing with point charges or a finite-sized charge distribution, we usually define $V = 0$ to be at a point infinitely far from the charges. However, if the charge distribution itself extends to infinity, some other nearby point must be selected as the reference point.

- The electric potential at some point P due to a continuous distribution of charge can be evaluated by dividing the charge distribution into infinitesimal elements of charge dq located at a distance r from the point P. You then treat this element as a point charge, so that the potential at P due to the element is $dV = \dfrac{k_e dq}{r}$.

The total potential at P is obtained by integrating dV over the entire charge distribution. In performing the integration for most problems, it is necessary to express dq and r in terms of a single variable. In order to simplify the integration, it is important to give careful consideration of the geometry involved in the problem.

- Another method that can be used to obtain the potential due to a finite continuous charge distribution is to start with the definition of the potential difference given by Equation 20.3, $\Delta V = -\displaystyle\int_A^B \vec{E} \cdot d\vec{s}$. If $\vec{E}$ is known or can be obtained easily (say from Gauss's law), then the line integral can be evaluated.

- When you know the electric potential at a point, it is possible to obtain the electric field at that point by remembering that the components of the electric field are equal to the negative of the derivatives of the potential with respect to the respective coordinates.

- Note that V is used to denote the value of the electric potential at a point while ΔV is used to denote the potential difference between two specified points. If the electric potential has values V_A and V_B at points A and B, the electric potential difference in moving from point A to point B will be $\Delta V = V_B - V_A$.

In practice, a variety of phrases are used to describe the potential difference between two points, the most common being "voltage." A voltage applied to a device or across a device has the same meaning as the potential difference across the device. For example, if we say that the voltage across a certain capacitor is 12 volts, we mean that the potential difference between the plates of the capacitor is 12 volts.

Problem-Solving Hints for Capacitance

- When analyzing a series-parallel combination of capacitors to determine the equivalent capacitance, you should make a sequence of circuit diagrams, which show the successive steps in the simplification of the circuit. At each step combine and calculate the equivalent capacitance of those capacitors that are in either simple-series or simple-parallel combination with each other. Continue the sequence of simplifying steps until the final circuit contains a single capacitor, the value of which is the equivalent capacitance of the original circuit. At each step, you know two of the three quantities: Q, ΔV, and C. You will be able to determine the remaining quantity using the equation $Q = C\Delta V$.

- When calculating capacitance, be careful with your choice of units. To calculate capacitance in farads, make sure that distances are in meters and use the SI value of ϵ_0. When checking consistency of units, remember that the units for the electric field are newtons per coulomb (N/C) or, equivalently, volts per meter (V/m).

- When two or more *unequal* capacitors are connected in series, they carry the same charge, but their potential differences are not the same. The capacitances add as reciprocals, and the equivalent capacitance of the combination is always less than the smallest individual capacitor.

- When two or more capacitors are connected in parallel, the potential differences across them are the same. The charge on each capacitor is proportional to its capacitance and the capacitances add directly to give the equivalent capacitance of the parallel combination.

- A dielectric increases capacitance by the factor κ (the dielectric constant) because induced surface charges on the dielectric reduce the electric field inside the material from E to $\dfrac{E}{\kappa}$.

- Be careful about problems in which you may be connecting or disconnecting a battery to a capacitor. It is important to note whether modifications to the capacitor are being made while the capacitor is connected to the battery or after it is disconnected. If the capacitor remains connected to the battery, the voltage across the capacitor necessarily remains the same (equal to the battery voltage), and the charge is proportional to the capacitance, however it may be modified (i.e. by insertion of a dielectric). On the other hand, if you disconnect the capacitor from the battery *before* making any modifications to the capacitor, then its charge remains the same. In this case, as you vary the capacitance, the voltage across the plates changes in inverse proportion to capacitance, according to $\Delta V = \dfrac{Q}{C}$.

REVIEW CHECKLIST

- Understand that each point in the vicinity of a charge distribution can be characterized by a scalar quantity called the electric potential, V. The values of this potential function over the region (a scalar) are related to the values of the electrostatic field, $\vec{E}$, over the region (a vector field).

- Calculate the electric potential difference between any two points in a uniform electric field, and the electric potential difference between any two points in the vicinity of a group of point charges.

- Calculate the electric potential energy associated with a group of point charges.

- Obtain an expression for the electric field (a vector quantity) over a region of space if the scalar electric potential function for the region is known.

- Calculate the work done by an external force in moving a charge q between any two points in an electric field when (a) an expression giving the field as a function of position is known, or when (b) the charge distribution (either point charges or a continuous distribution of charge) giving rise to the field is known.

- Determine the equivalent capacitance of a network of capacitors in series-parallel combination and calculate the final charge on each capacitor and the potential difference across each when a known potential is applied across the combination.

- Calculate the capacitance, potential difference, and stored energy of a capacitor that is partially or completely filled with a dielectric.

ANSWER TO AN OBJECTIVE QUESTION

19. (i) What happens to the magnitude of the charge on each plate of a capacitor if the potential difference between the conductors is doubled? (a) It becomes four times larger. (b) It becomes two times larger. (c) It is unchanged. (d) It becomes one-half as large. (e) It becomes one-fourth as large. **(ii)** If the potential difference across a capacitor is doubled, what happens to the energy stored? Choose from the same possibilities as in part (i).

Answer (i) (b) The equation $Q = C\Delta V$ describes how the charge is directly proportional to the applied potential difference. **(ii)** (a) Since $U = C(\Delta V)^2/2$, doubling ΔV will quadruple the stored energy.

ANSWERS TO SELECTED CONCEPTUAL QUESTIONS

7. When charged particles are separated by an infinite distance, the electric potential energy of the pair is zero. When the particles are brought close, the electric potential energy of a pair with the same sign is positive, whereas the electric potential energy of a pair with opposite signs is negative. Give a physical explanation of this statement.

Answer You may remember from the chapter on gravitational potential energy that potential energy of a system is defined to be positive when positive work must have been performed by an external agent to change the system from an initial configuration, to which we assign a zero value of potential energy, to a final configuration. For example, the system of a flag and the Earth has positive potential energy when the flag is raised if we define zero potential energy as the configuration with the flag at the ground, since positive work must be done by an external force in order to lift it from the ground to the top of the pole.

When assembling particles with like charges from an infinite separation, a configuration for which we have defined the potential energy as having a value of zero, it takes work to move them closer together to some distance r; therefore energy is being stored in the system of charges, and the potential energy is positive.

When assembling two particles with opposite charges from an infinite separation, the particles tend to accelerate toward each other, and thus energy is released as they approach a separation at distance r. Therefore, the potential energy of a pair of unlike charges is negative.

□   □   □   □

11. If you were asked to design a capacitor in which small size and large capacitance were required, what would be the two most important factors in your design?

Answer You should use a dielectric-filled capacitor whose dielectric constant is very large. Furthermore, you should make the dielectric as thin as you can, keeping in mind that the maximum allowed voltage should not produce dielectric breakdown of the insulator.

□   □   □   □

SOLUTIONS TO SELECTED END-OF-CHAPTER PROBLEMS

3. (a) Calculate the speed of a proton that is accelerated from rest through an electric potential difference of 120 V. (b) Calculate the speed of an electron that is accelerated through the same potential difference.

Solution

Conceptualize: Since 120 V is only a modest potential difference, we might expect that the final speed of the particles will be substantially less than the speed of light. We should also expect the speed of the electron to be significantly greater than the proton because, with $m_e << m_p$, an equal force on both particles will result in a much greater acceleration for the electron.

Categorize: Conservation of energy of the proton-field system can be applied to this problem to find the final speed from the kinetic energy of the particles. (Review this work-energy theory of motion from Chapter 6 if necessary.)

Analyze:

(a) For the particle-field system, energy is conserved as the proton moves from high to low potential, which can be defined for this problem as moving from 120 V down to 0 V:

$$K_i + U_i = K_f + U_f \qquad \rightarrow \qquad 0 + qV_i = \frac{1}{2}mv_p^2 + 0$$

The initial energy is $qV_i = (1.60 \times 10^{-19} \text{ C})(120 \text{ V})\left(\dfrac{1 \text{ J/C}}{1 \text{ V}}\right) = 1.92 \times 10^{-17}$ J.

The final speed is given by

$$v_p = \left(\frac{2qV_i}{m}\right)^{1/2} = \left[\frac{2\ (1.60 \times 10^{-19}\ \text{C})(120\ \text{J/C})}{1.67 \times 10^{-27}\ \text{kg}}\right]^{1/2} = 1.52 \times 10^5 \text{ m/s} \qquad \blacksquare$$

(b) The electron will gain speed in moving the other way, from $V_i = 0$ to $V_f = 120$ V. Here $K_i + U_i = K_f + U_f$ becomes

$$0 + 0 = \frac{1}{2}mv_e^2 + qV_f = \frac{1}{2}mv_e^2 - eV_f$$

so $\quad v_e = \left(\dfrac{-2qV_f}{m}\right)^{1/2} = \left[\dfrac{-2\ (-1.60 \times 10^{-19}\ \text{C})(120\ \text{J/C})}{9.11 \times 10^{-31}\ \text{kg}}\right]^{1/2} = 6.49 \times 10^6 \text{ m/s.} \quad \blacksquare$

Finalize: Both of these speeds are significantly less than the speed of light, as expected, which also means that we were justified in not using the relativistic kinetic energy formula. (For precision to three significant digits, the relativistic formula is only needed if v is greater than about $0.1c$.)

Both the electron and the proton here are described as having energy of 120 electron volts. Look back at the problem solution to see that each really does have kinetic energy $(e)(120\text{ V}) = 120$ eV. One eV is just a unit, so the energy of any object can be quoted as a number of eV's. But for a beam of charged particles, this specification tells just how high to turn the power supply knob for the accelerating voltage.

---

5. An electron moving parallel to the x axis has an initial speed of 3.70×10^6 m/s at the origin. Its speed is reduced to 1.40×10^5 m/s at the point $x = 2.00$ cm. (a) Calculate the electric potential difference between the origin and that point. (b) Which point is at the higher potential?

Solution

Conceptualize: A coasting skateboarder slows when moving uphill. We assume the electron is moving in vacuum. It slows because it is coasting up to higher electric potential energy. With negative charge, it must be moving to a location with a more negative potential. The starting point is at higher voltage.

Categorize: We use the energy version of the isolated system model to equate the energy of the electron-field system when the electron is at $x = 0$ to the energy when the electron is at $x = 2.00$ cm. The unknown will be the difference in potential $V_f - V_i$.

Analyze: Thus, $K_i + U_i = K_f + U_f$ becomes

$$\frac{1}{2}mv_i^2 + qV_i = \tfrac{1}{2}mv_f^2 + qV_f$$

or $\quad \dfrac{1}{2}m(v_i^2 - v_f^2) = q\,(V_f - V_i),$

so $\quad (V_f - V_i) = \dfrac{m\left(v_i^2 - v_f^2\right)}{2q}.$

(a) Noting that the electron's charge is negative, and evaluating the potential difference, we have

$$V_f - V_i = \frac{(9.11 \times 10^{-31} \text{ kg})\left[(3.70 \times 10^6 \text{ m/s})^2 - (1.40 \times 10^5 \text{ m/s})^2\right]}{2(-1.60 \times 10^{-19} \text{ C})}$$

$$= -38.9 \text{ V}$$ ■

(b) The negative sign means that the 2.00-cm location is lower in potential than the origin. ■

Finalize: Our calculation agrees with our qualitative argument that the origin is the point of higher potential. A particle with positive charge would slow in free flight toward higher voltage, but the negative electron slows as it moves into lower potential. The 2.00-cm distance was unnecessary information for this problem. If the field were uniform, we could use the 2.00-cm distance to find the x component of the field from $\Delta V = -E_x d$.

---

8. Show that the amount of work required to assemble four identical charged particles of magnitude Q at the corners of a square of side s is $5.41 k_e Q^2/s$.

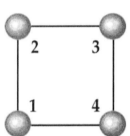

Solution

Conceptualize: Work is required to push the charges together against their mutual repulsion. It is reasonable that its magnitude increases if Q increases and if s decreases—if the charges are squeezed closer together.

Categorize: The work required equals the sum of the potential energies for all pairs of charges. No energy is involved in placing q_4 at a given position in empty space. When q_3 is brought from far away and placed close to q_4, the system potential energy can be expressed as $q_3 V_4$, where V_4 is the potential at the position of q_3 established by charge q_4. When q_2 is brought into the system, it interacts with two other charges, so we have two additional terms $q_2 V_3$ and $q_2 V_4$ in the total potential energy. Finally, when we bring the fourth charge q_1 into the system, it interacts with three other charges, giving us three more energy terms.

Analyze: Thus, the complete expression for the energy is

$$U = q_1 V_2 + q_1 V_3 + q_1 V_4 + q_2 V_3 + q_2 V_4 + q_3 V_4$$

$$U = \frac{q_1 k_e q_2}{r_{12}} + \frac{q_1 k_e q_3}{r_{13}} + \frac{q_1 k_e q_4}{r_{14}} + \frac{q_2 k_e q_3}{r_{23}} + \frac{q_2 k_e q_4}{r_{24}} + \frac{q_3 k_e q_4}{r_{34}}$$

$$U = \frac{Qk_eQ}{s} + \frac{Qk_eQ}{s\sqrt{2}} + \frac{Qk_eQ}{s} + \frac{Qk_eQ}{s} + \frac{Qk_eQ}{s\sqrt{2}} + \frac{Qk_eQ}{s}$$

Evaluating,

$$U = \frac{k_eQ^2}{s}\left(4 + \frac{2}{\sqrt{2}}\right) = 5.41k_e\frac{Q^2}{s}$$ ∎

Finalize: In describing the four charged particles, we never multiply all four or any three together. We think only of pairs, and we think of all of the pairs. That is where the 5.41 factor comes from. The most generally useful things you learn about electricity may be that a volt is a joule per coulomb and the associated definition about electric potential energy being $U = qV$.

11. The three charged particles in Figure P20.11 are at the vertices of an isosceles triangle (where $d = 2.00$ cm). Taking $q = 7.00$ μC, calculate the electric potential at point A, at the midpoint of the base.

Solution

Conceptualize: Being close to two negative charges, the point A probably has a negative potential of some kilovolts.

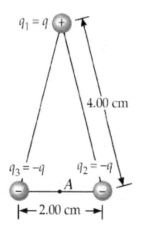

Figure P20.11

Categorize: Each charge creates potential at A on its own. We will add up the contributions.

Analyze: Define $q_1 = q$ and $q_2 = q_3 = -q$.

The charged particles are at distances from A of

$$r_1 = \sqrt{(0.040\ 0\ \text{m})^2 - (0.010\ 0\ \text{m})^2} = 3.87 \times 10^{-2}\ \text{m}$$

and $\qquad r_2 = r_3 = 0.010\ 0$ m.

The potential at point A is

$$V = \frac{k_eq_1}{r_1} + \frac{k_eq_2}{r_2} + \frac{k_eq_3}{r_3} = k_e(q)\left(\frac{1}{r_1} + \frac{-1}{r_2} + \frac{-1}{r_3}\right)$$

Substituting,

$$V = \left(8.99 \times 10^9 \, \frac{\text{N} \cdot \text{m}^2}{\text{C}^2}\right)(7.00 \times 10^{-6} \, \text{C})$$

$$\times \left(\frac{1}{0.038 \, 7 \, \text{m}} - \frac{1}{0.010 \, 0 \, \text{m}} - \frac{1}{0.010 \, 0 \, \text{m}}\right)$$

$$V = -11.0 \times 10^6 \, \text{V} \quad\blacksquare$$

Finalize: A microcoulomb is a large charge for a tabletop object. The net potential is in megavolts rather than in kilovolts. The two negative charges dominate the more distant positive charge in creating the net potential. For review and contrast, let us do this related calculation: Calculate the electric field vector at the same point due to the three charged particles. The separate fields of the two negative charges are in opposite directions and add to zero. The net field is

$$\vec{E} = \frac{k_e q_1}{r_1^2} \hat{r}_1 = \frac{(8.99 \times 10^9 \, \text{N} \cdot \text{m}^2/\text{C}^2)(7.00 \times 10^{-6} \, \text{C})}{(0.040 \, 0 \, \text{m})^2 - (0.010 \, 0 \, \text{m})^2} \text{ down}$$

$$\vec{E} = (42.0 \times 10^6 \, \text{N/C})(-\hat{j})$$

20. At a certain distance from a charged particle, the magnitude of the electric field is 500 V/m and the electric potential is –3.00 kV. (a) What is the distance to the particle? (b) What is the magnitude of the charge?

Solution

Conceptualize: The particle possesses its charge just at its own location. It creates both vector field and scalar potential throughout the surrounding empty space. The charge must be negative to create negative voltage. We expect less than a microcoulomb and a distance of several centimeters for a tabletop or vacuum-tube situation.

Categorize: We will use these equations: At a distance r from a charged particle, the voltage is $V = \frac{k_e q}{r}$ and the field magnitude is $E = \frac{k_e |q|}{r^2}$.

Analyze: Combining them, $E = \frac{r|V|}{r^2} = \frac{|V|}{r}$:

(a) $r = \dfrac{|V|}{E} = \dfrac{3\ 000\ \text{V}}{500\ \text{N/C}} = 6.00 \dfrac{\text{N} \cdot \text{m/C}}{\text{N/C}} = 6.00\ \text{m}$ ∎

(b) $q = \dfrac{rV}{k_e} = \dfrac{(6.00\ \text{m})(-3\ 000\ \text{V})}{8.99 \times 10^9\ \text{N} \cdot \text{m}^2 / \text{C}^2} = -2.00\ \mu\text{C}$ ∎

Finalize: Charge and distance are a bit larger than our guesses. Recall that a particle need not be small. A big ball can carry this rather large charge. We know that the vector field is pointing toward the negative charge.

---

23. Over a certain region of space, the electric potential is $V = 5x - 3x^2 y + 2yz^2$. (a) Find the expressions for the x, y, and z components of the electric field over this region. (b) What is the magnitude of the field at the point P that has coordinates $(1.00, 0, -2.00)$ m?

Solution

Conceptualize: The given equation tells a voltage value for every point in three-dimensional space. The downward slope of the potential from point to point is the electric field we will compute.

Categorize: First, we find the x, y, and z components of the field from the partial derivatives of the potential function with respect to x, y, and z. Then, we evaluate them at point P. We assume that V is given in volts, as a function of distances in meters.

Analyze:

(a) The components of the field are

$$E_x = -\frac{\partial V}{\partial x} = -\frac{\partial}{\partial x}\left(5x - 3x^2 y + 2yz^2\right) = -5 + 6xy + 0$$

and

$$E_y = -\frac{\partial V}{\partial y} = 3x^2 - 2z^2 \quad \text{and} \quad E_z = -\frac{\partial V}{\partial z} = -4yz$$ ∎

(b) At point P, $E_x = -5 + 6(1.00\ \text{m})(0\ \text{m}) = -5.00\ \text{N/C}$.

At point P, $E_y = 3(1.00\ \text{m})^2 - 2(-2.00\ \text{m})^2 = -5.00\ \text{N/C}$.

At point P, $E_z = -4(0 \text{ m})(-2.00 \text{ m}) = 0 \text{ N/C}$.

At P, the field's magnitude is

$$E = \sqrt{(-5.00 \text{ N/C})^2 + (-5.00 \text{ N/C})^2 + 0^2} = 7.07 \text{ N/C}$$ ∎

Finalize: It is not hard to take a partial derivative after you learn the rule to think of the other variables as constants. It would be cumbersome to draw a graph of V as a function of the three spatial variables. Picture the field at the particular point in question as a weak field in the xy plane, at 225° to the x axis.

---

26. A rod of length L (Fig. P20.26) lies along the x axis with its left end at the origin. It has a nonuniform charge density $\lambda = \alpha x$, where α is a positive constant. (a) What are the units of α? (b) Calculate the electric potential at A.

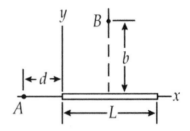

Figure P20.26

Solution

Conceptualize: The potential will be proportional to α. It will decrease as d increases.

Categorize: We think just of units in part (a). In part (b) we must do an integral, adding up infinitely many infinitesimal contributions to the total potential at A from all the elements of the rod.

Analyze:

(a) As a linear charge density, λ has units of C/m. So $\alpha = \lambda/x$ must have units of C/m^2.   ∎

(b) Consider a small segment of the rod at location x and of length dx. The amount of charge on it is $\lambda \, dx = (\alpha x) \, dx$. Its distance from A is $d + x$, so its contribution to the electric potential at A is

$$dV = k_e \frac{dq}{r} = k_e \frac{\alpha x \, dx}{d + x}$$

Relative to $V = 0$ infinitely far away, to find the potential at A we must integrate these contributions for the whole rod, from $x = 0$ to $x = L$. Then

$$V = \int\limits_{\text{all } q} dV = \int_0^L \frac{k_e \alpha x}{d + x} dx$$

To perform the integral, make a change of variables to $u = d + x$, $du = dx$, $u(\text{at } x = 0) = d$, and $u(\text{at } x = L) = d + L$:

$$V = \int_d^{d+L} \frac{k_e \alpha(u - d)}{u} du = k_e \alpha \int_d^{d+L} du - k_e \alpha d \int_d^{d+L} \left(\frac{1}{u}\right) du$$

$$V = k_e \alpha u \big|_d^{d+L} - k_e \alpha d \ln u \big|_d^{d+L} = k_e \alpha(d + L - d) - k_e \alpha d [\ln(d + L) - \ln d]$$

$$V = k_e \alpha L - k_e \alpha d \ln\left(\frac{d + L}{d}\right) \qquad \blacksquare$$

Finalize: At the start or in the middle of the calculation, it is good to keep track of symbols: the unknown is V. The values k_e, α, d, and L are known and constant. While we are doing the integral x and u are variables, and will not appear in the answer. We have the answer when the unknown is expressed in terms of the d, L, and α mentioned in the problem and the universal constant k_e.

---

32. A spherical conductor has a radius of 14.0 cm and charge of 26.0 μC. Calculate the electric field and the electric potential at (a) $r = 10.0$ cm, (b) $r = 20.0$ cm, and (c) $r = 14.0$ cm from the center.

Solution

Conceptualize: A metal sphere carries a charge. We review Chapter 19 to identify or calculate the field that the charge creates inside the sphere, at its surface, and outside. The potential follows related rules that are different in detail.

Categorize: For points on the surface and outside, the sphere of charge behaves like a charged particle at its center, both for creating field and potential. But ...

Analyze:

(a) ... inside a conductor when charges are not moving, the electric field is zero and the potential is uniform, the same as on the surface. We have $\vec{E} = 0$. $\qquad \blacksquare$

$$V = \frac{k_e q}{R} = \frac{\left(8.99 \times 10^9 \text{ N} \cdot \text{m}^2/\text{C}^2\right)\left(26.0 \times 10^{-6} \text{ C}\right)}{0.140 \text{ m}} = 1.67 \times 10^6 \text{ V} \quad \blacksquare$$

(b) The sphere behaves like a charged particle at its center when you stand outside.

$$\vec{E} = \frac{k_e q}{r^2}\hat{\mathbf{r}} = \frac{\left(8.99 \times 10^9 \text{ N} \cdot \text{m}^2/\text{C}^2\right)\left(26.0 \times 10^{-6} \text{ C}\right)}{\left(0.200 \text{ m}\right)^2}\hat{\mathbf{r}}$$

$$= \left(5.84 \times 10^6 \text{ N/C}\right)\hat{\mathbf{r}} \quad \blacksquare$$

$$V = \frac{k_e q}{r} = \frac{\left(8.99 \times 10^9 \text{ N} \cdot \text{m}^2/\text{C}^2\right)\left(26.0 \times 10^{-6} \text{ C}\right)}{0.200 \text{ m}}$$

$$= 1.17 \times 10^6 \text{ V} \quad \blacksquare$$

(c) $\vec{E} = \dfrac{k_e q}{r^2}\hat{\mathbf{r}} = \dfrac{\left(8.99 \times 10^9 \text{ N} \cdot \text{m}^2/\text{C}^2\right)\left(26.0 \times 10^{-6} \text{ C}\right)}{\left(0.140 \text{ m}\right)^2}\hat{\mathbf{r}} = \left(11.9 \times 10^6 \text{ N/C}\right)\hat{\mathbf{r}}$ $\quad \blacksquare$

$V = 1.67 \times 10^6$ V as in part (a) $\quad \blacksquare$

Finalize: The potential is a continuous function. It has the same value just below the surface and just above the surface. The field is discontinuous here. We have given in part (c) the field just outside the metal surface. It is zero just inside the surface.

---

35. An isolated charged conducting sphere of radius 12.0 cm creates an electric field of 4.90×10^4 N/C at a distance 21.0 cm from its center. (a) What is its surface charge density? (b) What is its capacitance?

Solution

Conceptualize: We can think of the sphere as a charge that creates a potential at exterior points just as a charged particle at its center would. We can also think of using a battery to put the sphere's surface at some potential, relative to zero volts at infinite distance. This action deposits some charge on the surface, which we can regard as stored on a capacitor plate.

Categorize: We use equations about a charge with spherical symmetry (an example is $V = Q/4\pi\epsilon_0 R$) to relate the charge to its exterior effects.

Analyze:

(a) The electric field outside a spherical charge distribution of radius R is $E = k_e Q/r^2$. Therefore, $Q = Er^2/k_e$. Since the surface charge density is $\sigma = Q/A$,

$$\sigma = \frac{Er^2}{k_e 4\pi R^2} = \frac{\left(4.90 \times 10^4 \text{ N/C}\right)(0.210 \text{ m})^2}{\left(8.99 \times 10^9 \text{ N} \cdot \text{m}^2/\text{C}^2\right)(4\pi)(0.120 \text{ m})^2} = 1.33 \ \mu\text{C/m}^2 \quad \blacksquare$$

(b) For an isolated charged sphere of radius R,

$$C = 4\pi\epsilon_0 R = (4\pi)(8.85 \times 10^{-12} \text{ C}^2/\text{N} \times \text{m}^2)(0.120 \text{ m}) = 13.3 \text{ pF} \quad \blacksquare$$

Finalize: This is a straightforward problem. The sphere is as big as a large classroom world globe. Its capacitance is small, in micromicrofarads.

------

39. A 50.0-m length of coaxial cable has an inner conductor that has a diameter of 2.58 mm and carries a charge of 8.10 μC. The surrounding conductor has an inner diameter of 7.27 mm and a charge of -8.10 μC. Assume the region between the conductors is air. (a)What is the capacitance of this cable? (b) What is the potential difference between the two conductors?

Solution

Conceptualize: We study parallel-plate and spherical capacitors. This problem is about a cylindrical capacitor, also briefly covered in the chapter text.

Categorize: We apply the definition of capacitance and the result proved in the chapter for this geometry.

Analyze:

(a) The capacitance of a cylindrical capacitor is

$$C = \frac{\ell}{2k_e \ln(b/a)} = \frac{50.0 \text{ m}}{2\left(8.99 \times 10^9 \text{ N} \cdot \text{m}^2/\text{C}^2\right)\ln(7.27/2.58)}$$

$$= 2.68 \times 10^{-9} \text{ F} \quad \blacksquare$$

(b) From the definition of capacitance, the potential difference between the cylindrical plates is

$$\Delta V = \frac{Q}{C} = \frac{8.10 \times 10^{-6} \text{ C}}{2.68 \times 10^{-9} \text{ F}} = 3.02 \text{ kV} \quad \blacksquare$$

Finalize: A person ordinarily buys a coaxial cable to carry electric current shielded from noise due to environmental electric fields. We have shown here that the cable itself possesses capacitance, which may be large enough in some applications to be important to the circuit of which it is part.

---

45. Four capacitors are connected as shown in Figure P20.45.

(a) Find the equivalent capacitance between points a and b.

(b) Calculate the charge on each capacitor, taking $\Delta V_{ab} = 15.0$ V.

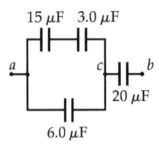

Figure P20.45

Solution

Conceptualize: To exterior devices, the section of electric circuit between a and b would behave like a single capacitor. When 15 volts is applied between a and b, each one of the actual capacitors will have some smaller voltage across it. We will see that the 20 microfarads stores the whole charge that passes points a and b, but the other actual capacitors store less.

Categorize: We find a set of capacitors in series or in parallel, replace them with a single equivalent capacitor, draw the new circuit, and repeat. The circuit diagrams will guide our analysis of how the charge divides up.

Analyze:

(a) We simplify the circuit of Figure P20.45 in three steps as shown in Figures (a), (b), and (c). First, the 15.0 μF and 3.00 μF in series are equivalent to

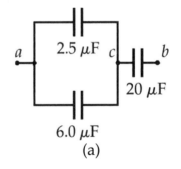

$$\frac{1}{(1/15.0\ \mu\text{F})+(1/3.00\ \mu\text{F})} = 2.50\ \mu\text{F}$$

Next, 2.50 μF combines in parallel with 6.00 μF, creating an equivalent capacitance of 8.50 μF. At last, 8.50 μF and 20.0 μF are in series, equivalent to

$$\frac{1}{(1/8.50\ \mu\text{F})+(1/20.00\ \mu\text{F})} = 5.96\ \mu\text{F} \quad\blacksquare$$

(b) We find the charge on each capacitor and the voltage across each by working backwards through solution figures (c)–(a), alternately applying $Q = C\Delta V$ and $\Delta V = Q/C$ to every capacitor, real or equivalent. For the 5.96-μF capacitor, we have

$$Q = C\Delta V = (5.96\ \mu\text{F})(15.0\ \text{V}) = 89.5\ \mu\text{C}$$

Thus, if a is higher in potential than b, just 89.5 μC flows between the wires and the plates to charge the capacitors in each picture. In (b) we have, for the 8.5-μF capacitor,

$$\Delta V_{ac} = \frac{Q}{C} = \frac{89.5\ \mu\text{C}}{8.50\ \mu\text{F}} = 10.5\ \text{V}$$

and for the 20.0 μF in (b), (a), and the original circuit, we have $Q_{20} = 89.5\ \mu\text{C}$. ∎

$$\Delta V_{cb} = \frac{Q}{C} = \frac{89.5\ \mu\text{C}}{20.0\ \mu\text{F}} = 4.47\ \text{V}$$

Next, the circuit in diagram (a) is equivalent to that in (b), so $\Delta V_{cb} = 4.47$ V and $\Delta V_{ac} = 10.5$ V.

For the 2.50 μF, $\Delta V = 10.5$ V and $Q = C\Delta V = (2.50\ \mu\text{F})(10.5\ \text{V}) = 26.3\ \mu\text{C}$.

For the 6.00 μF, $\Delta V = 10.5$ V and $Q_6 = C\Delta V = (6.00\ \mu\text{F})(10.5\ \text{V}) = 63.2\ \mu\text{C}$. ∎

Now, 26.3 μC having flowed in the upper parallel branch in (a), back in the original circuit we have $Q_{15} = 26.3\ \mu\text{C}$ and $Q_3 = 26.3\ \mu\text{C}$. ∎

Finalize: Some students think this kind of problem is just about using the formulas for series and parallel capacitor combinations. Really to analyze circuits you need to develop some essential skills beyond using formulas. One skill is to *recognize* parallel and series connections. In the original diagram the 6-μF capacitor and the 15-μF capacitor are drawn on parallel lines, but they are not in parallel or in series. To be in parallel, capacitors must be connected "top end to top end and bottom end to bottom end," so that the same voltage must appear on both. To be in series, capacitors must have no junction between them, so that the same charge will necessarily appear on both. Another essential skill, after calculating the capacitance of an equivalent capacitor, is to *draw* it in a new circuit diagram. All the diagrams are necessary for letting you practice the next skill, applying $Q = C\ \Delta V$ and $\Delta V = Q/C$ to *find the charges and potential differences* for all the capacitors, real and equivalent, including those in the original circuit. To bring the problem to a logical conclusion we do this **Related Calculation**:

A problem on your next exam may also ask for the voltage across each of the original capacitors. The remaining answers are:

$$\Delta V_{15} = \frac{Q}{C} = \frac{26.3\ \mu C}{15.0\ \mu F} = 1.75\text{V} \quad \text{and} \quad \Delta V_3 = \frac{Q}{C} = \frac{26.3\ \mu C}{3.0\ \mu F} = 8.77\text{ V}$$

---

52. Consider the circuit shown in Figure P20.52, where $C_1 = 6.00\ \mu F$, $C_2 = 3.00\ \mu F$, and $\Delta V = 20.0$ V. Capacitor C_1 is first charged by closing switch S_1. Switch S_1 is then opened, and the charged capacitor is connected to the uncharged capacitor by closing S_2. Calculate (a) the initial charge acquired by C_1 and (b) the final charge on each capacitor.

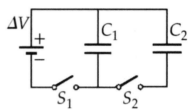

Figure P20.52

Solution

Conceptualize: The two switches are never closed at the same time. Capacitor 2 is never connected directly across the battery. The voltage across it will never be 20 V, but some smaller value after …

Categorize: … charge originally on capacitor 1 is shared with capacitor 2. We must visualize the sequence of operations. Then just the definition of capacitance, applied again and again, will give us the answers.

Analyze:

(a) When S_1 is closed, the charge on C_1 will be

$$Q_1 = C_1 \Delta V_1 = (6.00\ \mu F)(20.0\text{ V}) = 120\ \mu C \qquad \blacksquare$$

(b) When S_1 is opened and S_2 is closed, the total charge will remain constant and be shared by the two capacitors. We let primed symbols represent the new charges on the capacitors, in $Q_1' = 120\ \mu C - Q_2'$. The potential differences across the two capacitors will be equal.

$$\Delta V' = \frac{Q_1'}{C_1} = \frac{Q_2'}{C_2} \quad \text{or} \quad \frac{120\ \mu C - Q_2'}{6.00\ \mu F} = \frac{Q_2'}{3.00\ \mu F} \qquad \blacksquare$$

Then we do the algebra to find $Q_2' = 40.0\ \mu C$ and $\qquad \blacksquare$

$$Q_1' = 120\ \mu C - 40.0\ \mu C = 80.0\ \mu C \qquad \blacksquare$$

Finalize: It might help you to draw the circuit diagrams with switch 1 closed, then back to open, and then with just switch 2 closed. Note that the symbol for an open switch looks like an open door on a floor plan. An open door allows things to go through but an open switch does not allow charge to pass through it.

This problem may be the first you have seen to directly apply a universal law of nature, conservation of charge. This is the fundamental reason that the sum of the charges on the two capacitors in the final circuit must be 120 μC. Electrical potential energy is not conserved when the connection is made. Perhaps remarkably, the stored energy drops from 1 200 μJ to 800 μJ. The missing energy becomes internal energy in the wires joining the capacitors or leaves the circuit as electromagnetic radiation. If it were not so, most of the charge would bounce between one capacitor and the other, oscillating forever.

---

54. A parallel-plate capacitor has a charge Q and plates of area A. What force acts on one plate to attract it toward the other plate? Because the electric field between the plates is $E = Q/A\epsilon_0$, you might think that the force is $F = EQ = Q^2/A\epsilon_0$. This conclusion is wrong, because the field E includes contributions from both plates, and the field created by the positive plate cannot exert any force on the positive plate. Show that the force exerted on each plate is actually $F = Q^2/2\epsilon_0 A$. *Suggestion:* Let $C = \epsilon_0 A/x$ for an arbitrary plate separation x and note that the work done in separating the two charged plates is $W = \int F dx$.

Solution

Conceptualize: The force *on* one plate is exerted *by* the other, through the electric field created by that single other plate:

$$E = \frac{\sigma}{2\epsilon_0} = \frac{Q}{2A\epsilon_0}$$

The force on each plate is: $\quad F = (Q_{\text{self}})(E_{\text{other}}) = \dfrac{Q^2}{2A\epsilon_0}$

Categorize: To prove this, we follow the hint, and calculate the work done in separating the plates, which equals the potential energy stored in the charged capacitor:

$$U = \frac{1}{2}\frac{Q^2}{C} = \int F dx$$

Analyze: Now from the fundamental theorem of calculus, $dU = F dx$,

and
$$F = \frac{d}{dx}U = \frac{d}{dx}\left(\frac{Q^2}{2C}\right) = \frac{1}{2}\frac{d}{dx}\left(\frac{Q^2}{A\epsilon_0/x}\right).$$

Performing the differentiation, $F = \frac{1}{2}\frac{d}{dx}\left(\frac{Q^2 x}{A\epsilon_0}\right) = \frac{1}{2}\left(\frac{Q^2}{A\epsilon_0}\right).$ ∎

Finalize: The force exerted on one charged plate by another is sometimes used in a machine shop to hold a workpiece stationary. Larger charge on both plates will imply a larger field exerting on a larger charge an extra-large force, so the proportionality to charge squared is reasonable.

---

64. A commercial capacitor is to be constructed as shown in Figure 20.64. This particular capacitor is made from two strips of aluminum separated by a strip of paraffin-coated paper. Each strip of foil and paper is 7.00 cm wide. The foil is 0.004 00 mm thick, and the paper is 0.025 0 mm thick and has a dielectric constant of 3.70. What length should the strips have, if a capacitance of 9.50×10^{-8} F is desired before the capacitor is rolled up? (Adding a second strip of paper and rolling the capacitor effectively doubles its capacitance, by allowing charge storage on both sides of each strip of foil.)

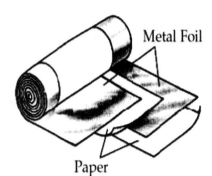

Metal Foil

Paper

Figure P20.64

Solution

Conceptualize: Ninety-five nanofarads is a reasonable capacitance for a physically compact component from the electronics store. The thinness of the dielectric—the closeness of the plates—is a primary reason for how large its capacitance is, so the long dimension of the foil plates may be well under a meter.

Categorize: We use just the equation giving the capacitance of a parallel-plate capacitor with a dielectric between its plates.

Analyze: Our diagram exaggerates how the strips can be offset to avoid contact between the two foils. It shows how a second paper strip can be used to roll the capacitor into a convenient cylindrical shape with electrical contacts at the two ends. We suppose that the overlapping width of the two metallic strips is still $w = 7.00$ cm. Then for the area of the plates we have $A = \ell w$ in $C = \kappa \epsilon_0 A/d = \kappa \epsilon_0 \ell w/d$. Solving the equation gives

$$\ell = \frac{Cd}{\kappa\epsilon_0 w} = \frac{(9.50\times10^{-8}\ \text{F})(2.50\times10^{-5}\ \text{m})}{3.70(8.85\times10^{-12}\ \text{C}^2/\text{N}\cdot\text{m}^2)(0.070\ 0\ \text{m})} = 1.04\ \text{m} \quad\blacksquare$$

Finalize: Oops! It's a bit over a meter, but the capacitor can still be smaller than a roll of coins. The distance between positive and negative charges is the thickness of one sheet of paper because these charges sit on adjacent surfaces of the metal foils. The electric field between the top and bottom surfaces of one piece of foil is zero. The thickness of the foil was given in the problem but it does not affect the capacitance.

———————————————

67. Lightning can be studied with a Van de Graaff generator, which consists of a spherical dome on which charge is continuously deposited by a moving belt. Charge can be added until the electric field at the surface of the dome becomes equal to the dielectric strength of air. Any more charge leaks off in sparks as shown in Figure P20.67. Assume the dome has a diameter of 30.0 cm and is surrounded by dry air with a "breakdown" electric field of 3.00×10^6 V/m. (a) What is the maximum potential of the dome? (b) What is the maximum charge on the dome?

Solution

Conceptualize: Van de Graaff generators produce voltages that can make your hair stand on end, somewhere on the order of about 100 kV. With these high voltages, the maximum charge on the dome is probably more than typical tabletop charged-particle values of about $1\ \mu$C.

The maximum potential and charge will be limited by the electric field strength at which the air surrounding the dome will ionize. This critical value is determined by the **dielectric strength** of air which, from Table 20.1, is $E_{\text{critical}} = 3 \times 10^6$ V/m. An electric field stronger than this will cause the air to act like a conductor instead of an insulator. This process is called dielectric breakdown and may be seen as a spark.

Categorize: From the maximum allowed electric field, we can find the charge and potential that would create this situation. Since we are only given the diameter of the dome, we will assume that the conductor is spherical, which allows us to use the electric field and potential equations for a spherical conductor. With these equations, it will be easier to do part (b) first and use the result for part (a).

Analyze: At the surface of a spherical conductor with total charge Q,

(b) $|\vec{E}| = \dfrac{k_e Q}{r^2}$:

$$Q = \frac{Er^2}{k_e} = \frac{\left(3.00 \times 10^6 \text{ V/m}\right)(0.150 \text{ m})^2}{8.99 \times 10^9 \text{ N} \cdot \text{m}^2/\text{C}^2}\left(1 \text{ N} \cdot \text{m/V} \cdot \text{C}\right) = 7.51 \ \mu\text{C} \qquad \blacksquare$$

(a) $V = \dfrac{k_e Q}{r} = \dfrac{\left(8.99 \times 10^9 \text{ N} \cdot \text{m}^2/\text{C}^2\right)\left(7.51 \times 10^{-6} \text{ C}\right)}{0.150 \text{ m}} = 450 \text{ kV} \qquad \blacksquare$

Finalize: These calculated results seem reasonable based on our predictions. The voltage is about 4 000 times larger than the 120 V of common electrical outlets, but the charge is similar in magnitude to many of the static charge problems we have solved earlier. This implies that many of the charge configurations mentioned in other problems would have to be in a vacuum because the electric field near these charged particles would be strong enough to cause sparking in air. For example, a charged ball with $Q = 1 \ \mu\text{C}$ and $r = 1$ mm would have an electric field near its surface of

$$E = \frac{k_e Q}{r^2} = \frac{\left(9 \times 10^9 \text{ N} \cdot \text{m}^2/\text{C}^2\right)\left(1 \times 10^{-6} \text{ C}\right)}{\left(0.001 \text{ m}\right)^2} = 9 \times 10^9 \text{ V/m}$$

which is well beyond the dielectric breakdown strength of air!

---

73. The liquid-drop model of the atomic nucleus suggests high-energy oscillations of certain nuclei can split the nucleus into two unequal fragments plus a few neutrons. The fission products acquire kinetic energy from their mutual Coulomb repulsion. Assume the charge is distributed uniformly throughout the volume of each spherical fragment and, immediately before separating, each fragment is at rest and their surfaces are in contact. The electrons surrounding the nucleus can be ignored. Calculate the electric potential energy (in electron volts) of two spherical fragments from a uranium nucleus having the following charges and radii: $38e$ and 5.50×10^{-15} m, $54e$ and 6.20×10^{-15} m.

Solution

Conceptualize: The answer should be on the order of a hundred million electron volts. The nuclear force ordinarily holds a nucleus together. No one knows how to write down an F = formula for the nuclear force. It is a little remarkable that we can compute a good estimate for the energy released by nuclear fission just from knowing about the electric force …

Categorize: … or really about electric potential energy. The problem is equivalent to finding the potential energy of a charged particle with $q = 38e$ at a distance 11.7×10^{-15} m (the distance between their centers when in contact) from a charged particle of $54e$.

Analyze: The two spheres of charge have together electric potential energy

$$U = qV = k_e \frac{q_1 q_2}{r_{12}}$$

$$= (8.99 \times 10^9 \text{ J} \cdot \text{m/C}^2) \frac{(38)(1.60 \times 10^{-19} \text{ C})(54)(1.60 \times 10^{-19} \text{ C})}{\left(5.50 \times 10^{-15} \text{ m} + 6.20 \times 10^{-15} \text{ m}\right)}$$

$$U = 4.04 \times 10^{-11} \text{ J} = 253 \text{ MeV} \qquad \blacksquare$$

Finalize: The analysis gives a good account of the energy released by uranium fission. There are no electrons involved here, but the proton charge is the same magnitude as the electron charge, and it is remarkably convenient to use the electron volt as a unit of energy in atoms, nuclei, and many laboratory processes.

---

81. A parallel-plate capacitor is constructed using a dielectric material whose dielectric constant is 3.00 and whose dielectric strength is 2.00×10^8 V/m. The desired capacitance is 0.250 μF, and the capacitor must withstand a maximum potential difference of 4.00 kV. Find the minimum area of the capacitor plates.

Solution

Conceptualize: This is a good design problem. We must combine the requirement that the electric field not be too large with the requirement that the capacitance be large enough.

Categorize: Algebra of simultaneous equations is the natural way to combine the requirements. Here the equations will be simple.

Analyze: The dielectric strength is $E_{max} = 2.00 \times 10^8$ V/m $= \dfrac{\Delta V_{max}}{d}$,

so we have for the distance between plates $d = \dfrac{\Delta V_{max}}{E_{max}}$.

Now to also satisfy $C = \dfrac{\kappa\epsilon_0 A}{d} = 0.250 \times 10^{-6}$ F with $\kappa = 3.00$, we combine by substitution to solve for the plate area:

$$A = \frac{Cd}{\kappa\epsilon_0} = \frac{C\Delta V_{max}}{\kappa\epsilon_0 E_{max}}$$

$$= \frac{(0.250 \times 10^{-6} \text{ F})(4\,000 \text{ V})}{(3.00)(8.85 \times 10^{-12} \text{ F/m})(2.00 \times 10^8 \text{ V/m})} = 0.188 \text{ m}^2 \qquad \blacksquare$$

Finalize: Along the way we have shown that the requirements are not mutually exclusive, and that there is only one possibility for the plate area. Neither of these facts might be obvious.

————————

82. An electric dipole is located along the y axis as shown in Figure P20.82. The magnitude of its electric dipole moment is defined as $p = 2aq$. (a) At a point P, which is far from the dipole ($r \gg a$), show that the electric potential is

$$V = \frac{k_e p \cos\theta}{r^2}$$

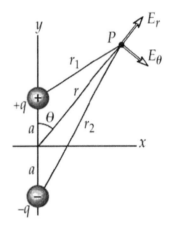

Figure P20.82

(b) Calculate the radial component E_r and the perpendicular component E_θ of the associated electric field. Note that $E_\theta = -(1/r)(\partial V/\partial\theta)$. Do these results seem reasonable for (c) $\theta = 90°$ and $0°$? (d) For $r = 0$? (e) For the dipole arrangement shown, express V in terms of Cartesian coordinates using $r = (x^2 + y^2)^{1/2}$ and

$$\cos\theta = \frac{y}{\left(x^2 + y^2\right)^{1/2}}$$

(f) Using these results and again taking $r \gg a$, calculate the field components E_x and E_y.

Solution

Conceptualize: We know how a charged particle (a monopole) creates potential around it. This is about how a dipole creates potential.

Categorize: We add up the potentials that the two charged particles separately create. Then we take partial derivatives to find field components.

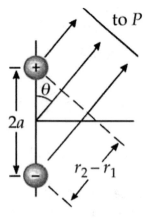

Analyze:

(a) The total potential is

$$V = \frac{k_e q}{r_1} - \frac{k_e q}{r_2} = \frac{k_e q}{r_1 r_2}(r_2 - r_1)$$

From the diagram, for $r \gg a$ we have $r_2 - r_1 \approx 2a \cos\theta$ and $r_1 \approx r \approx r_2$, so

$$V \approx \frac{k_e q}{(r)(r)}(2a \cos\theta) = \frac{k_e p \cos\theta}{r^2}$$ ■

(b) $E_r = -\dfrac{\partial V}{\partial r} = -\dfrac{\partial}{\partial r}\left(\dfrac{k_e p \cos\theta}{r^2}\right) = \dfrac{2k_e p \cos\theta}{r^3} = E_r$ ■

In spherical coordinates,

$$E_\theta = -\frac{1}{r}\left(\frac{\partial V}{\partial\theta}\right) = -\frac{1}{r}\frac{\partial}{\partial\theta}\left(\frac{k_e p \cos\theta}{r^2}\right) = \frac{k_e p \sin\theta}{r^3} = E_\theta$$ ■

(c) The values at $\theta = 0$, $\theta = 90°$, and $\theta = 180°$ correspond to an electric field that points outward at $\theta = 0$ from the positive charge on top, loops around to point downward at $\theta = 90°$, and continues around to point upward toward the negative charge from below at $\theta = 180°$. This is reasonable. ■

(d) Taking $r = 0$ makes no sense, because we assumed that r is much larger than a in the derivation. The field is not infinite at $r = 0$, at the point halfway between the charges, although the distant-field equations suggest that it goes to infinity. ■

(e) For $r = \sqrt{x^2 + y^2}$,

$$\cos\theta = \frac{y}{\sqrt{x^2 + y^2}} \quad \text{and} \quad V = \frac{k_e p y}{\left(x^2 + y^2\right)^{3/2}}$$ ■

(f) $\quad E_x = -\dfrac{\partial V}{\partial x} = -\dfrac{\partial}{\partial x}\left(\dfrac{k_e py}{\left(x^2 + y^2\right)^{3/2}}\right) = \dfrac{3k_e pxy}{\left(x^2 + y^2\right)^{5/2}} = E_x$ $\qquad$ ■

$\quad E_y = -\dfrac{\partial V}{\partial y} = -\dfrac{\partial}{\partial y}\left(\dfrac{k_e py}{\left(x^2 + y^2\right)^{3/2}}\right) = \dfrac{k_e p\left(2y^2 - x^2\right)}{\left(x^2 + y^2\right)^{5/2}} = E_y$ $\qquad$ ■

Finalize: In part (a) we can argue that the proportionality to $\cos\theta$ is reasonable because for $\theta = 90°$ we are on the x axis, equidistant from the positive and negative charges. Here the total potential is zero, as the cosine function describes. If we take the special case $y = 0$ in part (f) of this problem and interchange the labels x and y, we should get the result derived in part B of Example 19.4. With these assignments we get $E_y = 0$ and $|E_x| = k_e p(-y^2)/y^5| = k_e 2qa/y^3$, which does agree with the result of the Example.

If we take the special case $x = 0$ in part (f) and interchange x and y, we get the results $E_y = 0$ and $|E_x| = 4k_e qa / x^3$, agreeing with Example 19.4.

Chapter 21
Current and Direct Current Circuits

Section 21.1   Electric Current

The direction of conventional current is designated as the direction of motion of positive charge. *In an ordinary metal conductor, the direction of current will be opposite the direction of flow of electrons (which are the charge carriers in this case).*

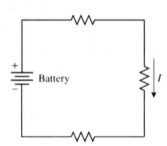

A simple dc circuit showing the direction of conventional current.

Section 21.2   Resistance and Ohm's Law

For ohmic materials, the ratio of the current density to the electric field (that causes the current) is equal to a constant σ, the **conductivity** of the material. The reciprocal of the conductivity is called the **resistivity**, ρ. Each ohmic material has a characteristic resistivity, which depends only on the properties of the specific material and is a function of temperature. Ohmic materials obey Ohm's law: R is constant in $\Delta V = IR$. Resistance has the SI units volts per ampere, called ohms (Ω); if a potential difference of 1 V across a conductor produces a current of 1 A, the resistance of the conductor is 1 Ω.

Section 21.4   A Structural Model for Electrical Conduction

In the classical model of electrical conduction in a metal, electrons are treated like molecules in a gas and, in the absence of an electric field, have a zero average velocity.

Under the influence of an electric field, the electrons move along a direction opposite the direction of the applied field with a drift

148

velocity which is proportional to the average time between collisions with atoms of the metal.

The current is proportional to the magnitude of the drift velocity and to the number of free electrons per unit volume.

Section 21.6   Sources of EMF

A source of emf (for example a battery or generator) maintains a constant current in a closed circuit. The emf, $\mathcal{E}$, of the source is the work done per unit charge in increasing the electric potential of the circulating charges. *One joule of work is required to move one coulomb of charge between two points that differ in potential by one volt.*

The **terminal voltage** (ΔV between the positive and negative terminals of a battery or other source) is equal to the emf of the battery when the current in the circuit is zero. In this case the terminal voltage is called the **open circuit voltage**. When the battery is delivering current to a circuit, the terminal voltage (closed circuit voltage) is less than the emf due to the **internal resistance** of the battery.

Section 21.7   Resistors in Series and in Parallel

For a group of resistors connected in parallel:

- The reciprocal of the equivalent resistance of the group is the algebraic sum of the reciprocals of the individual resistors, as seen in Equation 21.28, $\dfrac{1}{R_{eq}} = \dfrac{1}{R_1} + \dfrac{1}{R_2} + \dfrac{1}{R_3} + \cdots$.

- The potential differences across the individual resistors have the same value.

- The total current associated with the parallel group is the sum of the currents in the individual resistors.

For a group of resistors connected in series:

- The equivalent resistance of the series combination is the algebraic sum of the individual resistances, as seen in Equation 21.26, $R_{eq} = R_1 + R_2 + R_3 + \cdots$.

- The current is the same for each resistor in the group.

- The total potential difference across the group of resistors equals the sum of the potential differences across the individual resistors.

Section 21.8   Kirchhoff's Rules

- **Junction Rule (conservation of charge)**: With currents entering a junction counted as positive and currents leaving the junction counted as negative, the total current accumulating at any junction must be equal to zero. A junction is defined to be any point in the circuit where the current can split. *To generate independent equations, the junction rule may be used a number of times which is one fewer than the number of junction points in the circuit.*

- **Loop Rule (conservation of energy)**: The algebraic sum of the potential differences around any closed loop in a circuit must be zero. *Each application of the loop rule should result in an equation that contains a new circuit element (resistor or battery), or a new current.*

Overall, in order to solve a circuit problem using Kirchhoff's Rules, you must have as many independent equations as the total number of unknown quantities (currents, resistances, or emfs) in the circuit.

Section 21.9   *RC* Circuits

Consider a battery, resistor, and an uncharged capacitor in a circuit as shown in the figure at the right. When the switch is thrown to position *a* the battery, resistor, and capacitor will be in simple series; the capacitor will begin charging.

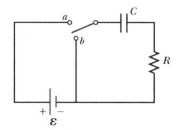

A circuit for charging and discharging a capacitor

The charge on the capacitor will increase (*at a decreasing rate*) from zero to a maximum value as the current decreases exponentially. (See Figure 21.27 in the textbook.)

When the maximum charge on the capacitor is reached, the potential difference across the capacitor will equal that of the battery. When the switch in the circuit is moved from position *a* to position *b* after the capacitor has been charged (leaving the capacitor and resistor in series) the initially charged capacitor will begin discharging. The charge on the capacitor, current in the circuit, and the voltage across the resistor will all decrease exponentially with time. The maximum charge ($Q = C\mathcal{E}$) will be accumulated on the capacitor only if the charging time is allowed to go to infinity. Charges are transferred from one plate to the other along the connecting wires; *they do not move across the gap between plates.*

EQUATIONS AND CONCEPTS

Electric current, I, is defined as the rate at which charge moves through a cross section of a conductor. The direction of conventional current is opposite the direction of the flow of electrons in the circuit. The SI unit of current is the **ampere** (A). *Under the action of an electric field, electric charges will move through gases, liquids, and solid conductors.*

$$I \equiv \lim_{t \to 0} \frac{\Delta Q}{\Delta t} = \frac{dQ}{dt} \qquad (21.2)$$

$$1\ \text{A} = 1\ \text{C/s} \qquad (21.3)$$

The **drift velocity**, v_d (velocity of the charge carriers in a conductor), is actually an average value of the individual velocities.

$$I_{\text{ave}} = \frac{\Delta Q}{\Delta t} = nqv_d A \qquad (21.4)$$

n = the number of mobile charge carriers per unit volume of conductor

The **current density**, J, in a conductor is defined as the current per unit cross-sectional area of the conductor.

$$J \equiv \frac{I}{A} = nqv_d \qquad (21.5)$$

The **definition of resistance** is expressed in Equation 21.6. According to Ohm's law, R is understood to be independent of ΔV for ohmic materials, which have a linear current-voltage relationship over a wide range of applied voltages.

$$R \equiv \frac{\Delta V}{I} \qquad (21.6)$$

The **resistance** of a given conductor of uniform cross section depends on the length, cross-sectional area, and a characteristic property of the material of which the conductor is made. The parameter, ρ, is the **resistivity** of the material from which the conductor is made. The resistivity is the inverse of the **conductivity** and has units of ohm-meters. The unit of resistance is the ohm (Ω). *The value of the resistivity of a given material depends on the electronic structure of the material and on the temperature.*

$$R = \rho \frac{\ell}{A} \qquad (21.8)$$

$$1\,\Omega = 1\ \text{V/A}$$

The **resistivity**, and therefore the resistance, of a conductor varies with temperature. Over a limited range of temperatures, this variation is approximately linear. *Semiconductors (for example, carbon) are characterized by a negative temperature coefficient of resistivity and in these materials the resistance decreases as the temperature increases.*

$$\rho = \rho_0 \left[1 + \alpha \left(T - T_0 \right) \right] \qquad (21.10)$$

$$R = R_0 \left[1 + \alpha \left(T - T_0 \right) \right] \qquad (21.12)$$

$\alpha =$ temperature coefficient of resistivity

T_0 is a reference temperature (usually taken to be 20.0°C).

Power will be supplied to a resistor or other current-carrying devices when a potential difference is maintained between the terminals of the circuit element. These quantities can be related, as in Equation 21.20, and the SI unit of power is the watt (W). When the device is purely resistive, the power delivered can be expressed in alternative forms. *Note that watt is a unit of power; kilowatt-hour is a unit of energy.*

$$P = I\Delta V \qquad (21.20)$$

$$P = I^2 R = \frac{\left(\Delta V \right)^2}{R} \qquad (21.21)$$

$$1\,\text{W} = 1\ \text{J/s} \qquad \text{(power)}$$
$$1\,\text{kWh} = 3.6 \times 10^6\ \text{J} \qquad \text{(energy)}$$

The **terminal voltage**, ΔV, of a battery will be less than the emf when the battery is providing a current to an external circuit. This is due to **internal resistance** of the battery.

$$\Delta V = \mathcal{E} - Ir \qquad (21.22)$$

The **current**, I, delivered by a battery in a simple dc circuit depends on the value

$$\mathcal{E} = IR + Ir \qquad (21.23)$$

of the emf of the source ($\mathcal{E}$) the total load resistance in the circuit (R) and the internal resistance of the source (r).

Resistors in series are connected so that pairs of adjacent resistors share only one common circuit point; there is a common current through each resistor in the group.

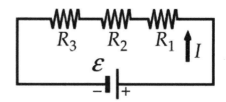

Resistors in series

A series combination of resistors has total or equivalent resistance equal to the sum of the resistances of the individual resistors.

$$R_{eq} = R_1 + R_2 + R_3 + \ldots \qquad (21.26)$$

(Series combination)

Resistors in parallel are connected so that each resistor in the parallel group has two circuit points in common with each of the other resistors; there is a common potential difference across each resistor in the group.

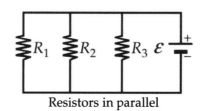

Resistors in parallel

A parallel combination of resistors has a total or equivalent resistance, which is the inverse of the sum of the inverses of the individual resistances.

$$\frac{1}{R_{eq}} = \frac{1}{R_1} + \frac{1}{R_2} + \frac{1}{R_3} + \ldots \qquad (21.28)$$

(Parallel combination)

Kirchhoff's rules can be used to analyze multiloop circuits which cannot be reduced to a single loop by using Ohm's law and the rules for combining resistors in series and parallel. Review the procedures for applying Kirchhoff's rules as outlined in *Suggestions, Skills, and Strategies*.

Junction rule: The sum of the currents entering any junction must be equal to the sum of the currents leaving that junction.

$$\sum_{junction} I = 0 \qquad (21.29)$$

Loop rule: The algebraic sum of the potential differences across all circuit elements around *any closed loop* in a circuit must equal zero.

$$\sum_{\substack{\text{closed} \\ \text{loop}}} \Delta V = 0 \qquad (21.30)$$

Charging a capacitor in series with a resistor involves a parameter called the **time constant** of the circuit (τ) which determines the manner in which the charge on the capacitor varies with time. The charge on the capacitor increases from zero to 63.2% of its maximum value in a time interval equal to one time constant. See the figure "Charging capacitor" below. Also, during one time constant, the charging current decreases from its initial maximum value of

$$I_0 = \frac{\varepsilon}{R} \text{ to } 36.8\% \text{ of } I_0.$$

$$I(t) = \frac{\varepsilon}{R} e^{-t/RC} \qquad (21.35)$$

$$q(t) = Q\left(1 - e^{-t/RC}\right) \qquad (21.34)$$

$$\tau = RC \qquad (21.36)$$

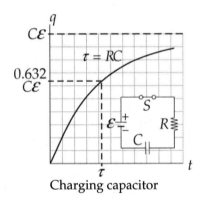

Charging capacitor

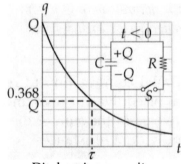

Discharging capacitor

When a **capacitor is discharged** through a resistor, the charge and current decrease exponentially with time. (See the figure "Discharging capacitor" above.) *The negative sign in Equation 21.39 indicates the direction of the current, which is opposite the direction during the charging process.*

$$q(t) = Q e^{-t/RC} \qquad (21.38)$$

$$I(t) = -\frac{Q}{RC} e^{-t/RC} \qquad (21.39)$$

$$\frac{Q}{RC} = I_i \qquad \text{(initial current)}$$

SUGGESTIONS, SKILLS, AND STRATEGIES

A Problem-Solving Strategy for Resistors

- Be careful with your choice of units. To calculate the resistance of a device in ohms using Equation 21.8, $R = \rho \dfrac{\ell}{A}$, make sure that dimensions are in meters and use the SI value of ρ.

- When two or more resistors are in series (connected end-to-end) they each carry the same current, but the potential differences across them are not equal (unless the resistors are equal). The resistors add directly to give the equivalent resistance of the series combination. *The equivalent resistance will always be **greater than** any of the individual resistances.*

- When two or more resistors are connected in parallel, the potential differences across them are the same. Since the current is inversely proportional to the resistance, the currents through them are not equal (unless the resistances are equal). The equivalent resistance of a parallel combination of resistors is found through reciprocal addition. *The equivalent resistance will always be **less than** the resistance of the smallest individual resistor.*

- A multiloop circuit consisting of resistors in series and parallel can often be reduced to a simple circuit containing only one resistor. Examine the circuit and replace any resistors that are in simple series or simple parallel, using Equations 21.26, $R_{eq} = R_1 + R_2 + R_3 + \cdots$, and Equation 21.28, $\dfrac{1}{R_{eq}} = \dfrac{1}{R_1} + \dfrac{1}{R_2} + \dfrac{1}{R_3} + \cdots$, and sketch the resulting circuit. Repeat this process (sketching the new circuit after each step in the simplification) until a single equivalent resistance is found.

- If the current through, or the potential difference across, a resistor in a multiloop circuit is to be determined, start with the final equivalent circuit found in the step above, and gradually work your way back through the circuits, using $V = IR$ and the properties of resistors in series and parallel at each step. Use the series of diagrams made during the simplification of the original circuit.

A Strategy for Using Kirchhoff's Rules

- First, draw the circuit diagram and assign labels and symbols to all the known and unknown quantities. *You must assign directions to the currents in each part of the circuit.* Do not be alarmed if your assumption for the direction of a current is wrong; the resulting value will be negative, but its magnitude will be correct. Although the assignment of current directions is arbitrary, you must stick with your guess throughout the problem as you apply Kirchhoff's rules. *A resulting current with a negative value has a direction opposite that which you assumed.*

- Apply the junction rule to any junction point in the circuit. The junction rule may be applied as many times as a new current (one not used in a previous application) appears in the resulting equation. *In general, the number of times the junction rule can be used is one fewer than the number of junction points in the circuit.*

- Now apply Kirchhoff's loop rule to as many loops in the circuit as are needed to solve for the unknowns. Remember you must have as many equations as there are unknowns (I's, R's, and $\mathcal{E}$'s). In order to apply this rule, you must correctly identify the change in potential as you cross each element in traversing the closed loop. Watch out for signs!

- Convenient rules which you may use to determine the change in potential as you cross a resistor or source of emf in traversing a circuit loop are illustrated below.

The potential decreases (changes by $-IR$) when a resistor is traversed in the direction of the current.

$$V_b - V_a = -IR$$

The potential increases (changes by $+IR$) when a resistor is traversed opposite the direction of the current.

$$V_b - V_a = +IR$$

The potential increases when a source of emf is traversed in the direction of the emf (from $-$ to $+$).

$$V_b - V_a = \mathcal{E}$$

The potential decreases when a source of emf is traversed opposite the direction of the emf (from $+$ to $-$).

$$V_b - V_a = -\mathcal{E}$$

- Finally, you must solve the equations simultaneously for the unknown quantities. Be careful of signs in your algebraic steps, and check your numerical answers for consistency. Do your values for the currents agree with the junction point and loop rules?

As an illustration of the use of Kirchhoff's rules, consider a three-loop circuit, which has the general form shown in the figure at left below. In this illustration, the actual circuit elements (R's and $\mathcal{E}$'s) are not shown but are assumed known.

There are six possible different values of I in the circuit; therefore you will need six independent equations to solve for the six values of I. There are four junction points in the circuit (at points a, d, f, and h). The first rule applied at any three of these points will yield three equations. The circuit can be thought of as a group of three "blocks" or meshes as shown in the figure at right below. Kirchhoff's second law, when applied to each of these loops (*abcda*, *ahfga*, and *defhd*), will yield three additional equations.

You can then solve the total of six equations simultaneously for the six values of $I_1, I_2, I_3, I_4, I_5,$ and I_6. You can, of course, expect that the sum of the changes in potential difference around any other closed loop in the circuit will be zero (for example, *abcdefga* or *ahfedcba*); *however the equations found by applying Kirchhoff's second rule to these additional loops will not be independent of the six equations found previously.*

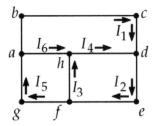

Assignment of currents
in a multiloop circuit

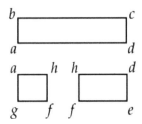

Three possible loops
for the circuit at left

REVIEW CHECKLIST

- Define electric current in terms of the rate of charge flow; and define the ampere as the unit of current. Calculate electron drift velocity, and the quantity of charge passing a point in a given time interval in a specified current-carrying conductor.

- Make calculations of the variation of resistance with temperature, which involves the concept of the temperature coefficient of resistivity.

- Determine the power delivered to a resistor.

- Calculate the equivalent resistance of a group of resistors in parallel, series, or series-parallel combination.

- Apply Kirchhoff's rules to solve multiloop circuits; that is, find the currents and the potential difference between any two points.

- Calculate the charging (discharging) current $I(t)$ and the accumulated (residual) charge $q(t)$ during charging (discharging) of a capacitor in an *RC* circuit.

ANSWER TO AN OBJECTIVE QUESTION

14. The terminals of a battery are connected across two resistors in series. The resistances of the resistors are not the same. Which of the following statements are correct? Choose all that are correct. (a) The resistor with the smaller resistance carries more current than the other resistor. (b) The resistor with the larger resistance carries less current than the other resistor. (c) The current in each resistor is the same. (d) The potential difference across each resistor is the same. (e) The potential difference is greatest across the resistor closest to the positive terminal.

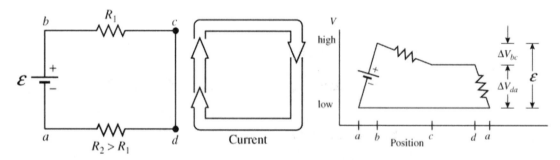

Figure OQ21.14

Answer You should learn the terminology and the standard symbols, so that from the two-sentence description of the circuit in the question, you can draw a circuit diagram, as in the first panel of our diagram here. And get into the habit of reasoning from the circuit diagram. The one correct answer is (c). There are no junctions in the circuit, so the current is the same everywhere, as in the second panel of the diagram, and statements (a) and (b) are false. If the second resistor has (say) twice the resistance of the first, the potential difference across it will be twice as large as across the first resistor. This relationship is described by $\Delta V = IR$, is shown in the third panel of the diagram, and makes statement (d) false. The *potential* is greatest at the positive terminal of the battery and at the end of the resistor connected to it; but the potential *difference* will be greater across the resistor of larger resistance, so statement (e) is false. The first panel of the diagram is the circuit schematic as it is usually drawn. It implies the information shown in the second and third panels of the diagram. Diagrams like these are not usually drawn, but you should learn to visualize them. The second panel shows the path of the current. The third panel is a graph of electric potential versus position in the circuit.

ANSWERS TO SELECTED CONCEPTUAL QUESTIONS

6. Use the atomic theory of matter to explain why the resistance of a material should increase as its temperature increases.

Answer As the temperature increases, the amplitude of atomic vibrations increases. This makes it more likely that the drifting electrons will be scattered by atomic vibrations, and makes it more difficult for charged particles to participate in organized motion inside the conductor.

This answer applies to the situation in metallic conduction, and generally to situations where the number of charge carriers is constant. But note that in a semiconductor the number of charge carriers may change with temperature, to make the resistance behave differently.

☐   ☐   ☐   ☐

8. (a) What advantage does 120-V operation offer over 240 V? (b) What disadvantages does it have?

Answer (a) Both 120-V and 240-V lines can deliver injurious or lethal shocks, but there is a somewhat better factor of safety with the lower voltage. To say it a different way, the insulation on a 120-V wire can be thinner. (b) On the other hand, a 240-V line carries less current to operate a device with the same power, so the conductor itself can be thinner.

☐   ☐   ☐   ☐

10. If charges flow very slowly through a metal, why does it not require several hours for a light to come on when you throw a switch?

Answer Individual electrons move with a small average velocity through the conductor, but as soon as the voltage is applied, electrons all along the conductor start to move. Actually the current does not exist "immediately" when the switch is turned on, but rather the establishment of the electric field throughout the conductor is limited by the speed of light.

☐   ☐   ☐   ☐

15. Why is it possible for a bird to sit on a high-voltage wire without being electrocuted?

Answer The resistance of the short segment of wire between the bird's feet is so small that the potential difference $\Delta V = IR$ between the feet is negligible. In order for the bird to be electrocuted, a much larger potential difference would be required.

It could also be said like this: A small amount of current does go through the bird's legs and body, according to the rule of parallel resistors. However, since the resistance of the bird's feet is much higher than that of the wire between them, the amount of current that exists in the bird is not enough to harm it.

☐   ☐   ☐   ☐

SOLUTIONS TO SELECTED END-OF-CHAPTER PROBLEMS

2.   Suppose the current in a conductor decreases exponentially with time according to the equation $I(t) = I_0 e^{-t/\tau}$, where I_0 is the initial current (at $t = 0$) and τ is a constant having dimensions of time. Consider a fixed observation point within the conductor. (a) How much charge passes this point between $t = 0$ and $t = \tau$? (b) How much charge passes this point between $t = 0$ and $t = 10\tau$? (c) **What If?** How much charge passes this point between $t = 0$ and $t = \infty$?

Solution

Conceptualize: The exponential function of Euler's number to the negative power t/τ describes a current starting with its maximum value and decreasing to 36.9% of its previous value in every time interval τ thereafter. The answer to (b) must be larger than the answer to (a), but less than ten times larger. It is not obvious that the answer to (c) is finite.

Categorize: If the current were constant, we would just multiply its value by a time interval to find the charge passing in that time. Here the current is always changing, but we can use its definition to find the charge by integration.

Analyze: From $I = \dfrac{dQ}{dt}$, we have $dQ = I\,dt$.

From this, we derive the general integral $Q = \int dQ = \int I \, dt$.

In all three cases, define an end-time, T: $\quad Q = \int_0^T I_0 e^{-t/\tau} dt$.

Integrating from time $t = 0$ to time $t = T$, $\quad Q = \int_0^T (-I_0 \tau) e^{-t/\tau} \left(-\dfrac{dt}{\tau} \right)$.

We perform the integral and set $Q = 0$ at $t = 0$ to obtain

$$Q = -I_0 \tau \left(e^{-T/\tau} - e^0 \right) = I_0 \tau \left(1 - e^{-T/\tau} \right)$$

(a) If $T = \tau$, $\quad Q = I_0 \, \tau \, (1 - e^{-1}) = 0.632 \, 1 \, I_0 t$. ∎

(b) If $T = 10\tau$, $\quad Q = I_0 \, \tau \, (1 - e^{-10}) = 0.999 \, 95 \, I_0 t$. ∎

(c) If $T = \infty$, $\quad Q = I_0 \tau \, (1 - e^{-\infty}) = I_0 \tau$. ∎

Finalize: The current decreases steeply enough in time that the total charge it transports is finite. In fact, a large majority of the charge is transported in the first "time constant" τ and nearly all of the charge in a time of 10τ. We will see in the next chapter that a capacitor discharging through a resistor produces a current with just this pattern of change in time.

———————————————

5. The electron beam emerging from a certain high-energy electron accelerator has a circular cross section of radius 1.00 mm. (a) The beam current is 8.00 μA. Find the current density in the beam assuming it is uniform throughout. (b) The speed of the electrons is so close to the speed of light that their speed can be taken as 300 Mm/s with negligible error. Find the electron density in the beam. (c) Over what time interval does Avogadro's number of electrons emerge from the accelerator?

Solution

Conceptualize: The current density can be several amps per square meter, as for a small current in a metal wire. In a metal wire the electron density is high and the speed is low, but here the speed is high and the electron density is very low. Avogadro's number is so large that a long time will be required for that number of electrons to go past.

Categorize: This current is not in a wire, but we can use the same ideas of electron density and current density.

Analyze:

(a) The current density is

$$J = \frac{I}{A} = \frac{8.00 \times 10^{-6} \text{ A}}{\pi \left(1.00 \times 10^{-3} \text{ m}\right)^2} = 2.55 \text{ A/m}^2$$ ∎

(b) From $J = nev_d$, we have

$$n = \frac{J}{ev_d} = \frac{2.55 \text{ A/m}^2}{\left(1.60 \times 10^{-19} \text{ C}\right)\left(3.00 \times 10^8 \text{ m/s}\right)}$$

$$n = 5.31 \times 10^{10} \text{ m}^{-3}$$ ∎

(c) From $I = \Delta Q / \Delta t$, we have

$$\Delta t = \frac{\Delta Q}{I} = \frac{N_A e}{I} = \frac{\left(6.02 \times 10^{23}\right)\left(1.60 \times 10^{-19} \text{ C}\right)}{8.00 \times 10^{-6} \text{ A}}$$

$$\Delta t = 1.20 \times 10^{10} \text{ s (or about 382 years!)}$$ ∎

Finalize: In answer (b), tens of billions of electrons per cubic meter counts as a low density, compared to the density in matter. The problem mentions a high-energy accelerator, but it could be about the electron beam in a television picture tube. An electron-beam tube, a chemical battery, a fuel cell, metal wires, and an electrochemical plating tank can all be in the same electric circuit, carrying the same current, described everywhere by $nqv_d A$, with wildly different values for the factors in this expression.

---

9. An aluminum wire with a diameter of 0.100 mm has a uniform electric field of 0.200 V/m imposed along its entire length. The temperature of the wire is 50.0°C. Assume one free electron per atom. (a) Use the information in Table 21.1 and determine the resistivity of aluminum at this temperature. (b) What is the current density in the wire? (c) What is the total current in the wire? (d) What is the drift speed of the conduction electrons? (e) What potential difference must exist between the ends of a 2.00-m length of the wire to produce the stated electric field?

Solution

Conceptualize: Aluminum is a very good conductor but this wire is quite thin. We expect a current of some milliamps driven by the (0.2 V/m)(2 m) = 0.4 V potential difference. We expect a drift speed in meters per hour.

Categorize: We will use the definition of the temperature coefficient of resistivity. Knowing that conductivity = 1/resistivity, we can find the current density directly from the applied field.

Analyze:

(a)   The resistivity is computed from $\rho = \rho_0\left[1 + \alpha(T - T_0)\right]$:

$$\rho = (2.82 \times 10^{-8}\ \Omega \cdot m)\left[1 + (3.90 \times 10^{-3}\ °C^{-1})(30.0°C)\right]$$
$$= 3.15 \times 10^{-8}\ \Omega \cdot m \qquad \blacksquare$$

(b)   The current density is

$$J = \sigma E = \frac{E}{\rho} = \left(\frac{0.200\ \text{V/m}}{3.15 \times 10^{-8}\ \Omega \cdot m}\right)\left(\frac{1\ \Omega \cdot A}{V}\right) = 6.35 \times 10^6\ \text{A/m}^2 \qquad \blacksquare$$

(c)   The current density is related to the current by $J = \dfrac{I}{A} = \dfrac{I}{\pi r^2}$.

$$I = J\pi r^2 = (6.35 \times 10^6\ \text{A/m}^2)[\pi (5.00 \times 10^{-5}\ m)^2] = 49.9\ \text{mA} \qquad \blacksquare$$

(d)   The mass density gives the number-density of free electrons; we assume that each atom donates one conduction electron:

$$n = \left(\frac{2.70 \times 10^3\,\text{kg}}{m^3}\right)\left(\frac{1\ \text{mol}}{26.98\ \text{g}}\right)\left(\frac{10^3\,\text{g}}{\text{kg}}\right)\left(\frac{6.02 \times 10^{23}\ \text{free } e^-}{1\ \text{mol}}\right)$$
$$= 6.02 \times 10^{28}\ e^-/m^3$$

Now $J = nqv_d$ gives the drift speed as

$$v_d = \frac{J}{nq} = \frac{6.35 \times 10^6\ \text{A/m}^2}{(6.02 \times 10^{28}\ e^-/m^3)(-1.60 \times 10^{-19}\ \text{C/}e^-)}$$
$$= -6.59 \times 10^{-4}\ \text{m/s} \qquad \blacksquare$$

The sign indicates that the electrons drift opposite to the field and current.

(e) The applied voltage is $\Delta V = E\ell = (0.200 \text{ V/m})(2.00 \text{ m}) = 0.400 \text{ V}$. ∎

Finalize: It would be natural to compute the resistance of the two-meter strand as 8.02 Ω, but our set of equations is overcomplete and makes that identification optional. Observe that a "typical tabletop" current density in a metal wire is in megamperes per square meter. Aluminum and copper, quite different in mass density, have very similar densities of conduction electrons, on the order of 10^{29} per cubic meter.

═══════════

17. If the magnitude of the drift velocity of free electrons in a copper wire is 7.84×10^{-4} m/s, what is the electric field in the conductor?

Solution

Conceptualize: For electrostatic cases, we learned that the electric field inside a conductor is always zero. On the other hand, if there is a current, a non-zero electric field must be maintained by a battery or other source to make the charges flow. Therefore, we might expect the electric field to be small, but definitely **not** zero.

Categorize: The drift velocity of the electrons can be used to find the current density, which can be used with the definition of conductivity to find the electric field inside the conductor.

Analyze: We choose to find first the electron density in copper. Its mass density is 8.92 g/cm^3 and from a periodic table its molar mass is 63.5 g/mol, so the number density of atoms is

$$n = \left(8.92 \frac{\text{g}}{\text{cm}^3} \right) \left(\frac{1 \text{ mol}}{63.5 \text{ g}} \right) \left(\frac{6.02 \times 10^{23} \text{ atoms}}{1 \text{ mol}} \right) \left(\frac{1 \times 10^6 \text{ cm}^3}{1 \text{ m}^3} \right)$$

$$= 8.46 \times 10^{28} / \text{m}^3$$

We assume that each atom contributes one conduction electron. Then the density of conduction electrons is this same number.

The current density in this wire is then

$$J = nqv_d = (8.46 \times 10^{28} \text{ e}^-/\text{m}^3)(1.60 \times 10^{-19} \text{ C/e}^-)(7.84 \times 10^{-4} \text{ m/s})$$

$$J = 1.06 \times 10^7 \text{ A/m}^2$$

The definition of conductivity can be stated as

$$J = \sigma E = E/\rho$$

where $\rho = 1.7 \times 10^{-8} \ \Omega \cdot m$ for copper,

so then $E = \rho J = (1.70 \times 10^{-8} \ \Omega \cdot m)(1.06 \times 10^{7} \ A/m^2) = 0.180 \ V/m.$ ∎

Finalize: This electric field is certainly smaller than typical static values outside charged objects. The direction of the electric field should be along the length of the conductor, for otherwise the electrons would be forced to leave the wire! In fact, excess charges arrange themselves on the surface of the wire, notably at bends in it, to create an electric field that "steers" the free electrons to flow along the length of the wire from low to high potential (opposite the direction of a positive test charge). It is also interesting to note that when the electric field is being established it travels at the speed of light; but the drift velocity of the electrons is literally a "snail's pace"!

---

21. A certain toaster has a heating element made of Nichrome wire. When the toaster is first connected to a 120-V source (and the wire is at a temperature of 20.0°C), the initial current is 1.80 A. The current decreases as the heating element warms up. When the toaster reaches its final operating temperature, the current is 1.53 A. (a) Find the power delivered to the toaster when it is at its operating temperature. (b) What is the final temperature of the heating element?

Solution

Conceptualize: Most toasters are rated at about 1 000 W (usually stamped on the bottom of the unit), so we might expect this one to have a similar power rating. The temperature of the heating element should be hot enough to toast bread but low enough that the nickel-chromium alloy element does not melt. (The melting point of nickel is 1 455°C, and chromium melts at 1 907°C.)

Categorize: The power can be calculated directly by multiplying the current and the voltage. The temperature can be found from the linear resistivity equation for Nichrome, with $\alpha = 0.4 \times 10^{-3} \ °C^{-1}$ from Table 21.1.

Analyze:

(a) The power at the operating point is

$$P = \Delta VI = (120 \ V)(1.53 \ A) = 184 \ W$$ ∎

(b) The resistance at 20.0°C is

$$R_0 = \frac{\Delta V}{I} = \frac{120 \text{ V}}{1.80 \text{ A}} = 66.7 \text{ }\Omega$$

At the operating temperature,

$$R = \frac{120 \text{ V}}{1.53 \text{ A}} = 78.4 \text{ }\Omega$$

Neglecting thermal expansion, we have

$$R = \frac{\rho \ell}{A} = \frac{\rho_0 \left[1 + \alpha(T - T_0)\right]\ell}{A} = R_0 \left[1 + \alpha(T - T_0)\right]$$

$$T = T_0 + \frac{R/R_0 - 1}{\alpha} = 20.0°C + \frac{78.4 \text{ }\Omega/66.7 \text{ }\Omega - 1}{0.4 \times 10^{-3}°C^{-1}} = 461°C \qquad \blacksquare$$

Finalize: Although this toaster appears to use significantly less power than most, the temperature seems high enough to toast a piece of bread in a reasonable amount of time. The absolute temperature of the filament in a typical 1 000-W toaster would be much less than five times higher because Stefan's radiation law (Equation 17.36) tells us that (assuming all power is lost through radiation) $T \propto \sqrt[4]{P}$, so that the temperature might be about 850°C. In either case, the operating temperature is well below the melting point of the heating element.

═══════════════════════

27. Assuming the cost of energy from the electric company is \$0.110/kWh, compute the cost per day of operating a lamp that draws a current of 1.70 A from a 110-V line.

Solution

Conceptualize: We estimate cheaper than a fancy cup of coffee.

Categorize: This problem is about identifying what we buy from the electric company. It is not charge, current, field, potential, potential difference, resistance, or power, but …

Analyze: … energy. The power of the lamp is $P = I\,\Delta V = U/\Delta t$, where U is the energy transformed. Then the energy you buy, in standard units, is

$$U = \Delta V I \Delta t$$

$$= (110 \text{ V})(1.70 \text{ A})(1 \text{ day})\left(\frac{24 \text{ h}}{1 \text{ day}}\right)\left(\frac{3\,600 \text{ s}}{\text{h}}\right)\left(\frac{1 \text{ J}}{\text{V}\cdot\text{C}}\right)\left(\frac{1 \text{ C}}{\text{A}\cdot\text{s}}\right)$$

$$= 16.2 \text{ MJ}$$

In kilowatt hours, the energy is

$$U = \Delta VI\Delta t$$

$$= (110 \text{ V})(1.70 \text{ A})(1 \text{ day})\left(\frac{24 \text{ h}}{1 \text{ day}}\right)\left(\frac{1 \text{ J}}{\text{V} \cdot \text{C}}\right)\left(\frac{1 \text{ C}}{\text{A} \cdot \text{s}}\right)\left(\frac{\text{W} \cdot \text{s}}{\text{J}}\right)$$

$$= 4.49 \text{ kWh}$$

So operating the lamp costs (4.49 kWh)($0.110/kWh) = 49.4 cents.   ∎

Finalize: This is a 187-W bulb at 110 V, so it might be sold in the hardware store as a 200-W bulb at 120 V. The good news is that it is not expensive to operate. If your goal is saving energy and money, the bad news is that it does not help a lot to avoid using an electric pencil sharpener or corn popper. The heavy hitters in household energy use may be the furnace, air conditioner, water heater, and refrigerator; but your car can dwarf them all.

---

31. An all-electric car (not a hybrid) is designed to run from a bank of 12.0-V batteries with total energy storage of 2.00×10^7 J. If the electric motor draws 8.00 kW as the car moves at a steady speed of 20.0 m/s, (a) what is the current delivered to the motor? (b) How far can the car travel before it is "out of juice"?

Solution

Conceptualize: We guess a current on the order of a hundred amperes. A practical car should travel on the order of a hundred km on one charge.

Categorize: We need not know details about the car's construction to follow the electrically transmitted input and the mechanical output of the motor.

Analyze:

(a) Since $P = I\Delta V$, we have $I = \dfrac{P}{\Delta V} = \dfrac{8.00 \times 10^3 \text{ W}}{12.0 \text{ V}} = 667$ A.   ∎

(b) From $P = U/\Delta t$, the time the car runs is

$$\Delta t = \frac{U}{P} = \left(\frac{2.00 \times 10^7 \text{ J}}{8.00 \times 10^3 \text{ W}}\right)\left(\frac{1 \text{ W} \cdot \text{s}}{\text{J}}\right) = 2.50 \times 10^3 \text{ s}$$

So it moves a distance of

$$\Delta x = v\Delta t = (20.0 \text{ m/s})(2.50 \times 10^3 \text{ s}) = 50.0 \text{ km}$$ ∎

Finalize: A fifty-kilometer range can be sufficient for commuting and local deliveries, but this car does not sound very practical. A current of hundreds of amperes is hard to handle. Connecting some of the batteries in series to run the motor at higher voltage could let it run on less current.

═══════════════

33. A battery has an emf of 15.0 V. The terminal voltage of the battery is 11.6 V when it is delivering 20.0 W of power to an external load resistor R. (a) What is the value of R? (b) What is the internal resistance of the battery?

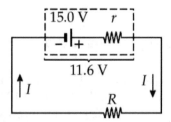

Solution

Conceptualize: The internal resistance of a battery usually is less than 1 Ω with physically larger batteries having less resistance due to the larger anode and cathode areas. The voltage of this battery drops significantly (23%) when the load resistance is added, so a sizable amount of current must be drawn from the battery. If we assume that the internal resistance is about 1 Ω, then the current must be about 3 A to give the 3.4-V drop across the battery's internal resistance. If this is true, then the load resistance must be about $R \approx 12 \text{ V}/3 \text{ A} = 4 \Omega$.

Categorize: We can find R precisely by using the power delivered to the load resistor when the voltage is 11.6 V. Then we can find the internal resistance of the battery by summing the electric potential differences around the circuit.

Analyze:

(a) Combining Joule's law, $P = I\Delta V$, and the definition of resistance, $\Delta V = IR$, gives

$$R = \frac{(\Delta V)^2}{P} = \frac{(11.6 \text{ V})^2}{20.0 \text{ W}} = 6.73 \ \Omega \qquad \blacksquare$$

(b) The electromotive force of the battery must equal the voltage drops across the resistances: $\mathcal{E} = IR + Ir$, where $I = \Delta V/R$.

$$r = \frac{(\mathcal{E} - IR)}{I} = \frac{(\mathcal{E} - \Delta V)R}{\Delta V} = \frac{(15.0 \text{ V} - 11.6 \text{ V})(6.73 \ \Omega)}{11.6 \text{ V}} = 1.97 \ \Omega \qquad \blacksquare$$

Finalize: The resistance of the battery is larger than 1 Ω, but it is reasonable for an old battery or for a battery consisting of several small electric cells in series. The load resistance agrees reasonably well with our prediction, despite the fact that the battery's internal resistance was about twice as large as we assumed.

Note that in our initial guess we did not consider the power delivered to the load resistance; however, this datum is required for an accurate solution.

═══════════

39. Consider the circuit shown in Figure P21.39. Find (a) the current in the 20.0-Ω resistor and (b) the potential difference between points a and b.

Solution

Conceptualize: The current will be from b to a through the 20-Ω and then the 5-Ω resistors, perhaps about half an amp. Then b will be at higher potential than a, perhaps by about 10 V.

10 Ω 25 V

a | 10 Ω | b

5.0 Ω

5.0 Ω

20 Ω

Figure P21.39

Categorize: The problem might as well ask for the current in every resistor, because we must analyze the whole circuit to find the current in any one resistor. This circuit has only one power supply, so we can try ideas of equivalent resistances for series and parallel combinations.

Analyze: If we turn the given diagram on its side, we find that it is the same as Figure (a). The 20.0-Ω and 5.00-Ω resistors are in series, so the first reduction is as shown in (b).

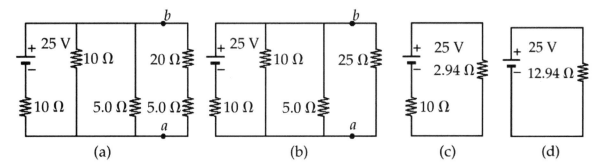

(a) (b) (c) (d)

Next, since the 10.0-Ω, 5.00-Ω, and 25.0-Ω resistors are then in parallel, we can compute their equivalent resistance as

$$R_{eq} = \frac{1}{(1/10.0\ \Omega) + (1/5.00\ \Omega) + (1/25.0\ \Omega)} = 2.94\ \Omega$$

This is shown in Figure (c), which in turn reduces to the circuit shown in (d).

Next, we work backwards through the diagrams, applying $I = \Delta V / R$ and $\Delta V = IR$ alternately to each resistor, real or equivalent. The 12.94-Ω resistor is connected across 25.0 V, so the current through the voltage source in every diagram is

$$I = \frac{\Delta V}{R} = \frac{25.0 \text{ V}}{12.94 \text{ }\Omega} = 1.93 \text{ A}$$

In Figure (c), this 1.93 A goes through the 2.94-Ω equivalent resistor to give a voltage drop across this resistor of

$$\Delta V = IR = (1.93 \text{ A})(2.94 \text{ }\Omega) = 5.68 \text{ V}$$

From Figure (b), we see that this voltage drop is the same across ΔV_{ab}, the 10-Ω resistor, and the 5.00-Ω resistor.

(b) It happens that the answer to part (b) emerges first, as $\Delta V_{ab} = 5.68$ V.   ∎

(a) Since the current through the 20-Ω resistor is also the current through the 25.0-Ω line *ab*, this current is

$$I = \frac{\Delta V_{ab}}{R_{ab}} = \frac{5.68 \text{ V}}{25.0 \text{ }\Omega} = 0.227 \text{ A}$$ ∎

Finalize: Each step in the solution is small, but you may be surprised at how many steps are required. We could not use $\Delta V = IR$ in the original circuit because there was not a single resistor for which we knew either ΔV or I. It really was necessary to draw all of the diagrams (a), (b), (c), and (d). Both answers are smaller than our guesses. Most of the battery current goes through the 5-Ω resistor in the center of the original diagram.

---

43. Calculate the power delivered to each resistor in the circuit shown in Figure P21.43.

Solution

Conceptualize: If the current in the battery is a few amperes, the resistor powers will add up to about 50 watts. The 4-Ω resistor will have twice the power of the 2-Ω resistor since these two carry the same current. Three times more power will be delivered to the 1-Ω resistor compared to the 3-Ω resistor, because the same potential difference is applied to both and the 1-Ω resistor carries more current.

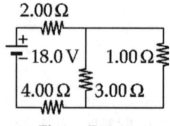

Figure P21.43

Categorize: The problem might just as well ask for the current in each resistor, as a step toward finding the power delivered to each. In turn, to find the current in each resistor, we . . .

Analyze: . . . find the resistance seen by the battery. The given circuit reduces as shown in Figure (a), since

$$\frac{1}{(1/1.00\ \Omega) + (1/3.00\ \Omega)} = 0.750\ \Omega$$

In Figure (b), $I = 18.0\ \text{V}/6.75\ \Omega = 2.67\ \text{A}$.

This is also the current in (a), so the 2.00-Ω and 4.00-Ω resistors convert powers

$$P_2 = I\Delta V = I^2R = (2.67\ \text{A})^2(2.00\ \Omega) = 14.2\ \text{W}$$

and $\quad P_4 = I^2R = (2.67\ \text{A})^2(4.00\ \Omega) = 28.4\ \text{W}.$

The voltage across the 0.750-Ω resistor in Figure (a), and across both the 3.00-Ω and the 1.00-Ω resistors in Figure P21.43, is

$$\Delta V = IR = (2.67\ \text{A})(0.750\ \Omega) = 2.00\ \text{V}$$

Then for the 3.00-Ω resistor,

$$I = \frac{\Delta V}{R} = \frac{2.00\ \text{V}}{3.00\ \Omega}$$

and the power is

$$P_3 = I\Delta V = \left(\frac{2.00\ \text{V}}{3.00\ \Omega}\right)(2.00\ \text{V}) = 1.33\ \text{W}$$

For the 1.00-Ω resistor,

$$I = \frac{2.00\ \text{V}}{1.00\ \Omega} \quad \text{and} \quad P_1 = \left(\frac{2.00\ \text{V}}{1.00\ \Omega}\right)(2.00\ \text{V}) = 4.00\ \text{W}$$

Finalize: The total power of the resistors, 14.2 W + 28.4 W + 1.33 W + 4.00 W = 48 W agrees with our prediction and with the battery power (18 V)(2.67A) = 48.0 W. Every small step in the solution was necessary.

47. The circuit shown in Figure P21.47 is connected for 2.00 min. (a) Determine the current in each branch of the circuit. (b) Find the energy delivered by each battery. (c) Find the energy delivered to each resistor. (d) Identify the type of energy storage transformation that occurs in the operation of the circuit. (e) Find the total amount of energy transformed into internal energy in the resistors.

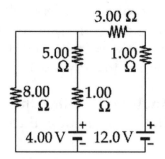

Solution

Conceptualize: We estimate a current of about 1 A in the 12-V battery and in the 8-Ω resistor, and a much smaller current in the 4-V battery. Then the total energy conversion might be about (25 J/s)(120 s) = 3 000 J.

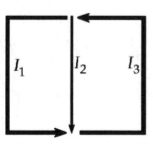

Figure P21.47

Categorize: Many students might start their estimates, or even their solutions, with thinking about 12 V/4 Ω = 3 A and 4 V/6 Ω = 0.7 A, in the circuit diagram. This is quite unproductive—those voltages do not go with those resistances. You may wish to just carefully follow the steps of the Kirchhoff's-rules method. Or you may want to visualize a mountain landscape. The bottom line (wire) in the circuit diagram is at a uniform low potential which we naturally take as zero volts. The highest peak in the landscape is the positive pole of the 12-V battery. From this point, charge rolls downhill through the resistors and perhaps backward—from positive pole to negative pole—through the 4-V battery. After we find the three currents in part (a), it will in principle be easy to multiply voltage and current to find the power converted by each battery and resistor. Multiplying each of the seven numbers for power by 120 s will give the energy converted while the circuit is connected.

Analyze for part (a): We use the method of Kirchhoff's rules. First, we arbitrarily assign current directions and names, as shown in the second figure. The current rule then says that

$$I_3 - I_1 - I_2 = 0 \quad \text{or} \quad I_3 = I_1 + I_2 \qquad \text{[1]}$$

By the voltage rule, clockwise around the left-hand loop,

$$+ I_1(8.00 \ \Omega) - I_2(5.00 \ \Omega) - I_2(1.00 \ \Omega) - 4.00 \ \text{V} = 0 \qquad \text{[2]}$$

Clockwise around the right-hand loop, combining the 1- and 5-Ω resistors and the 1- and 3-Ω resistors, the voltage rule reads

$$+ 4.00 \text{ V} + I_2(6.00 \ \Omega) + I_3(4.00 \ \Omega) - 12.0 \text{ V} = 0 \qquad\qquad [3]$$

We check that we have three independent equations in three unknowns. To solve, we substitute $(I_1 + I_2)$ for I_3, and reduce our three equations to the two equations

$$(8.00 \ \Omega)I_1 - (6.00 \ \Omega)I_2 - 4.00 \text{ V} = 0$$

and $\qquad 4.00 \text{ V} + (6.00 \ \Omega)I_2 + (4.00 \ \Omega)(I_1 + I_2) - 12.0 \text{ V} = 0.$

Solving the first of these equations for I_2 gives

$$I_2 = \frac{(8.00 \ \Omega)I_1 - 4.00 \text{ V}}{6.00 \ \Omega}$$

and rearranging the second of our pair of equations gives

$$I_1 + \frac{10.00 \ \Omega}{4.00 \ \Omega} I_2 = \frac{8.00 \text{ V}}{4.00 \ \Omega}$$

We substitute for I_2 to get down to one equation in one unknown:

$$I_1 + 3.33 I_1 - 1.67 \text{ V}/\Omega = 2.00 \text{ V}/\Omega$$

We solve for the current I_1, which is down the 8-Ω resistor:

$$I_1 = 0.846 \text{ V}/\Omega = 0.846 \text{ A} \qquad\qquad \blacksquare$$

If we were solving the three simultaneous equations by determinants or by calculator matrix inversion, we would have to do much more work to find I_2 and I_3. Our method of solving by substitution is more generally useful, and makes it easy to work out the remaining answers after the first. We substitute again, putting the value for I_1 into equations we already solved for the other unknowns. Thus,

$$I_2 = \frac{(8.00 \ \Omega)(0.846 \text{ A}) - 4.00 \text{ V}}{6.00 \ \Omega}$$
$$= 0.462 \text{ A down in the middle branch} \qquad\qquad \blacksquare$$

and

$$I_3 = 0.846 \text{ A} + 0.462 \text{ A} = 1.31 \text{ A up in the right-hand branch.} \qquad \blacksquare$$

Finalize for part (a): Trust us: there is no method that is really simpler. When resistors are in series, like the 3-Ω and 1-Ω resistors, their equivalent resistance will automatically appear in the loop equation. No resistors are in parallel with each other. At the first analysis step, you might naturally assume that the 4-V battery pushes current out of its positive terminal to go upward rather than downward in the middle branch. Get some really good practice by writing out an entirely separate solution proceeding from this assumption. Prove to yourself that it reaches the same physical answers as the solution here.

Analyze for parts (b) through (e):

(b) The power converted by a battery is $P = \Delta VI = U/\Delta t$, so the energy U converted is $U = (\Delta V)I\Delta t$. For the 4.00-V battery, we have

$$U = (\Delta V)I\Delta t = (4.00 \text{ V})(-0.462 \text{ A})(120 \text{ s}) = -222 \text{ J} \qquad \blacksquare$$

We have counted the current as negative because it is opposite in direction to the direction of increasing potential in this battery. For the 12.0-V battery,

$$U = (\Delta V)I\Delta t = (12.0 \text{ V})(1.31 \text{ A})(120 \text{ s}) = 1.88 \text{ kJ} \qquad \blacksquare$$

(c) For a resistor (and only for a resistor) $\Delta V = IR$, so the energy converted $U = (\Delta V)I\Delta t$ becomes $U = I^2R\Delta t$. The energy delivered to the 8.00-Ω resistor is

$$U = I^2R\Delta t = (0.846 \text{ A})^2(8.00 \ \Omega)(120 \text{ s}) = 687 \text{ J} \qquad \blacksquare$$

For the 5.00-Ω resistor, $U = (0.462 \text{ A})^2 (5.00 \ \Omega)(120 \text{ s}) = 128 \text{ J}.$ $\qquad \blacksquare$

For the 1.00-Ω resistor in the center branch,

$$U = (0.462 \text{ A})^2(1.00 \ \Omega)(120 \text{ s}) = 25.6 \text{ J} \qquad \blacksquare$$

For the 3.00-Ω resistor, $U = (1.31 \text{ A})^2(3.00 \ \Omega)(120 \text{ s}) = 616 \text{ J}.$ $\qquad \blacksquare$

For the 1.00-Ω resistor in the right-hand branch,

$$U = (1.31 \text{ A})^2(1.00 \ \Omega)(120 \text{ s}) = 205 \text{ J} \qquad \blacksquare$$

(d) As the circuit carries current, the 12.0-V battery converts chemical energy into electrically transmitted energy. The 4.00-V battery is converting +222 J from electrically transmitted energy into chemical energy. All of the resistors are converting electrically transmitted energy into internal energy. The net transformation, in terms of energy storage, is from chemical energy into internal energy. $\qquad \blacksquare$

(e) The amount of energy transformed is altogether

$$687 \text{ J} + 128 \text{ J} + 25.6 \text{ J} + 616 \text{ J} + 205 \text{ J} = 1.66 \text{ kJ} \qquad \blacksquare$$

Finalize for parts (b) through (e): Part (e) was about the internal energy created in just the resistors. We can count the same energy as the batteries convert it from chemical into electrically transmitted:

$$+1.88 \text{ kJ} - 222 \text{ J} = 1.66 \text{ kJ}$$

Make sure that you do not think that the total energy converted is 1.66 kJ plus another 1.66 kJ. A hungry child does not have twice as much lunch money if he counts it at ten o'clock and again at eleven. Rather, the "net accumulation" of electrically transmitted energy is nonexistent, according to 1.66 kJ − 1.66 kJ = 0. Electrically transmitted energy is not stored energy and does not accumulate. Again, notice that $P = \Delta VI$ is true for both batteries and resistors, but $P = I^2R$ is true only for a resistor. A resistor can transform energy only from electrically transmitted into internal, but a battery with emf can transform energy either way between chemical and electrically transmitted. To be more complete, we add that a charging or discharging capacitor converts energy between electric potential energy and electrically transmitted energy. Later in the course we will see that an inductor converts energy between energy stored in its magnetic field and electrically transmitted energy.

---

49. Taking $R = 1.00 \text{ k}\Omega$ and $\mathcal{E} = 250 \text{ V}$ in Figure P21.49, determine the direction and magnitude of the current in the horizontal wire between a and e.

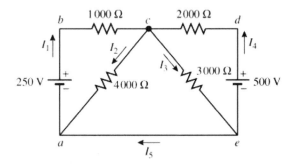

Figure P21.49

Solution

Conceptualize: Are you tempted to think that the wire ae carries no current, because no battery or resistor is in the branch that it forms? Or because it connects the low-potential terminals of two batteries? The negative terminal of one battery can be assigned the value zero volts, and then in this circuit the negative terminal of the other battery will also be at zero volts, but the zero-resistance wire will still carry a finite current. We can estimate its order of magnitude from the currents we might expect in the other branches, as a few hundred volts divided by a few thousand ohms, or on the order of a tenth of an amp.

Categorize: The 3 000-Ω and 4 000-Ω resistors are in parallel. But if we redrew an equivalent circuit diagram with a single equivalent resistance replacing them, the new circuit would not contain a branch with the current we are asked to find. For this reason, we choose to apply Kirchhoff's rules to the circuit as it stands. There are five branches, so there are five unknown currents. There are three junctions (one where four wires meet!) and we could think up many different trips around different large or small loops in the circuit. But we can minimize algebraic difficulties by carefully writing down a set of five equations before we think about solving them, and then doing just the algebra to solve for the one particular current required.

Analyze: We arbitrarily assign names I_1 through I_5 and possible directions to the currents in the different branches, as shown in our version of the diagram. We arbitrarily write junction equations just for junctions a and c. An equation for junction e would not be a mathematically independent equation.

$$\text{Junction } a: \quad -I_1 + I_2 + I_5 = 0$$

$$\text{Junction } c: \quad +I_1 - I_2 - I_3 + I_4 = 0$$

We need $5 - 2 = 3$ loop equations. We choose arbitrarily to write them for clockwise trips around the loops *abca*, *acea*, and *cdec*. (These smallest loops are called meshes. If we wrote any other loop equation, such as for the perimeter *abcdea*, it would just be a linear combination of the three equations we have chosen.)

$$\text{Loop } abca: \quad +250\text{ V} - I_1(1.00\text{ k}\Omega) - I_2(4.00\text{ k}\Omega) = 0$$

$$\text{Loop } acea: \quad +I_2(4.00\text{ k}\Omega) - I_3(3.00\text{ k}\Omega) = 0$$

$$\text{Loop } cdec: \quad +I_4(2.00\text{ k}\Omega) - 500\text{ V} + I_3(3.00\text{ k}\Omega) = 0$$

Our purpose is to solve for I_5. One sure-fire method is to eliminate other unknowns one by one. From the first equation, we substitute $I_1 = I_2 + I_5$ into each of the others, obtaining the four equations

$$I_2 + I_5 - I_2 - I_3 + I_4 = 0, \quad \text{or simplified} \quad I_3 = I_5 + I_4:$$

$$+250\text{ V} - I_2(1.00\text{ k}\Omega) - I_5(1.00\text{ k}\Omega) - I_2(4.00\text{ k}\Omega) = 0$$

$$+I_2(4.00\text{ k}\Omega) - I_3(3.00\text{ k}\Omega) = 0$$

$$+I_4(2.00\text{ k}\Omega) - 500\text{ V} + I_3(3.00\text{ k}\Omega) = 0$$

Next we make the free choice to substitute $I_3 = I_5 + I_4$ into each of the others, to get the three equations

$$+ 250 \text{ V} - I_2(5.00 \text{ k}\Omega) - I_5(1.00 \text{ k}\Omega) = 0$$

$$+ I_2(4.00 \text{ k}\Omega) - I_5(3.00 \text{ k}\Omega) - I_4(3.00 \text{ k}\Omega) = 0$$

$$+ I_4(2.00 \text{ k}\Omega) - 500 \text{ V} + I_5(3.00 \text{ k}\Omega) = I_4(3.00 \text{ k}\Omega) = 0$$

In just one more step, we can eliminate both the unknowns $I_2 = 0.050\ 0 \text{ A} - 0.200\ I_5$ and $I_4 = 0.100 \text{ A} - 0.600\ I_5$ (from the first and third equations) by substituting these expressions into the second equation:

$$4(0.050\ 0 \text{ A} - 0.200\ I_5) - 3\ I_5 - 3(0.100 \text{ A} - 0.600\ I_5) = 0$$

Clearing parentheses, $0.200 \text{ A} - 0.800\ I_5 - 3.00\ I_5 - 0.300 \text{ A} + 1.80\ I_5 = 0$.

Gathering like terms, $-2.00\ I_5 = 0.100 \text{ A}$.

And at last solving, $I_5 = -0.050\ 0 \text{ A}$. The fact that it is negative means that it is opposite to the direction we assumed.

This current is $+50.0$ mA from a to e. ∎

Finalize: We made a lot of arbitrary choices about which equations to write and what order to do the steps of algebra in. It is good practice to solve the problem again with a different set of choices—call the unknowns u, w, x, y, and z. You may need practice in working with sets of equations. At every stage, the number of equations must be equal to the number of remaining unknowns. Because we used the method of substitution, it would be easy to find all the other currents: equations like $I_4 = 0.100 \text{ A} - 0.600\ I_5$ would give us the answers directly.

═══════════

53. Consider a series RC circuit as in Figure P21.53 for which $R = 1.00$ MΩ, $C = 5.00\ \mu$F, and $\mathcal{E} = 30.0$ V. Find (a) the time constant of the circuit and (b) the maximum charge on the capacitor after the switch is thrown closed. (c) Find the current in the resistor 10.0 s after the switch is thrown closed.

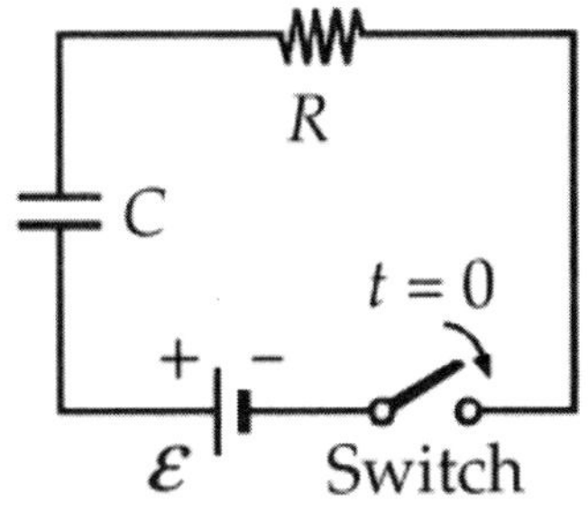

Figure P21.53

Solution

Conceptualize: The capacitor charges up after the switch is first thrown closed.

Categorize: The definition of the time constant, the definition of capacitance, and the equation for the time-dependent current in a charging-capacitor circuit will give the answers directly.

Analyze:

(a) The time constant is

$$\tau = RC = (1.00 \times 10^6 \ \Omega)(5.00 \times 10^{-6} \ F) = 5.00 \ \Omega \cdot F = 5.00 \ s \qquad \blacksquare$$

(b) After a long time interval, the capacitor is "charged to thirty volts," separating charges of

$$Q = C\Delta V = (5.00 \times 10^{-6} \ F)(30.0 \ V) = 150 \ \mu C \qquad \blacksquare$$

(c) $I = I_i e^{-t/\tau}$, where $I_i = \dfrac{\mathcal{E}}{R}$ and $\tau = RC = 5.00$ s.

$$I = \frac{\mathcal{E}}{R} e^{-t/\tau} = \left(\frac{30.0 \ V}{1.00 \times 10^6 \ \Omega} \right) e^{-10.0 \ s / 5.00 \ s}$$

$$I = 4.06 \times 10^{-6} \ A = 4.06 \ \mu A \qquad \blacksquare$$

Finalize: The capacitor starts to charge instantly, at time zero, when the switch is closed. It takes an infinite time to finish charging. Still it has a well defined time constant of five seconds. We can think of this as a fairly long time interval, associated with the large resistance in the circuit. Learn to do the proof that RC has units of time:

$$\Omega \cdot F = \frac{V}{A} \frac{C}{V} = \frac{C}{C/s} = s$$

---

59. The circuit in Figure P21.59 has been connected for a long time. (a) What is the potential difference across the capacitor? (b) If the battery is disconnected from the circuit, over what time interval does the capacitor discharge to one tenth of its initial voltage?

Solution

Conceptualize: The network of resistors contains voltage dividers so that the voltage across the capacitor will be a few volts, less than 10 V. The capacitor will discharge through an equivalent resistance on the order of a few ohms. With a microfarad capacitance, the time scale of its discharge is in microseconds.

Categorize: We must use equivalent-resistance ideas in two quite different circuits to do the separate parts (a) and (b).

Analyze:

(a) After a long time interval, the capacitor branch will carry negligible current. The current is as shown in Figure (a). To find the voltage at point *a*, we first find the current, using the voltage rule:

$$10.0 \text{ V} - (1.00 \ \Omega)I_2 - (4.00 \ \Omega)\, I_2 = 0$$

$$I_2 = 2.00 \text{ A}$$

$$V_a - V_0 = (4.00 \ \Omega)I_2 = 8.00 \text{ V}$$

Similarly for the right-hand branch,

$$10.0 \text{ V} - (8.00 \ \Omega)I_3 - (2.00 \ \Omega)I_3 = 0$$

$$I_3 = 1.00 \text{ A}$$

At point *b*, $V_b - V_0 = (2.00 \ \Omega)I_3 = 2.00$ V.

Thus, the voltage across the capacitor is

$$V_a - V_b = 8.00 \text{ V} - 2.00 \text{ V} - 6.00 \text{ V}$$

(b) We suppose the battery is pulled out leaving an open circuit. We are left with Figure (b), which can be reduced to equivalent circuits (c) and (d). From (d), we can see that the capacitor discharges through a 3.60-Ω equivalent resistance.

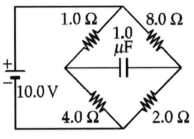

Figure P21.59

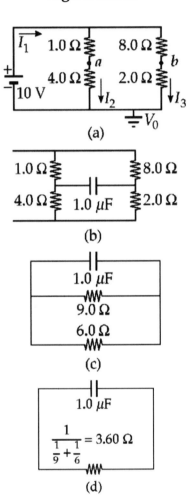

(a)

(b)

(c)

(d)

Starting with $q = Qe^{-t/RC}$, we multiply both sides by *C*, which gives

$$qC = QCe^{-t/RC} \quad \text{and} \quad \Delta V = \Delta V_i e^{-t/RC}.$$

We proceed to solve for *t*:

$$V/V_i = e^{-t/RC} \quad \text{or} \quad V_i/V = e^{+t/RC}$$

Take natural logarithms of both sides:

$$\ln(V_i/V) = +t/RC$$

$$t = RC \ln(V_i/V) = (3.60 \ \Omega)(1.00 \ \mu F) \ln(V_i/0.100V_i)$$

$$t = 3.60 \ \mu s \ \ln(10) = 3.60 \ \mu s \ (2.30) = 8.29 \ \mu s \qquad \blacksquare$$

Finalize: Our estimates were good. Every circuit diagram was necessary to draw.

---

75. An electric heater is rated at 1.50×10^3 W, a toaster at 750 W, and an electric grill at 1.00×10^3 W. The three appliances are connected to a common 120-V household circuit. (a) How much current does each draw? (b) If the circuit is protected with a 25.0-A circuit breaker, will the circuit breaker be tripped in this situation? Explain your answer.

Solution

Conceptualize: This is a very practical problem. An electric appliance is marked with the power that it converts, and just this information lets the householder figure out whether he or she can safely plug it in. Each appliance here will carry current of several amps. If all are turned on at once, they will likely pop the 25-A circuit breaker.

Categorize: We use just $P = I\Delta V$ to find the current in each appliance, and then use addition to find the total current.

Analyze:

(a)   Heater: $I = \dfrac{P}{\Delta V} = \left(\dfrac{1\,500 \ \text{W}}{120 \ \text{V}}\right)\left(\dfrac{1 \ \text{J/s}}{1 \ \text{W}}\right)\left(\dfrac{1 \ \text{V}}{1 \ \text{J/C}}\right)\left(\dfrac{1 \ \text{A}}{1 \ \text{C/s}}\right) = 12.5 \ \text{A} \qquad \blacksquare$

Toaster: $I = \dfrac{750 \ \text{W}}{120 \ \text{V}} = 6.25 \ \text{A} \qquad \blacksquare$

Grill: $I = \dfrac{1\,000 \ \text{W}}{120 \ \text{V}} = 8.33 \ \text{A} \qquad \blacksquare$

(b)   Together in parallel they pass current 12.5 A + 6.25 A + 8.33 A = 27.1 A, so a 25.0-A circuit breaker will trip to switch all three off if the householder turns all three on at once. $\qquad \blacksquare$

Finalize: It is instructive to draw a circuit diagram for this case. The appliances are resistors in parallel, and that makes the addition of their currents meaningful.

––––––––––––––––––

76. An experiment is conducted to measure the electrical resistivity of Nichrome in the form of wires with different lengths and cross-sectional areas. For one set of measurements, a student uses 30-gauge wire, which has a cross-sectional area of 7.30×10^{-8} m^2. The student measures the potential difference across the wire and the current in the wire with a voltmeter and an ammeter, respectively. (a) For each set of measurements given in the table taken on wires of three different lengths, calculate the resistance of the wires and the corresponding values of the resistivity. (b) What is the average value of the resistivity? (c) Explain how this value compares with the value given in Table 21.1.

L (m)	ΔV (V)	I (A)	R (Ω)	ρ ($\Omega \cdot$ m)
0.540	5.22	0.72		
1.028	5.82	0.414		
1.543	5.94	0.281		

Solution

Conceptualize: With material and cross-sectional area constant, we expect resistance to be proportional to length. It should be about twice as large for the second sample compared to the first, and three times as large for the third. The resistivity should be about the same for all three. The amount of scatter will suggest the experimental uncertainty.

Categorize: For each row in the table, we will first find the resistance from $R = \Delta V / I$, and then find the resistivity from $R = \rho \ell / A$ or $\rho = RA / \ell$.

Analyze:

(a) In the first row,

$$R = \frac{\Delta V}{I} = \frac{5.22 \text{ V}}{0.72 \text{ A}} = 7.25 \ \Omega$$

and $$\rho = \frac{RA}{\ell} = \frac{(7.25 \ \Omega)(7.30 \times 10^{-8} \text{ m}^2)}{0.540 \text{ m}} = 9.80 \times 10^{-7} \ \Omega \cdot \text{m}.$$

Applying this method to each entry, we obtain:

L (m)	R (Ω)	ρ (Ω · m)
0.540	7.25	9.80×10^{-7}
1.028	14.1	9.98×10^{-7}
1.543	21.1	1.00×10^{-6}

■

(b) Thus the average resistivity is $\rho = 9.93 \times 10^{-7}$ Ω · m.   ■

(c) This differs from the tabulated 1.00×10^{-6} Ω by 0.7%. The difference is accounted for by the experimental uncertainty, which we may estimate as $(1.00 - 0.98)/1.00 = 2\%$.   ■

Finalize: We have estimated the *amount* of the experimental uncertainty. If we looked at the apparatus and procedure, we might form hypotheses about the *causes* of the uncertainty, and these could suggest ways to reduce the uncertainty.

Chapter 22
Magnetic Forces
and Magnetic Fields

Section 22.2   The Magnetic Field

Particles with charge q, moving with speed $\vec{v}$ in a magnetic field $\vec{B}$, experience a magnetic force $\vec{F}_B$. Properties of the magnetic force are:

- The magnetic force is proportional to the charge q and speed v of the particle.

- The magnitude and direction of the magnetic force depend on the angle between the velocity vector of the particle and the magnetic field vector.

- When a charged particle moves in a direction parallel to the magnetic field vector, the magnetic force $\vec{F}_B$ on the charge is zero.

- The magnetic force acts in a direction perpendicular to both $\vec{v}$ and $\vec{B}$; that is, $\vec{F}_B$ is perpendicular to the plane formed by $\vec{v}$ and $\vec{B}$.

- The magnetic force on a positive charge is in the direction opposite to the force on a negative charge moving in the same direction.

- If the velocity vector makes an angle θ with the magnetic field, the magnitude of the magnetic force is proportional to $\sin\theta$.

There are several important differences between electric and magnetic forces:

- The electric force is always along the direction of the electric field, whereas the magnetic force is perpendicular to the magnetic field.

- The electric force acts on a charged particle independent of the particle's velocity, whereas the magnetic force acts on a charged particle only when the particle is in motion.

- The electric force does work in displacing a charged particle, whereas the magnetic force associated with a steady magnetic field does no work when a particle is displaced.

Graphical methods have been adopted to represent magnetic field lines in the plane of the page and perpendicular to the plane of the page (pointing into or out of the page).

Lines lying in the plane of the page are shown by an arrow in the plane of the page.

Lines directed out of the page are shown by dots representing tips of the arrows coming outward.

Lines directed into the page are shown by crosses representing the feathers of arrows going inward.

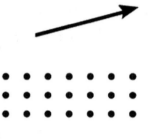

Section 22.7   The Biot-Savart Law

The Biot-Savart law states that if a wire carries a steady current I, the magnetic field $d\vec{\mathbf{B}}$ at a point P associated with an element $d\vec{\mathbf{s}}$ has the following properties:

- The vector $d\vec{\mathbf{B}}$ is perpendicular both to $d\vec{\mathbf{s}}$ (which is in the direction of the current) and to the unit vector $\hat{\mathbf{r}}$ (directed from the current element to the point P). **See Figure 22.22.**

- The magnitude of $d\vec{\mathbf{B}}$ is inversely proportional to r^2, where r is the distance from the element to the point P.

- The magnitude of $d\vec{\mathbf{B}}$ is proportional to the current and to the length of the element $d\vec{\mathbf{s}}$.

- The magnitude of $d\vec{\mathbf{B}}$ is proportional to sin θ, where θ is the angle between the vectors $d\vec{\mathbf{s}}$ and $\hat{\mathbf{r}}$.

Section 22.8   The Magnetic Force Between Two Parallel Conductors

Parallel conductors carrying currents in the same direction attract each other, and parallel conductors carrying currents in opposite directions repel each other.

The force between two parallel wires each carrying a current is used to define the ampere as follows:
If two long, parallel wires 1 m apart in a vacuum carry the same current and the force per unit length on each wire is 2×10^{-7} N/m, then the current is defined to be 1 A.

Section 22.9   Ampère's Law

The direction of the magnetic field due to a current in a conductor is given by the right-hand rule: (See Figure 22.23.)

If the wire is grasped in the right hand with the thumb in the direction of the current, the fingers will wrap (or curl) in the direction of $\vec{\mathbf{B}}$.

Ampère's law is valid only for steady currents and is useful only in those cases where the current configuration has a high degree of symmetry.

EQUATIONS AND CONCEPTS

The **magnetic force** exerted on a positive electric charge moving in a magnetic field can be expressed as a vector cross product.

$$\vec{\mathbf{F}}_B = q\vec{\mathbf{v}} \times \vec{\mathbf{B}} \qquad (22.1)$$

$\vec{\mathbf{F}}_B$ is perpendicular to $\vec{\mathbf{v}}$ and $\vec{\mathbf{B}}$.

The **magnitude of the magnetic force** will be a maximum when the charge moves along a direction perpendicular to the direction of the magnetic field.

$$F_B = |q|\, v B \sin\theta \qquad (22.2)$$

A charge moving at an angle θ with respect to the magnetic field.

The **SI unit of the magnetic field** is the tesla (T) or weber per square meter (Wb/m²).

$$1\,T = 1\,\frac{N}{C \cdot m/s} = 1\frac{N}{A \cdot m} = 1\frac{W}{m^2}$$

The **direction of the magnetic force on a charged particle moving in a magnetic field** can be determined by applying the right-hand rule. Hold your right hand open and point your fingers along the direction of $\vec{B}$ while pointing your thumb along the direction of $\vec{v}$. The magnetic force on a *positive charge* will be directed out of the palm of your hand.

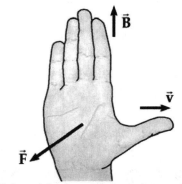

Using right-hand rule to determine direction of force on a charged particle moving in amagnetic field.

Motion of a charged particle entering a uniform magnetic field with the velocity vector initially perpendicular to the field has the following characteristics:

- The **path** of the particle will be circular and in a plane perpendicular to the direction of the field. The direction of rotation of the particle will be as determined by the right-hand rule. The magnetic force exerted on the particle is directed toward the center of the circular path.

$$r = \frac{mv}{qB} \qquad (22.3)$$

- The **radius** of the circular path will be proportional to the linear momentum of the charged particle.

$$\omega = \frac{qB}{m} \qquad (22.4)$$

- The **angular frequency** (or cyclotron frequency) of the particle will be proportional to the ratio of charge to mass. *Note that the frequency, and hence the period of rotation, does not depend on the radius of the path.*

$$T = \frac{2\pi m}{qB} \qquad (22.5)$$

If the **initial velocity is not perpendicular** to the field, the particle will move in a helical or spiral path. The axis of the spiral will be parallel to the magnetic field, and the "pitch" (distance between adjacent coils) of the helix will depend on the component of velocity parallel to the field.

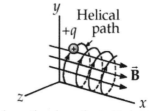

Particle path when the initial velocity is not perpendicular to the field.

The **Lorentz force** is the total force experienced by a charged particle moving in a region where both a magnetic field and an electric field are present.

$$\vec{F} = q\vec{E} + q\vec{v} \times \vec{B} \qquad (22.6)$$

The force on a straight segment of current-carrying conductor in a uniform external magnetic field depends on the angle between the direction of the current and the direction of the field. The **direction of the force is determined** by the right-hand rule. In the figure at right, the *external magnetic field* is directed out of the page in the region above the conductor.

$$\vec{F}_B = I\vec{L} \times \vec{B} \qquad (22.10)$$

$$F_B = BI\ell \sin\theta$$

Force on a current carrying conductor.

For a wire of arbitrary shape the magnetic force is found by integrating over the length of the wire. In these equations the direction of $d\vec{s}$ is that of the current. *The magnetic force on a curved current-carrying wire in a uniform magnetic field is equal to that of a straight wire connecting the end points and carrying the same current.*

$$\vec{F}_B = I \int_a^b d\vec{s} \times \vec{B} \qquad (22.12)$$

The net magnetic force on a closed current loop in a uniform field is zero.

Applications of the motion of charged particles in a magnetic field include:

- **Velocity Selector:** When a beam of charged particles is directed into a region where uniform electric and magnetic fields are perpendicular to each other and to the initial direction of the particle beam, only those particles with a velocity $v = \dfrac{E}{B}$ will emerge undeflected. The force due to the electric field (qE) is equal in magnitude and directed opposite the force due to the magnetic field (qvB).

$$v = \frac{E}{B} \qquad (22.7)$$

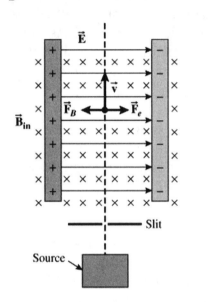

Velocity selector

- **Mass Spectrometer:** If an ion beam, after passing through a velocity selector, is directed perpendicularly into a second uniform magnetic field (B_0), the ratio of charge-to-mass of the isotopic species can be determined by measuring the radius of curvature of the beam in the second field.

$$\frac{m}{q} = \frac{rB_0B}{E} \qquad (22.8)$$

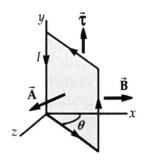

Mass spectrometer

$$K = \frac{1}{2}mv^2 = \frac{q^2B^2R^2}{2m} \qquad (22.9)$$

- **Cyclotron:** The maximum kinetic energy acquired by an ion in a cyclotron depends on the radius of the "dees" and the intensity of the magnetic field. This relationship holds until the ion reaches relativistic energies (~20 MeV). *This is true for ions of proton mass or greater, but is not true for electrons.*

A **net torque** is exerted on a conducting loop carrying a current when placed in an external magnetic field. The magnitude of vector $\vec{\mathbf{A}}$ is numerically equal to the area of the loop and its direction is perpendicular to the area. *When the fingers of the right hand are curled around the loop in the direction of the current, the thumb points in the direction of $\vec{\mathbf{A}}$.*

$$\vec{\tau} = I\vec{\mathbf{A}} \times \vec{\mathbf{B}} \qquad (22.14)$$

Rectangular current loop
in a uniform field

The **magnetic dipole moment, $\vec{\mu}$,** of a loop can be used to state the torque on the loop. The direction of $\vec{\mu}$ is the same as the direction of $\vec{\mathbf{A}}$.

$$\vec{\tau} = \vec{\mu} \times \vec{\mathbf{B}} \qquad (22.16)$$

$$\vec{\mu} = I\vec{\mathbf{A}} \qquad (22.15)$$

The **magnitude of the torque** will depend on the angle between the direction of the magnetic field and the direction of the normal (or perpendicular) to the plane of the loop and will be maximum when the magnetic field is parallel to the plane of the loop.

$$\tau = IAB\sin\theta$$

$$\tau_{max} = IAB \qquad (22.13)$$

The **direction of rotation of the loop** is in the direction of decreasing values of τ (i.e. vector $\vec{A}$ rotates toward the direction of the magnetic field). When $\vec{A}$ becomes parallel to $\vec{B}$ the torque on the loop will be zero. The loop shown in the figure will rotate counterclockwise as seen from above. *The direction of the vector torque $\vec{\tau}$ is indicated by the thumb of the right hand when the fingers curl $\vec{A}$ into the direction of $\vec{B}$.*

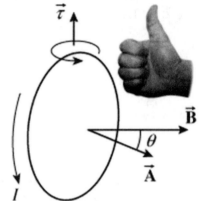

The direction of rotation of a current loop in a magnetic field

The **Biot–Savart law** gives the magnetic field at a point in space due to an element of conductor $d\vec{s}$, which carries a current I and is at a distance r away from the point. *The vector $d\vec{B}$ is perpendicular to both $d\vec{s}$ (which points in the direction of the current) and to the unit vector $\hat{r}$ (which points from the current element toward the point where the field is to be determined).* The **total magnetic field** at a point in the vicinity of a current of finite length is found by integrating the Biot–Savart law expression over the entire current distribution. *Remember, the integrand is a vector quantity.*

$$d\vec{B} = \frac{\mu_0}{4\pi}\frac{I\,d\vec{s}\times\hat{r}}{r^2} \qquad (22.20)$$

$$\mu_0 = 4\pi\times10^{-7}\ \text{T}\cdot\text{m/A}$$

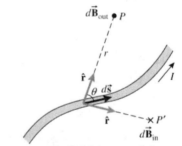

The magnetic field is out of the page at point P and into the page at point P'.

The **magnitude of the magnetic field** due to several important geometric arrangements of a current-carrying conductor can be calculated by use of the Biot-Savart law.
Some examples are:

B at a distance r from a **long, straight conductor**, carrying a current I.

$$B = \frac{\mu_0 I}{2\pi r} \qquad (22.21)$$

B at the **center of an arc** of radius a that subtends an angle θ (in radians) at the center of the arc.

$$B = \frac{\mu_0 I}{4\pi a}\theta$$

B_x on the axis of a **circular loop** of radius a and at a distance x from the plane of the loop.

$$B_x = \frac{\mu_0 I a^2}{2\left(a^2 + x^2\right)^{3/2}} \qquad (22.23)$$

$$B \approx \frac{\mu_0 I a^2}{2x^3} \quad (\text{for } x \gg a) \qquad (22.25)$$

B at the **center of a circular loop** of radius a.

$$B = \frac{\mu_0 I}{2a} \qquad (22.24)$$

The **magnetic force per unit length** between very long parallel conductors depends on the distance a between the conductors and the magnitudes of the two currents. Parallel conductors carrying currents in the same direction attract each other, and parallel conductors carrying currents in opposite directions repel each other. *The forces on the two conductors will be equal in magnitude regardless of the relative magnitude of the two currents.*

$$\frac{F}{\ell} = \frac{\mu_0 I_1 I_2}{2\pi a} \qquad (22.27)$$

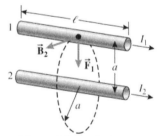

Magnet force between parallel conductors.

Ampère's law describes the creation of magnetic fields by continuous pathways. I is the total current passing through a surface bounded by the closed path.

$$\oint \vec{\mathbf{B}} \cdot d\vec{\mathbf{s}} = \mu_0 I \qquad (22.29)$$

B **inside a toroid** having N turns and at a distance r from the center of the toroid.

$$B = \frac{\mu_0 N I}{2\pi r} \qquad (22.31)$$

B near the center of a **solenoid** of n turns per unit length.

$$B = \mu_0 n I \qquad (22.32)$$

To determine the direction of the magnetic field:

Due to a long, straight conductor:
Hold the conductor in your right hand with your thumb pointing in the direction of the conventional current. Your fingers will then wrap around the wire in the direction of the magnetic field lines. *The magnetic field is tangent to the circular field lines at every point in the region around the conductor.*

Direction of the magnetic field due to a long, straight conductor

At the center of a current loop:
The magnetic field is perpendicular to the plane of the loop and directed in the sense given by the right-hand rule.

Direction of the magnetic field due to a current loop

Within a solenoid:
The magnetic field is parallel to the axis of the solenoid and pointing in the direction determined by applying the right-hand rule to one of the coils.

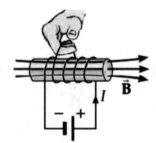

Direction of the magnetic field within a solenoid

SUGGESTIONS, SKILLS, AND STRATEGIES

The **direction of the magnetic force on a moving charge** can be determined by applying the right-hand rule. Two versions for application of the rule are shown in the figure below. The direction shown for $\vec{F}$ is the direction of force on a positive charge; *if the charge is negative the direction of the force will be reversed.*

Version 1

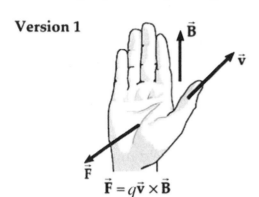

$$\vec{F} = q\vec{v} \times \vec{B}$$

Version 2

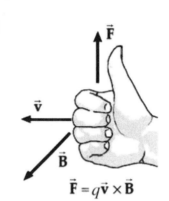

$$\vec{F} = q\vec{v} \times \vec{B}$$

Hold your open right hand with your thumb pointing in the direction of $\vec{v}$ (*the first named vector quantity*) and your fingers pointing in the direction of $\vec{B}$ (*the second named vector quantity*). By necessity, your fingers and thumb cannot stretch farther than 180°, so if you do this, your hand will be oriented properly. The force vector $\vec{F}$ now is directed out of the palm of your hand.

Orient your hand so that your fingers point in the direction of $\vec{v}$ (*the first named vector quantity*) and then curl your fingers to point in the direction of $\vec{B}$ (*the second named vector quantity*). Note that since your fingers cannot bend farther than 180°, you may have to flip your hand upside down to do this. Your thumb now points in the direction of $\vec{F}$, where $\vec{F}$ is perpendicular to both $\vec{v}$ and $\vec{B}$.

REVIEW CHECKLIST

- Use the defining equation for a magnetic field $\vec{B}$ to determine the magnitude and direction of the magnetic force exerted on an electric charge moving in a region where there is a magnetic field. You should understand clearly the important differences between the forces exerted on electric charges by electric fields and those forces exerted on moving electric charges by magnetic fields.

- Calculate the magnitude and direction of the magnetic force on a current-carrying conductor when placed in an external magnetic field.

- Determine the magnitude and direction of the torque exerted on a closed current loop in an external magnetic field. You should understand how to designate the direction of the area vector corresponding to a given current loop, and to incorporate the magnetic moment of the loop into the calculation of the torque on the loop.

- Use the Biot-Savart law to calculate the magnetic field at a specified point in the vicinity of a current element, and by integration find the total magnetic field due to a number of important geometric arrangements. Your use of the Biot-Savart law must include a clear understanding of the direction of the magnetic field contribution relative to the direction of the current element, which produces it and the direction of the vector, which locates the point at which the field is to be calculated.

- Use Ampère's law to calculate the magnetic field due to steady current configurations, which have a sufficiently high degree of symmetry such as a long, straight conductor, a long solenoid, and a toroidal coil.

ANSWER TO SELECTED OBJECTIVE QUESTIONS

2. What creates a magnetic field? Choose every correct answer. (a) A stationary object with electric charge, (b) a moving object with electric charge, (c) a stationary conductor carrying electric current, (d) a difference in electric potential, (e) a charged capacitor disconnected from a battery and at rest. *Note:* In Chapter 24, we will see that a changing electric field also creates a magnetic field.

Answer The correct choices are (b) and (c). By contrast, if the question asked what creates an electric field, definitely correct answers would be (a), (b), and (e); and a difference in potential (d) can only exist in a space that also contains an electric field. Notice that a moving charged object creates both an electric field and a magnetic field. Both a stationary and a moving current-carrying conductor create magnetic fields. But a typical conductor has zero net charge, and creates no electric field.

□   □   □   □

7. Answer each question yes or no. Assume the motions and currents mentioned are along the x axis and fields are in the y direction. (a) Does an electric field exert a force on a stationary charged object? (b) Does a magnetic field do so? (c) Does an electric field exert a force on a moving charged object? (d) Does a magnetic field do so? (e) Does an electric field exert a force on a straight current-carrying wire? (f) Does a magnetic field do so? (g) Does an electric field exert a force on a beam of moving electrons? (h) Does a magnetic field do so?

Answer   (a) Yes, the definition of the electric field, represented by the equation $\vec{\mathbf{F}}_e = q\vec{\mathbf{E}}$, says that the electric field exerts a force on an object with charge.

(b)   No, a stationary object has zero velocity and no magnetic force acts on it.

(c)   Yes, a moving object feels an electric force with the same strength and direction that it would feel if it were stationary.

(d)   Yes, the equation $\vec{\mathbf{F}}_B = q\vec{\mathbf{v}} \times \vec{\mathbf{B}}$, associated with the definition of the magnetic field, describes the magnetic force on a moving object with charge.

(e)   No. An ordinary current-carrying wire has zero net charge. A metallic wire has equal numbers of (moving) conduction electrons and (stationary) ion cores in each cubic millimeter. So it feels zero force in an electric field.

(f)   Yes. The moving electrons feel a magnetic force and the stationary positive charges feel no force, so the whole wire feels magnetic force $\vec{\mathbf{F}}_B = I\vec{\mathbf{L}} \times \vec{\mathbf{B}}$.

(g)   and (h) Yes and yes. A beam of electrons, as in a cathode ray tube (such as the picture tube in a boxy television receiver or computer monitor), possesses net charge and consists of moving charges, so it feels electric and magnetic forces at the same time.

Notice that the only kind of charge is electric charge. There is no such thing as a magnetic charge. It is an object with electric charge that feels an electric force if it is in an electric field, and that feels a magnetic force if it is moving in a magnetic field. Any force is a vector measured in newtons and might have the effect of causing acceleration of the object on which it acts, or the effect of counterbalancing some other force. An electric field is an utterly different thing from a magnetic field, produced by a different source, measured in different units, and having different effects.

□   □   □   □

ANSWERS TO SELECTED CONCEPTUAL QUESTIONS

1.   Two charged particles are projected in the same direction into a magnetic field perpendicular to their velocities. If the particles are deflected in opposite directions, what can you say about them?

Answer   We know the magnetic field is the same for both particles. The velocity vectors are in the same direction, but one force vector is the negative of the other. From $\vec{\mathbf{F}}_B = q\vec{\mathbf{v}} \times \vec{\mathbf{B}}$, we conclude that the forces are in opposite directions because the charges have opposite signs.

□   □   □   □

5. Is it possible to orient a current loop in a uniform magnetic field such that the loop does not tend to rotate? Explain.

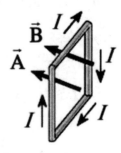

Answer Yes. If the magnetic field is perpendicular to the plane of the loop, the forces on opposite sides will be equal in magnitude, opposite in direction, and along the same lines of action, so they will produce no net torque about any axis.

□   □   □   □

9. How can a current loop be used to determine the presence of a magnetic field in a given region of space?

Answer The loop can be mounted free to rotate around an axis. The loop will rotate about this axis when placed in an external magnetic field at some arbitrary orientation. As the current through the loop is increased, the torque on it will increase.

□   □   □   □

SOLUTIONS TO SELECTED END-OF-CHAPTER PROBLEMS

2. Determine the initial direction of the deflection of charged particles as they enter the magnetic fields shown in Figure P22.2.

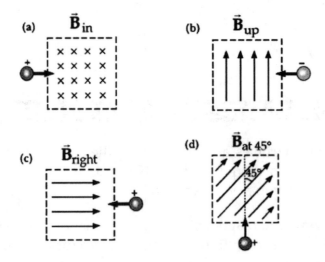

Figure P22.2

Solution

Conceptualize: The electric force exerted by an electric field on an electric charge is the same for any velocity of the charge. This problem is about how the magnetic force exerted by a magnetic field on an electric charge follows a quite different pattern.

Categorize: We must use the right-hand rule.

Analyze: Here we show both versions of the right-hand rule, open-handed and curled, for each cross product. Later on, we will draw just the curled-finger version. That version lets you work out any cross product, such as the torque on a magnetic moment, $|\vec{\tau}| = \vec{\mu} \times \vec{B}$, without further memorization. If you are already familiar with the open-handed version of the rule, you can use the set of figures to practice the curled-finger version.

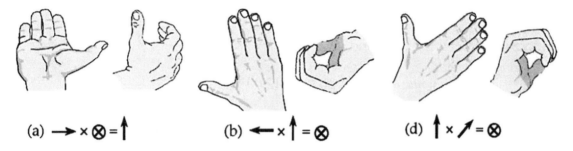

(a) $\longrightarrow \times \otimes = \uparrow$ (b) $\longleftarrow \times \uparrow = \otimes$ (d) $\uparrow \times \nearrow = \otimes$

(a)   By solution Figure (a), $\vec{v} \times \vec{B}$ is (right) $\times$ (away) = up.   ∎

(b)   By solution Figure (b), $\vec{v} \times \vec{B}$ is (left) $\times$ (up) = away.

    Since the charge is negative, $q\vec{v} \times \vec{B}$ is toward you.   ∎

(c) $\vec{v} \times \vec{B}$ is zero since the angle between $\vec{v}$ and $\vec{B}$ is 180° and sin 180° = 0.

    There is no deflection.   ∎

(d) $\vec{v} \times \vec{B}$ is (up) $\times$ (up and right), or away from you.   ∎

Finalize: The cross product has a direction, and it also has a size. In parts (a) and (b) the factor sin 90° = 1 makes magnitude of the magnetic force equal to $|qvB|$. In part (c) it is zero. In part (d), where sin 45° = 0.707, the magnitude of the force is 0.707 $|qvB|$.

7. A proton moves perpendicular to a uniform magnetic field $\vec{B}$ at a speed of 1.00×10^7 m/s and experiences an acceleration of 2.00×10^{13} m/s² in the positive x direction when its velocity is in the positive z direction. Determine the magnitude and direction of the field.

Solution

Conceptualize: The magnetic field exerts a magnetic force on the proton, and the force causes the acceleration. Ten million meters per second is a perfectly reasonable speed for a proton in a vacuum. If the computed field magnitude is anything between a microtesla and a tesla, then we can infer that 10^{13} m/s² is a reasonable acceleration.

Categorize: We will use the particle under acceleration model to find the net force on the proton, which is the magnetic force. Then the particle in a magnetic field model will tell us the field.

Analyze: By Newton's second law,

$$\Sigma F = ma = (1.67 \times 10^{-27} \text{ kg})(2.00 \times 10^{13} \text{ m/s}^2)$$

$$= 3.34 \times 10^{-14} \text{ N in the } +x \text{ direction.}$$

The magnetic force is $F = 3.34 \times 10^{-14}$ N $= qvB \sin 90°$.

Rearranging,

$$B = \frac{F}{qv} = \frac{3.34 \times 10^{-14} \text{ N}}{(1.60 \times 10^{-19} \text{ C})(1.00 \times 10^7 \text{ m/s})} = 2.09 \times 10^{-2} \text{ T} \qquad \blacksquare$$

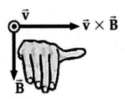

By the right-hand rule, $\vec{B}$ must be in the $-y$ direction. $\qquad \blacksquare$

This yields a force on the proton in the $+x$ direction when $\vec{v}$ points in the $+z$ direction.

Finalize: Twenty milliteslas is an easy field to produce with an electromagnet. In the student laboratory you may do an experiment like this, but with electrons rather than protons. The experimenter chooses to fire the particle at a ninety degree angle to the field. Then nature makes the direction of the force at ninety degrees to both the velocity and the field. Note that one tesla is one newton-second per coulomb-meter.

═══════════════════

10. A cosmic-ray proton in interstellar space has an energy of 10.0 MeV and executes a circular orbit having a radius equal to that of Mercury's orbit around the Sun (5.80×10^{10} m). What is the magnetic field in that region of space?

Solution

Conceptualize: A very big orbit implies that the magnetic field in this interstellar space is a tiny fraction of a tesla.

Categorize: Think of the proton as having accelerated through a potential difference $\Delta V = 10^7$ V. We use the energy version of the isolated system model, applied to the particle and the electric field that made it speed up from rest. Then we use the particle in a magnetic field model and the particle in uniform circular motion model.

Analyze: By conservation of energy for the proton-electric-field system in the process that set the proton moving, its kinetic energy is

$$E = \frac{1}{2}mv^2 = e\Delta V \quad \text{so its speed is} \quad v = \sqrt{\frac{2e\Delta V}{m}}.$$

Now Newton's second law for its circular motion in the magnetic field is

$$\sum F = ma: \quad \frac{mv^2}{R} = evB \sin 90°$$

$$B = \frac{mv}{eR} = \frac{m}{eR}\sqrt{\frac{2e\Delta V}{m}} = \frac{1}{R}\sqrt{\frac{2m\Delta V}{e}}$$

$$B = \left(\frac{1}{5.80 \times 10^{10} \text{ m}}\right)\sqrt{\frac{2\left(1.67 \times 10^{-27} \text{ kg}\right)\left(10^7 \text{ V}\right)}{1.60 \times 10^{-19} \text{ C}}} = 7.88 \times 10^{-12} \text{ T} \quad \blacksquare$$

Finalize: This is a reasonable answer. Some matter far away created this field that exists in a space empty of matter, and exists unchanged when the proton enters, to exert a force on the proton.

═══════════════════

17. The picture tube in an old black-and-white television uses magnetic deflection coils rather than electric deflection plates. Suppose an electron beam is accelerated through a 50.0-kV potential difference and then through a region of uniform magnetic field 1.00cm wide. The screen is located 10.0 cm from the center of the coils and is 50.0 cm wide. When the field is turned off, the electron beam hits the center of the screen. Ignoring relativistic corrections, what field magnitude is necessary to deflect the beam to the side of the screen?

Solution

Conceptualize: Electrons are the smallest-mass particles of any that have measured mass. They are the easiest particles to deflect. A field of some milliteslas should do the trick.

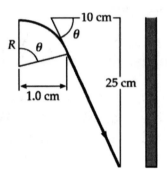

Categorize: We must do some geometry to quantify the angle of deflection and relate it to the radius of curvature of the electrons' path, an arc of a circle, when they are in the field.

Analyze: The beam is deflected by the angle

$$\theta = \tan^{-1}\left(\frac{25.0 \text{ cm}}{10.0 \text{ cm}}\right) = 68.2°$$

The two angles θ shown are equal because their sides are perpendicular, right side to right side and left side to left side. The radius of curvature of the electrons' path in the field is

$$R = \frac{1.00 \text{ cm}}{\sin 68.2°} = 1.077 \text{ cm}$$

Now $\frac{1}{2}mv^2 = |q|\Delta V,$

so $v = \sqrt{\frac{2|q|\Delta V}{m}} = \sqrt{\frac{2(1.60 \times 10^{-19} \text{ C})(50\,000 \text{ V})}{9.11 \times 10^{-31} \text{kg}}} = 1.33 \times 10^{8} \text{ m/s.}$

$\sum \vec{F} = m\vec{a}$ becomes $\frac{mv^2}{R} = |q|vB \sin 90°.$

$$B = \frac{mv}{|q|\,R} = \frac{(9.11 \times 10^{-31} \text{ kg})(1.33 \times 10^{8} \text{ m/s})}{(1.60 \times 10^{-19} \text{ C})(1.077 \times 10^{-2} \text{ m})} = 70.1 \text{ mT} \qquad \blacksquare$$

Finalize: The remarkably useful theorem about equal angles with perpendicular sides is in Appendix B.3 of the textbook, with a diagram that can help you to identify the right and left sides of the two angles marked θ in the diagram here. In the problem answer, several milliteslas is what we estimated. Do not endanger yourself, but if you look inside the case of an old television set you can see the magnetic coils, with many turns of fine copper wire, outside the throat of the picture tube. The speed is well over one-tenth of the speed of light, so using equations from Chapter 9 would give a noticeably more accurate answer, according to the theory of relativity.

---

18. A strong magnet is placed under a horizontal conducting ring of radius r that carries current I as shown in Figure P22.18. If the magnetic field $\vec{B}$ makes an angle θ with the vertical at the ring's location, what are (a) the magnitude and (b) the direction of the resultant magnetic force on the ring?

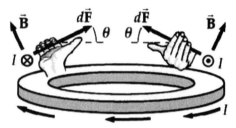

Figure P22.18

Solution

Conceptualize: You might have learned that a uniform electric field exerts zero force on an electric dipole, but a nonuniform electric field does exert a force on an electric dipole. This chapter points out that a uniform magnetic field exerts a torque but zero force on a magnetic dipole. This problem illustrates that a nonuniform magnetic field exerts a net force on a magnetic dipole. The force considered here is important in some simple applications.

Categorize: We evaluate in symbolic terms the magnetic force on one bit of the current loop, and then the force on the whole loop.

Analyze: (a) and (b) The magnetic force on each bit of ring is inward and upward, at an angle θ above the radial line, according to

$$\left| d\vec{F} \right| = I \left| d\vec{s} \times \vec{B} \right| = I\,ds\,B$$

The radially inward components tend to squeeze the ring, but cancel out as forces. The upward components $I\,ds\,B\sin\theta$ all add to

$$\vec{F} = I(2\pi r)\,B\sin\theta \text{ up} \qquad \blacksquare$$

Finalize: The magnetic moment of the ring is down. The current loop as a magnetic dipole has a downward-facing north pole, which is repelled by the upward-pointing (and upward-weakening) field coming out of the north pole of the external magnet shown in the textbook picture. This problem is a model for the force that one magnet exerts on another magnet.

---

21. A wire having a mass per unit length of 0.500 g/cm carries a 2.00-A current horizontally to the south. What are (a) the direction and (b) the magnitude of the minimum magnetic field needed to lift this wire vertically upward?

Solution

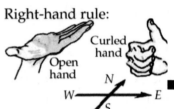

Right-hand rule:

Conceptualize: (a) Since I = 2.00 A south, $\vec{B}$ must be to the east to make $\vec{F}$ upward according to the right-hand rule for currents in a magnetic field.

As before, in viewing the diagrams get used to using the curled-hand version of the right-hand rule for later applications.

The magnitude of $\vec{B}$ should be significantly greater than the Earth's magnetic field (~50 μT), since we do not typically see electric currents making wires levitate.

Categorize: The force on a current-carrying wire in a magnetic field is ϕ, from which we can find $\vec{B}$.

Analyze: (b) With I to the south and $\vec{B}$ to the east, the force on the wire is $F_B = ILB \sin 90°$, which must counterbalance the weight of the wire, mg.

So, $B = \dfrac{F_B}{IL} = \dfrac{mg}{IL} = \dfrac{g}{I}\left(\dfrac{m}{L}\right) = \left(\dfrac{9.80 \text{ m/s}^2}{2.00 \text{ A}}\right)\left(0.500\dfrac{g}{cm}\right)\left(\dfrac{10^2 \text{ cm/m}}{10^3 \text{ g/kg}}\right) = 0.245$ T. ∎

Finalize: The required magnetic field is about 5 000 times stronger than the Earth's magnetic field. Thus it was reasonable to ignore the Earth's magnetic field in this problem. In other situations the Earth's field can have a significant effect. If the field were not straight to the east horizontally, it could lift the wire but it would have to be stronger, so we have successfully found the minimum field. In our description of the magnetic force on a wire carrying electric current, we attribute the vector direction to the length of the wire and not to the current. Current cannot be a vector because at a junction in an electric circuit currents add as scalars and not as vectors.

---

25. A rectangular coil consists of $N = 100$ closely wrapped turns and has dimensions $a = 0.400$ m and $b = 0.300$ m. The coil is hinged along the y axis, and its plane makes an angle $\theta = 30.0°$ with the x axis (Fig. P22.25). (a) What is the magnitude of the torque exerted on the coil by a uniform magnetic field $B = 0.800$ T directed in the positive x direction when the current is $I = 1.20$ A in the direction shown? (b) What is the expected direction of rotation of the coil?

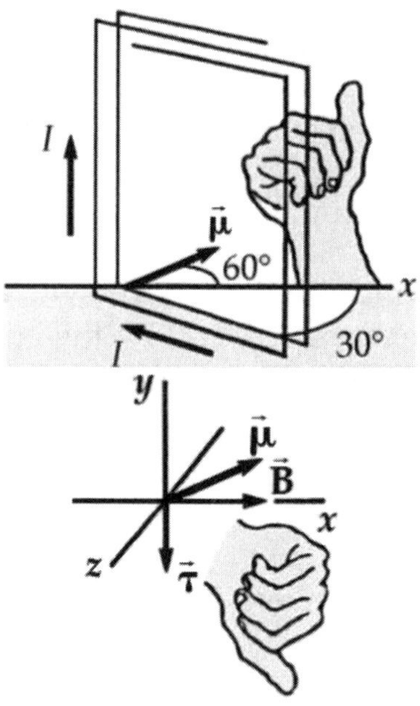

Figure P22.25

Solution

Conceptualize: A magnetic field twists a compass needle, exerting a torque tending to align the needle with the field. Here we have a model for that phenomenon.

Categorize: We use the result proved in the chapter for the torque on a loop of current in a magnetic field.

Analyze: The magnetic moment of the coil is $\mu = NIA$, perpendicular to its plane and making a $\phi = 60°$ angle with the x axis as shown above. The torque on the dipole is then

$$|\vec{\tau}| = \vec{\mu} \times \vec{\mathbf{B}} = NBAI \sin\phi \text{ down}$$

(a) having a magnitude

$$\begin{aligned}\tau &= NBAI \sin\phi \\ &= (100)(0.800\text{ T})(0.400\text{ m} \times 0.300\text{ m})(1.20\text{ A})\sin 60.0° \\ &= 9.98\text{ N}\cdot\text{m}\end{aligned}$$ ■

Note that ϕ is the angle between the magnetic moment and the $\vec{\mathbf{B}}$ field.

(b) We model the coil as a rigid body under a net torque; the coil will rotate so as to align the magnetic moment with the $\vec{\mathbf{B}}$ field. Looking down along the y axis, we will see the coil rotate in the clockwise direction. ■

Finalize: This is a big coil in a strong field, so it feels a rather large torque. This is a model for the functioning of an electric motor.

30. An infinitely long wire carrying a current I is bent at a right angle as shown in Figure P22.30. Determine the magnetic field at point P, located a distance x from the corner of the wire.

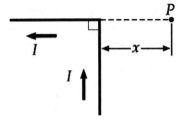

Solution

Conceptualize: We will add as vectors the field created at P by the current in the vertical

Figure P22.30

section of wire, and the field created at P by the horizontal section. The answer should be directly proportional to μ_0 and directly proportional to I, should decrease as x increases, and should contain some proportionality constant we cannot guess in advance.

Categorize: For the vertical section, we can use half of the field created by an infinitely long, straight wire. For the horizontal section this $\mu_0 I / 2\pi\, a$ result does not apply, because the point P is not perpendicularly off to the side of the horizontal section. Instead, we will reason from the Biot-Savart law.

Analyze: The vertical section of wire constitutes one half of an infinitely long, straight wire at distance x from P, so it creates a field equal to

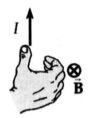

$$B = \frac{1}{2}\left(\frac{\mu_0 I}{2\pi x}\right)$$

Hold your right hand with extended thumb in the direction of the current; the field is away from you, into the paper.

For each bit of the horizontal section of wire $d\vec{\mathbf{s}}$ is to the left and $\hat{\mathbf{r}}$ is to the right, so $d\vec{\mathbf{s}} \times \hat{\mathbf{r}} = 0$. The horizontal current produces zero field at P. Thus,

$$\vec{\mathbf{B}} = \frac{\mu_0 I}{4\pi x} \qquad \text{into the paper.} \qquad\blacksquare$$

Finalize: The dependence on x turns out to be simply inverse proportionality. The proportionality constant is $1/4\pi$.

An electric current creating a magnetic field is a phenomenon different from any we have studied before, and the field direction is at least as important as the magnitude. (Make sure your light saber is pointing away from you when you switch it on!) The magnetic field is generally in the last direction you could guess. Get plenty of practice with the right-hand rules. In this problem reasoning about vector directions contributes to knowing the magnitude of the field.

33. One long wire carries current 30.0 A to the left along the *x* axis. A second long wire carries current 50.0 A to the right along the line ($y = 0.280$ m, $z = 0$). (a) Where in the plane of the two wires is the total magnetic field equal to zero? (b) A particle with a charge of -2.00 μC is moving with a velocity of $150\hat{\mathbf{i}}$ Mm/s along the line ($y = 0.100$ m, $z = 0$). Calculate the vector magnetic force acting on the particle. (c) **What If?** A uniform electric field is applied to allow this particle to pass through this region undeflected. Calculate the required vector electric field.

Solution

Conceptualize: In an experiment or demonstration, you may see current-carrying wires causing deflection of moving charged particles in a vacuum tube. We estimate that the magnetic fields of the two currents here may add to zero at a location some tens of centimeters below the 30-A wire. In part (b) the force should be a small fraction of a newton. Figuring out its direction will be as important as calculating its magnitude. In part (c) the strength of the electric field may be some kilonewtons per coulomb.

Categorize: In part (a), we must find a location, farther away from the 50-A current and closer to the 30-A current, where the fields which the currents create separately are equal in magnitude and opposite in direction. In part (b) we will compute the magnitude and identify the direction of the magnetic field each current creates at the location between them; then add the fields as vectors; and then find the magnetic force on the moving charge by working out a cross product. Part (c) reviews the definition of the electric field by asking us for an extra electric field that will exert a counterbalancing force on the same moving charge.

Analyze:

(a) Above the pair of wires, the field out of the page of the 50A current will be stronger than the ($-\hat{\mathbf{k}}$) field of the 30 A current, so they cannot add to zero. Between the wires, both produce fields into the page. The magnetic fields can only add to zero below the wires, at coordinate $y = -|y|$. Here we represent the total field as

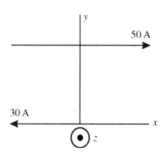

$$\vec{\mathbf{B}} = \frac{\mu_0 I}{2\pi r} \quad + \quad \frac{\mu_0 I}{2\pi r} \qquad \text{which requires } y \text{ to satisfy}$$

$$0 = \frac{\mu_0}{2\pi}\left[\frac{50.0 \text{ A}}{\left(|y| + 0.280 \text{ m}\right)}\left(-\hat{\mathbf{k}}\right) + \frac{30.0 \text{ A}}{|y|}\left(\hat{\mathbf{k}}\right)\right]$$

or

$50.0|y| = 30.0\left(|y| + 0.280 \text{ m}\right)$

$50.0(-y) = 30.0(0.280 \text{ m} - y)$

$-20.0y = 30.0(0.280 \text{ m})$

So the net field is zero all along the line at $y = -0.420$ m. ∎

(b) At $y = 0.100$ m the total field is $\vec{\mathbf{B}} = \dfrac{\mu_0 I}{2\pi r}$ ☜ $+ \dfrac{\mu_0 I}{2\pi r}$ ☜

$$\vec{\mathbf{B}} = \left(\frac{4\pi \times 10^{-7} \text{ T} \cdot \text{m/A}}{2\pi}\right)\left[\frac{50.0 \text{ A}}{(0.280 - 0.100) \text{ m}}\left(-\hat{\mathbf{k}}\right) + \frac{30.0 \text{ A}}{0.100 \text{ m}}\left(-\hat{\mathbf{k}}\right)\right]$$

$= 1.16 \times 10^{-4} \text{ T}\left(-\hat{\mathbf{k}}\right) = 116 \ \mu\text{T}$ into the plane of the paper.

The force on the particle is then

$$\vec{\mathbf{F}} = q\vec{\mathbf{v}} \times \vec{\mathbf{B}} = \left(-2.00 \times 10^{-6} \text{ C}\right)\left(150 \times 10^6 \text{ m/s}\right)\left(\hat{\mathbf{i}}\right)$$

$$\times\left(1.16 \times 10^{-4} \text{ N} \cdot \text{s/C} \cdot \text{m}\right)\left(-\hat{\mathbf{k}}\right)$$

$$= 3.47 \times 10^{-2} \text{ N}\left(-\hat{\mathbf{j}}\right)$$

$= 34.7$ mN downward toward the bottom of the page. ∎

(c) We require the electric force to be the same size but in the opposite direction to the magnetic force, according to

$$\vec{\mathbf{F}}_e = 3.47 \times 10^{-2}\text{N}\left(+\hat{\mathbf{j}}\right) = q\vec{\mathbf{E}} = \left(-2 \times 10^{-6}\text{C}\right)\vec{\mathbf{E}}$$

So $\vec{\mathbf{E}} = -1.73 \times 10^4 \hat{\mathbf{j}}$ N/C $= 17.3$ kN/C toward the bottom of the page. ∎

Finalize: In part (a), we might think of the zero-net-field location as remarkably far away, compared to where $+50 \ \mu$C and $-30 \ \mu$C charged particles would create zero net electric field. That extra distance happens because the magnetic field of a straight wire decreases as $1/r$ and not in the inverse-square-law pattern of $1/r^2$.

We could think of part (b) as requiring us to use a right hand three times, to find the magnetic fields of both currents and also the direction of the magnetic force. The field is never toward the wire or away from it, and never in the direction of the current or the opposite direction. The magnetic force exerted on the extra moving charge, in contrast, is toward the 30-A current and away from the 50-A current. The negative charge moving to the right constitutes an element of current to the left, which is attracted by the 30-A current in that same direction and repelled by the oppositely directed 50-A current. Notice that the forces that the two wires exert on each other have nothing to do with this problem at all.

Use part (c) to remind yourself of how different electric and magnetic forces are. The electric force does not depend on the velocity of the charge, but the magnetic force does. The electric force on a negative charge is directly opposite to the electric field direction, while the magnetic force is perpendicular to the magnetic field. The kilonewton-per-coulomb electric field and the microtesla magnetic field here exert forces of equal size.

---

41. (a) A conducting loop in the shape of a square of edge length $\ell = 0.400$ m carries a current $I = 10.0$ A as shown in Figure P22.41. Calculate the magnitude and direction of the magnetic field at the center of the square. (b) **What If?** If this conductor is reshaped to form a circular loop and carries the same current, what is the value of the magnetic field at the center?

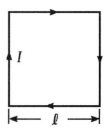

Figure P22.41

Solution

Conceptualize: As shown in Figure (a) below, the magnetic field at the center is directed into the page from the clockwise current. If we consider the sides of the square to be sections of four infinite wires, then we could expect the magnetic field at the center of the square to be a little less than four times the strength of the field at a point $\ell/2$ away from an infinite wire with current I.

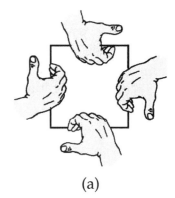

(a)

$$B < 4\frac{\mu_0 I}{2\pi a} = 4\frac{\left(4\pi \times 10^{-7}\ \text{T} \cdot \text{m/A}\right)(10.0\ \text{A})}{2\pi(0.200\ \text{m})}$$

$$B < 40.0\ \mu\text{T}$$

Forming the wire into a circle should not greatly change the magnetic field at the center since the average distance of the wire from the center will not be much different.

Categorize: Each side of the square is simply a section of a thin, straight conductor, so to solve part (a), we will derive the expression for the magnetic field of a finite-length wire from the Biot-Savart law. For part (b), the Biot-Savart law can also be used to derive the equation for the magnetic field at the center of a circular current loop as shown in Example 22.6.

Analyze:

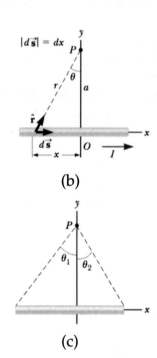

(b)

(c)

(a) First, we need to determine the magnetic field due to a finite-length wire. Let's start by considering a length element $d\vec{s}$ located a distance r from P, as shown in Figure (b). The direction of the magnetic field at point P due to the current in this element is out of the page because $d\vec{s} \times \hat{r}$ is out of the page. In fact, because *all* the current elements $I\,d\vec{s}$ lie in the plane of the page, they all produce a magnetic field directed out of the page at point P. Therefore, the direction of the magnetic field at point P is out of the page and we need only find the magnitude of the field. We place the origin at O and let point P be along the positive y axis, with $\hat{k}$ being a unit vector pointing out of the page. Evaluating the cross product in the Biot-Savart law gives

$$d\vec{s} \times \hat{r} = |d\vec{s} \times \hat{r}|\hat{k} = \left[dx \sin\left(\frac{\pi}{2} - \theta \right) \right]\hat{k} = (dx \cos\theta)\hat{k}$$

Substituting into Equation 22.20 gives

$$d\vec{B} = dB\hat{k} = \frac{\mu_0 I}{4\pi} \frac{dx \cos\theta}{r^2}\hat{k} \qquad [1]$$

From the geometry of Figure (b), we express r in terms of θ:

$$r = \frac{a}{\cos\theta} \qquad [2]$$

Notice that $\tan\theta = -x/a$ from the right triangle in Figure (b) (the negative sign is necessary because $d\vec{s}$ is located at a negative value of x) and solve for x:

$$x = -a\tan\theta$$

Then

$$dx = -a\sec^2\theta d\theta = -\frac{ad\theta}{\cos^2\theta} \qquad [3]$$

Substituting [2] and [3] into the magnitude of the field from equation [1]:

$$dB = -\frac{\mu_0 I}{4\pi}\left(\frac{ad\theta}{\cos^2\theta}\right)\left(\frac{\cos^2\theta}{a^2}\right)\cos\theta = -\frac{\mu_0 I}{4\pi a}\cos\theta d\theta \qquad [4]$$

We then integrate equation [4] over all length elements on the wire, where the subtending angles range from θ_1 to θ_2 as defined in Figure (c):

$$B = -\frac{\mu_0 I}{4\pi a}\int_{\theta_1}^{\theta_2}\cos\theta d\theta = \frac{\mu_0 I}{4\pi a}\left(\sin\theta_1 - \sin\theta_2\right) \qquad [5]$$

From Figure P22.41, we note that $\theta_1 = 45.0°$, $\theta_2 = -45.0°$, and $a = \dfrac{\ell}{2}$. Also, from Figure (a), each side produces a field into the page. The four sides altogether produce

$$B_{\text{center}} = 4B = 4\frac{\mu_0 I}{4\pi a}\left(\sin\theta_1 - \sin\theta_2\right)$$

$$= \frac{\mu_0 I}{\pi\ell/2}\left[\sin 45.0° - \sin\left(-45.0°\right)\right]$$

$$= \frac{2\mu_0 I}{\pi\ell}\left[\frac{2}{\sqrt{2}}\right] = \frac{2\sqrt{2}\mu_0 I}{\pi\ell}$$

$$B = \frac{2\sqrt{2}\left(4\pi\times 10^{-7}\ \text{T}\cdot\text{m/A}\right)(10.0\ \text{A})}{\pi(0.400\ \text{m})}$$

$$= 2\sqrt{2}\times 10^{-5}\ \text{T} = 28.3\ \mu\text{T}\ \text{ perpendicularly into the page} \qquad \blacksquare$$

(b) As in the first part of the problem, the direction of the magnetic field will be into the page. The new radius is found from the length of wire: $4\ell = 2\pi R$, so $R = 2\ell/\pi = 0.255$ m. Equation 22.24 gives the magnetic field at the center of a circular current loop:

$$B = \frac{\mu_0 I}{2R} = \frac{(4\pi\times 10^{-7}\ \text{T}\cdot\text{m/A})(10.0\ \text{A})}{2(0.255\ \text{m})} = 2.47\times 10^{-5}\ \text{T} = 24.7\ \mu\text{T} \qquad \blacksquare$$

Caution! If you use your calculator, it may not understand the keystrokes
$\boxed{4}\,\boxed{\times}\,\boxed{\pi}\,\boxed{\text{EXP}}\,\boxed{+/-}\,\boxed{7}$. To get the right answer, you may need to use
$\boxed{4}\,\boxed{\text{EXP}}\,\boxed{+/-}\,\boxed{7}\,\boxed{\times}\,\boxed{\pi}$.

Finalize: The magnetic field in part (a) is less than 40 μT as we predicted. Also, the magnetic fields from the square and circular loops are similar in magnitude, with the field from the circular loop being about 15% less than from the square loop.

Quick Tip: A simple way to use your right hand to find the magnetic field due to a current loop is to curl the fingers of your right hand in the direction of the current. Your extended thumb will then point in the direction of the magnetic field within the loop or solenoid. This is also the direction of the magnetic moment of the current loop.

---

43. In Figure P22.43, the current in the long, straight wire is I_1 = 5.00 A and the wire lies in the plane of the rectangular loop, which carries the current I_2 = 10.0 A. The dimensions in the figure are c = 0.100 m, a = 0.150 m, and ℓ = 0.450 m. Find the magnitude and direction of the net force exerted on the loop by the magnetic field created by the wire.

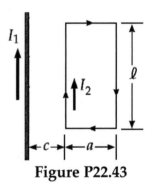

Figure P22.43

Solution

Conceptualize: There are forces in opposite directions on the loop, but we must remember that the magnetic field is stronger near the wire than it is farther away. By symmetry, the forces exerted on sides 2 and 4 (the horizontal segments of length a) are equal and opposite, and therefore add to zero. The magnetic field in the plane of the loop is directed into the page to the right of I_1. By the right-hand rule, $\vec{F} = I\vec{L} \times \vec{B}$ is directed toward the **left** for side 1 of the loop and a smaller force is directed toward the **right** for side 3. Therefore, we should expect the net force to be to the left, possibly in the μN range for the currents and distances given.

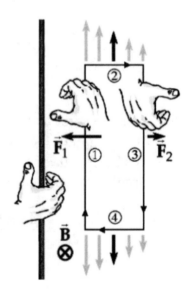

Categorize: The magnetic force between two parallel wires can be found from the first equation in Section 22.8, which can be applied to sides 1 and 3 of the loop to find the net force resulting from these opposing force vectors.

Analyze:

$$\vec{\mathbf{F}} = \vec{\mathbf{F}}_1 + \vec{\mathbf{F}}_2 = \frac{\mu_0 I_1 I_2 \ell}{2\pi}\left(\frac{1}{c+a} - \frac{1}{c}\right)\hat{\mathbf{i}} = \frac{\mu_0 I_1 I_2 \ell}{2\pi}\left(\frac{-a}{c(c+a)}\right)\hat{\mathbf{i}}$$

$$\vec{\mathbf{F}} = \frac{\left(4\pi \times 10^{-7}\ \text{N/A}^2\right)(5.00\ \text{A})(10.0\ \text{A})(0.450\ \text{m})}{2\pi}\left(\frac{-0.150\ \text{m}}{(0.100\ \text{m})(0.250\ \text{m})}\right)\hat{\mathbf{i}}$$

$$\vec{\mathbf{F}} = \left(-2.70 \times 10^{-5}\hat{\mathbf{i}}\right)\ \text{N} = 2.70 \times 10^{-5}\ \text{N toward the left} \qquad\blacksquare$$

Finalize: The net force is to the left and in the μN range as we expected. The symbolic representation of the net force on the loop shows that the net force would be zero if either current disappeared, if either dimension of the loop became very small ($a \to 0$ or $\ell \to 0$), or if the magnetic field were uniform ($c \to \infty$).

---

54. Four long, parallel conductors carry equal currents of $I = 5.00$ A. Figure P22.54 is an end view of the conductors. The current direction is into the page at points A and B and out of the page at C and D. Calculate (a) the magnitude and (b) the direction of the magnetic field at point P, located at the center of the square of edge length $\ell = 0.200$ m.

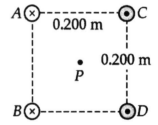

Figure P22.54

Solution

Conceptualize: To get a field in the millitesla range we would typically need a coil with many turns. Here we expect a microtesla field. The currents point out of the page and into the page, so the field will not. Rather, the field will lie in the plane of the page. If A and B were positive charges and C and D were negative, their net electric field would be to the right. But the total magnetic field will not be right or left.

Categorize: We find the field created at P by each wire, and do a vector addition of the four fields.

Analyze: Each wire is distant from P by (0.200 m) cos 45.0° = 0.141 m. Each wire produces a field at P of the same magnitude:

$$B = \frac{\mu_0 I}{2\pi a} = \frac{\left(2.00 \times 10^{-7} \text{ T} \cdot \text{m/A}\right)(5.00 \text{ A})}{0.141 \text{ m}} = 7.07 \ \mu\text{T}$$

Carrying currents away from you, the left-hand wires produce fields at P of 7.07 μT, in the following directions, determined by the right-hand rule:

A: to the bottom of the diagram and left, at 225°

B: to the bottom of the diagram and right, at 315°

Carrying currents toward you, the wires to the right also produce fields at P of 7.07 μT, in the following directions:

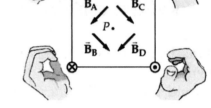

C: downward and to the right, at 315°

D: downward and to the left, at 225°

The diagram shows the four contributions to the total field at P. Vector addition shows that the total field is (a) 4(7.07 μT) sin 45.0° = 20.0 μT (b) toward the bottom of the page. ■

Finalize: The calculation would be significantly more complicated to find the field at other points away from the center of the array of wires. If you need to review vector addition by the component method from Chapter 1, do so. After that, you need to practice using your right hand to identify directions of magnetic fields. Describing the field as "clockwise" for *A* and *B* and "counterclockwise" for *C* and *D* is not definite enough. The field has a single direction at any particular point.

---

57. A long solenoid that has 1 000 turns uniformly distributed over a length of 0.400 m produces a magnetic field of magnitude 1.00×10^{-4} T at its center. What current is required in the windings for that to occur?

Solution

Conceptualize: We estimate on the order of one amp. The field is many microteslas. It is larger than fields typically produced by a single wire, mostly (we think) because of the many turns and not because the current is very large.

Categorize: We use just the equation derived in the chapter text for the field of a solenoid.

Analyze: The magnetic field at the center of a solenoid is $B = \mu_0 \frac{N}{\ell} I$.

So
$$I = \frac{B\ell}{\mu_0 N} = \frac{(1.00 \times 10^{-4} \text{ T} \cdot \text{A})(0.400 \text{ m})}{(4\pi \times 10^{-7} \text{ T} \cdot \text{m})(1\,000)} = 31.8 \text{ mA}.$$ ■

Finalize: The answer is thirty times smaller than our guess. This solenoid looks like a roll for wrapping paper. It could have any cross-sectional area, so long as its radius is small compared to its 40-cm length. Winding the wire onto the form would require a machine or else be quite tedious. Once the coil is made, the problem is perfectly realistic. The field could be displayed by iron filings or by an oscillating compass needle.

---

59. The magnetic moment of the Earth is approximately 8.00×10^{22} A · m². Imagine that the planetary magnetic field were caused by the complete magnetization of a huge iron deposit with density 7 900 kg/m³ and approximately 8.50×10^{28} iron atoms/m³. (a) How many unpaired electrons, each with a magnetic moment of 9.27×10^{-24} A · m², would participate? (b) At two unpaired electrons per iron atom, how many kilograms of iron would be present in the deposit?

Solution

Conceptualize: We know that much of the Earth is not iron, so if the situation described provides an accurate model, then the iron deposit must certainly be less than the mass of the Earth ($M_{Earth} = 5.97 \times 10^{24}$ kg). One mole of iron has a mass of 55.8 g and contributes $2(6.02 \times 10^{23})$ unpaired electrons, so we should expect the total number of unpaired electrons to be less than 10^{50}.

Categorize: The value 9.27×10^{-24} A · m² is equivalent to the Bohr magneton, 9.27×10^{-24} J/T (see Chapter 29). The Bohr magneton μ_B is the measured value for the magnetic moment of a single unpaired electron. Therefore, we can find the number of unpaired electrons by dividing the magnetic moment of the Earth by μ_B. We can then use the density of iron to find the mass of the iron atoms that each contribute two electrons.

Analyze:

(a) The Bohr magneton is

$$\mu_B = \left(9.27 \times 10^{-24} \frac{\text{J}}{\text{T}}\right)\left(\frac{\text{N} \cdot \text{m}}{1 \text{ J}}\right)\left(\frac{1 \text{ T}}{\text{N} \cdot \text{s/C} \cdot \text{m}}\right)\left(\frac{1 \text{ A}}{\text{C/s}}\right) = 9.27 \times 10^{-24} \text{ A} \cdot \text{m}^2$$

The number of unpaired electrons is

$$N = \frac{8.00 \times 10^{22} \text{ A} \cdot \text{m}^2}{9.27 \times 10^{-24} \text{ A} \cdot \text{m}^2} = 8.63 \times 10^{45} \text{ e}^- \qquad \blacksquare$$

(b) Each iron atom has two unpaired electrons, so the number of iron atoms required is

$$\frac{1}{2}N = \frac{1}{2}(8.63 \times 10^{45}) = 4.31 \times 10^{45} \text{ iron atoms}$$

Thus, $\quad M_{\text{Fe}} = \dfrac{(4.31 \times 10^{45} \text{ atoms})(7\,900 \text{ kg/m}^3)}{8.50 \times 10^{28} \text{ atoms/m}^3} = 4.01 \times 10^{20} \text{ kg.}$ $\qquad \blacksquare$

Finalize: The calculated answers seem reasonable based on the limits we expected. From the data in this problem, the iron deposit required to produce the magnetic moment would only be about 1/15 000 the mass of the Earth and would form a sphere 500 km in diameter. Although this is certainly a large amount of iron, it is much smaller than the inner core of the Earth, which is estimated to have a diameter of about 3 000 km.

---

72. Heart-lung machines and artificial kidney machines employ blood pumps. The blood is confined to an electrically insulating tube, cylindrical in practice but represented here for simplicity as a rectangle of width w and height h. Figure P22.72 shows a rectangular section of blood within the tube. Two electrodes fit into the top and the bottom of the tube. The potential difference between them establishes an electric current through the blood, with current density J over a section of length L shown in Figure P22.72. A perpendicular magnetic field exists in the same region. (a) Explain why this arrangement produces on the liquid a force that is directed along the length of the pipe. (b) Show that the section of liquid in the magnetic field experiences a pressure increase JLB. (c) After the blood leaves the pump, is it charged? (d) Is it carrying current? (e) Is it magnetized? (The same electromagnetic pump can be used for any fluid that conducts electricity, such as liquid sodium in a nuclear reactor.)

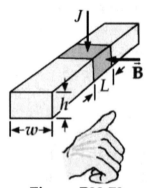

Figure P22.72

Solution

Conceptualize: Electricity is economically important because (for one thing) it is so easy to convert electrically transmitted energy into mechanical energy with high efficiency, using a magnetic force. This conversion happens in a spinning electric motor, and also in this electromagnetic pump. A mechanical pump could mangle blood cells. The simplicity of design makes this pump dependable. The blood is easily kept uncontaminated; the tube is simple to clean or cheap to replace.

Categorize: We think about the magnetic force acting on the electric current between the electrodes. We need not think about the motion of the blood.

Analyze:

(a) We define vector $\vec{\mathbf{h}}$ as downward, in the direction of the current. By the right-hand rule, the electric current carried by the material experiences a force $I\vec{\mathbf{h}} \times \vec{\mathbf{B}}$ along the pipe, into the page.   ■

(b) The blood, containing ions and electrons, flows along the pipe transporting no net charge. But inside the section of length L, electrons drift upward to constitute downward electric current $J(\text{area}) = JLw$. The current feels magnetic force $I\vec{\mathbf{h}} \times \vec{\mathbf{B}} = JLwhB \sin 90°$. This force along the pipe axis can make the fluid move, exerting pressure

$$\frac{F}{\text{area}} = \frac{JLwhB}{hw} = JLB$$ ■

The hand in the figure shows that the fluid moves away from you, into the page.

(c) Charge moves within the fluid inside the length L, but charge does not accumulate. The fluid is not charged after it leaves the pump. (d) It is not current-carrying and (e) it is not magnetized.   ■

Finalize: An electric force and a magnetic force can act on the same electric charge. Here an ion in the blood feels a vertical electric force making it participate in the vertical electric current, and then also a horizontal magnetic force. Make sure you know that a magnetic charge or a magnetic current does not exist.

78. A nonconducting ring of radius R is uniformly charged with a total positive charge q. The ring rotates at a constant angular speed ω about an axis through its center, perpendicular to the plane of the ring. What is the magnitude of the magnetic field on the axis of the ring a distance $R/2$ from its center?

Solution

Conceptualize: The "static" charge on the ring constitutes an electric current as the ring rotates. The current will be proportional to q and to ω. Then the field will be proportional to both of these quantities and to μ_0. It will decrease as R increases.

Categorize: We will find the current and then use the equation for the magnetic field on the axis of a current loop.

Analyze: The time interval required for the charge q to go past a point is the period of rotation,

$$T = \frac{\theta}{\omega} = \frac{2\pi}{\omega}$$

The current is $I = \dfrac{q}{T} = \dfrac{q}{2\pi / \omega} = \dfrac{q\omega}{2\pi}$.

The current has the shape of a flat compact circular coil with one turn, so the magnetic field it creates is

$$B = \frac{\mu_0 I R^2}{2\left(x^2 + R^2\right)^{3/2}}$$

In this case, the distance x is equal to $R/2$, so

$$B = \frac{\mu_0 q\omega R^2}{4\pi\left(R^2/4 + R^2\right)^{3/2}} = \frac{\mu_0 q\omega R^2}{4\pi\left(5R^2/4\right)^{3/2}} = \frac{2\mu_0 \, q\omega}{5\sqrt{5}\pi R} \qquad \blacksquare$$

On the side of the ring from which it is seen as turning counterclockwise, the magnetic field is along the axis away from its center. On the other side of the ring, the magnetic field is toward the center of the ring, in the same direction in space.

Finalize: The result has the predicted proportionalities. Its dependence on R turns out to be simply an inverse proportionality. We have found the magnetic field at $R/2$ from the center. It would be stronger at the center itself. The ring also creates an electric field, which is zero at the center and points away from the center, in opposite directions on opposite sides of the axis.

81. A very long, thin strip of metal of width w carries a current I along its length as shown in Figure P22.81. The current is distributed uniformly across the width of the strip. Find the magnetic field at point P in the diagram. Point P is in the plane of the strip at distance b away from its edge.

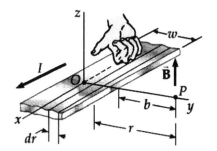

Figure P22.81 (modified)

Solution

Conceptualize: The field direction will not be parallel or antiparallel to the current. It will not be toward the strip or away from it. It must be in the $+z$ or $-z$ direction. Get out your right hand. Hold it with the extended thumb pointing toward you, in the direction of the current in the diagram. Then the field at a point directly to the right is straight up. ∎

The field magnitude should be proportional to μ_0 and proportional to I. It should decrease as b increases.

Categorize: We think of the strip as made of infinitely many filaments, each acting as a long, straight wire to create field at P. We identify the field dB due to one filament and then integrate to find the field of them all.

Analyze: Consider a long filament in the strip, which has width dr and is a distance r from point P. The magnetic field at a distance r from a long conductor is

$$B = \frac{\mu_0 I}{2\pi r}$$

Thus, the field due to the filament is $d\vec{\mathbf{B}} = \dfrac{\mu_0 dI}{2\pi r}\hat{\mathbf{k}}$, where $dI = I\left(\dfrac{dr}{w}\right)$,

so $\qquad \vec{\mathbf{B}} = \displaystyle\int_b^{b+w} \frac{\mu_0}{2\pi r}\left(I\frac{dr}{w}\right)\hat{\mathbf{k}} = \frac{\mu_0 I}{2\pi w}\int_b^{b+w}\frac{dr}{r}\hat{\mathbf{k}} = \frac{\mu_0 I}{2\pi w}\ln\left(1+\frac{w}{b}\right)\hat{\mathbf{k}}.$ ∎

Finalize: The field has the predicted proportionalities to μ_0 and I. For b large compared to w, the fact that the current is a strip rather than just a wire should be unimportant, so the field should be inversely proportional to b. And it is, because the condition $b \gg w$ makes $\ln(1 + w/b)$ take the form

$$w/b - (w/b)^2/2 + (w/b)^3/3 - \cdots \approx w/b$$

so the field becomes $\dfrac{\mu_0 I}{2\pi w}\dfrac{w}{b}\hat{\mathbf{k}} = \dfrac{\mu_0 I}{2\pi b}\hat{\mathbf{k}}$ as it should be for a thin wire.

Chapter 23
Faraday's Law and Inductance

NOTES FROM SELECTED CHAPTER SECTIONS

Section 23.1   Faraday's Law of Induction

The emf induced in a circuit is equal to the time rate of change of magnetic flux through the circuit.

An emf can be induced in the circuit in several ways:

- The magnitude of the magnetic field can change as a function of time.

- The area of the circuit can change with time.

- The direction of the magnetic field relative to the circuit can change with time.

- Any combination of the above can change.

Section 23.2   Motional emf

A potential difference, ΔV, will be maintained across a conductor moving so that it is "cutting lines of flux" in a magnetic field. The potential difference will be zero when the direction of motion of the conductor is in a plane parallel or antiparallel to the field direction. If the motion is reversed, the polarity of the potential difference will also be reversed.

Section 23.3   Lenz's Law

The polarity of the induced emf is such that it tends to produce a current that will create a magnetic flux to *oppose the change in flux through the circuit.*

Section 23.5   Inductance
Section 23.6   *RL* Circuits

The self-induced emf is always *proportional to the time rate of change of current* in the circuit. The inductance of a device (an inductor) depends on its geometry.

If a resistor and an inductor are connected in series to a battery, the current in the circuit will reach an equilibrium value (ε/R) after a time that is long compared to the time constant of the circuit $(\tau = L/R)$.

Section 23.7   Energy Stored in a Magnetic Field

In an *RL* circuit the rate at which energy is supplied by the battery equals the sum of the rate at which energy is delivered to the resistor and the rate at which energy is stored in the inductor. *The energy density is proportional to the square of the magnetic field.*

EQUATIONS AND CONCEPTS

The **total magnetic flux** threading a circuit is the integral of the normal component of the magnetic field over the area bounded by the circuit.

$$\Phi_B = \int \vec{\mathbf{B}} \cdot d\vec{\mathbf{A}} \qquad (23.1)$$

$d\vec{\mathbf{A}}$ is an element of area perpendicular to the surface.

The **magnetic flux through a plane area bounded by a circuit** depends on the angle between the direction of the magnetic field and the direction perpendicular to the surface area. *The maximum flux through the area occurs when the magnetic field is perpendicular to the plane of the surface area.*

$$\Phi_B = BA\cos\theta$$

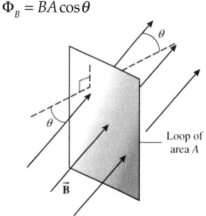

A uniform magnetic field through a conducting loop of area A

The unit of magnetic flux is the weber (Wb):
$1\ \text{Wb} = 1\ \text{T}\cdot\text{m}^2$

Faraday's law of induction states that the emf induced in a circuit equals the rate of change of magnetic flux through the circuit. *The minus sign is included to indicate the polarity of the induced emf, which can be found by use of Lenz's law.*

$$\mathcal{E} = -\frac{d\Phi_B}{dt} \quad \text{(single loop)} \quad (23.2)$$

An **emf will be induced in a circuit** when any one or more of the following change with time:

- magnitude of $\vec{B}$
- area enclosed by the circuit
- angle θ between $\vec{B}$ and the normal

$$\mathcal{E} = -\frac{d}{dt}\left(BA\cos\theta\right) \quad \text{(single loop)}$$

$$(23.4)$$

Lenz's law states that the polarity of the induced emf in a loop creates a current which produces a magnetic field that *opposes the change in the flux through the loop.* Consider the example illustrated in the figure below. In the coil on the left, the magnetic flux increases to the right as the N-pole of the magnet approaches. The resulting induced current has a direction around the loop which creates a magnetic field (and associated flux) directed toward the left as shown the loop on the right. *The induced current (and the associated magnetic field) tends to keep the magnetic flux through the loop coil from changing.*

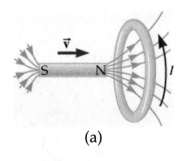

(a)

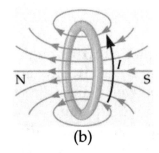

(b)

(a) N-pole of a bar magnet approaches a loop; this increases the flux through the loop toward the right. The change in flux through the loop creates an induced current in the direction shown by the arrow.

(b) The induced current produces a magnetic field directed toward the left. The overall effect is to *oppose any change* in the flux through the loop.

A **motional emf** is induced in a conductor of length ℓ moving with speed v perpendicular to a magnetic field (so that the conductor is "cutting" lines of magnetic flux). *Recall the significance of the negative sign as required by Lenz's law.*
The right-hand rule predicts that the magnetic force on an electron in the conductor will be directed downward as shown in the figure.

$$\mathcal{E} = -B\ell v \qquad (23.5)$$

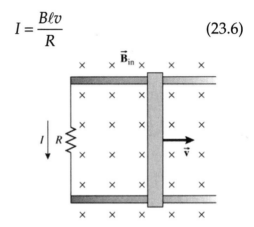

An illustration of motional emf

An **induced current** will exist in a conductor moving in a magnetic field if it is part of a complete circuit. Confirm that Lenz's law correctly predicts the direction of the induced current as shown in the figure. *Remember that the direction of conventional current is opposite the direction of electron flow in the circuit.*

$$I = \frac{B\ell v}{R} \qquad (23.6)$$

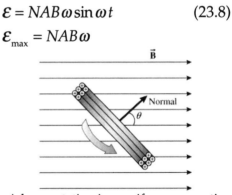

The direction of I is determined by Lenz's law.

A **sinusoidally varying emf** is produced when a conducting loop of N turns and cross-sectional area A rotates with a constant angular velocity in a magnetic field. *For a given loop, the maximum value of the induced emf will be proportional to the angular velocity of the loop.*

$$\mathcal{E} = NAB\omega \sin\omega t \qquad (23.8)$$
$$\mathcal{E}_{max} = NAB\omega$$

A loop rotating in a uniform magnetic field. The normal (perpendicular to the plane of the loop) makes an angle θ relative to the field.

Faraday's law in a more general form can be written as the integral of the electric field around a closed path. In this form the electric field is a nonconservative field that is generated by a *changing magnetic field*.

$$\oint \vec{E} \cdot d\vec{s} = -\frac{d\Phi_B}{dt} \qquad (23.9)$$

A **self-induced emf** is present in a coil when the current in the coil changes in time. The inductance (L) is a measure of the opposition of the coil to *a change in the current*.
Recall that resistance opposes the current in a circuit.

$$\varepsilon_L = -L\frac{dI}{dt} \qquad (23.10)$$

Inductance of a given device (coil, solenoid, toroid, coaxial cable, or other conducting device) can be calculated if the flux and current are known (Eq. 23.11) or as the ratio of the magnitude of the induced emf divided by the time rate of change of current in the circuit (Eq. 23.12). The SI unit of inductance is the henry.

$$L = \frac{N\Phi_B}{I} \qquad (23.11)$$

$$L = -\frac{\varepsilon_L}{dI/dt} \qquad (23.12)$$

$$1\,\text{H} = 1\,\frac{\text{V} \cdot \text{s}}{\text{A}} = 1\,\Omega \cdot \text{s}$$

Increasing current in an *RL* circuit (which contains a battery, resistor, and inductor) produces a *back emf*. If the switch in the series circuit shown below is closed in position 1 at time $t = 0$, current in the circuit will increase in a characteristic fashion toward a maximum value of $\left(\dfrac{\varepsilon}{R}\right)$. This is shown in the graph to the right.

$$I = \frac{\varepsilon}{R}(1 - e^{-t/\tau}) \qquad (23.14)$$

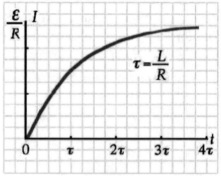

Increase of current (from zero) in an *RL* circuit when the switch at left is thrown to position 1.

The **time constant** of the circuit (τ) is the time required for the current to reach 63.2% of its maximum value.

$$\tau = \frac{L}{R} \qquad (23.15)$$

Let the switch in the circuit be at position 1 with the current at its maximum value, $\left(\dfrac{\varepsilon}{R}\right)$. If the switch is now thrown to position 2 (as shown below) at $t = 0$, the current will decay exponentially with time. The graph at right shows the manner in which the current decays.

$$I = \frac{\varepsilon}{R} e^{-t/\tau} = I_i e^{-t/\tau} \qquad (23.18)$$

$$\frac{\varepsilon}{R} = I_i$$

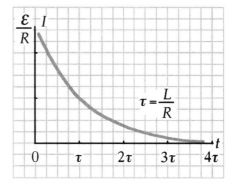

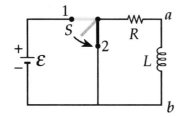

Exponential decay of current in an *RL* circuit when the switch at left is thrown to position 2.

The **energy stored** in the magnetic field of an inductor is proportional to the square of the current in the inductor.

$$U = \frac{1}{2} L I^2 \qquad \text{(energy)}$$

The **energy density** (u_B) is the energy per unit volume.

$$u_B = \frac{B^2}{2\mu_0} \qquad \text{(energy density)}$$

SUGGESTIONS, SKILLS, AND STRATEGIES

Instantaneous and Average Induced Emf

It is important to distinguish clearly between the **instantaneous value** of emf induced in a circuit and the **average value** of the emf induced in the circuit over a finite time interval.

To calculate the **average induced emf**, it is often useful to write Equation 23.3 as

$$\varepsilon_{av} = -N \left(\frac{d\Phi_B}{dt} \right)_{av} = -N \frac{\Delta \Phi_B}{\Delta t} \qquad \text{or} \qquad \varepsilon_{av} = -N \left(\frac{\Phi_{B,f} - \Phi_{B,i}}{\Delta t} \right)$$

where the subscripts *i* and *f* refer to the magnetic flux through the circuit at the beginning and end of the time interval Δ*t*. For a circuit (or coil) in a single plane, $\Phi_B = BA\cos\theta$, where *θ* is the angle between the direction of the normal to the plane of the circuit (conducting loop) and the direction of the magnetic field.

Equation 23.4 can be used to calculate the **instantaneous value of an induced emf**. For a multiple turn coil, the induced emf is

$$\varepsilon = -N\frac{d}{dt}\left(BA\cos\theta\right)$$

where in a particular case *B*, *A*, *θ*, or any combination of those parameters can be time dependent while the others remain constant. The expression resulting from the differentiation is then evaluated using the values of *B*, *A*, and *θ* corresponding to the specified value.

AN EXAMPLE OF LENZ'S LAW

Consider two single-turn, concentric coils **lying in the plane of the paper** as shown in the figure below. The outside coil is part of a circuit containing a resistor (*R*), a battery (*ε*), and switch (*S*). The inner coil is not part of the circuit. When the switch is moved from "**open**" to "**closed**" the direction of the induced current in the inner coil can be predicted by Lenz's law. Consider the following steps:

(1)   When the switch is in the "**open**" position as shown, there will be no current in the circuit.

(2)   When the switch is moved to the "**closed**" position, there will be a clockwise current in the outside coil.

(3)   Magnetic field lines due to current in the outside coil will be directed into the page through the area enclosed by the coil. **Use the right-hand rule (first application) to confirm this.** Magnetic flux into the page will penetrate the entire area enclosed by the outside coil **including the area of the inner coil.**

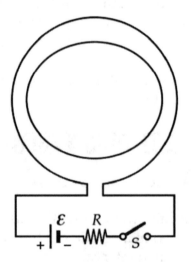

Two single-turn, concentric coils in the plane of the paper

(4)   By Faraday's law, the increasing flux produces an induced emf (and current) in the inner coil.

(5)   Lenz's law requires that the induced current have a direction, which will tend to maintain the initial flux condition (**which in this case was zero**). By using the **right-hand rule (second application)**, you should be able to determine that the direction of the induced current in the inner coil must be counterclockwise, contributing to a flux out of the page. Try this!

As a second example you should follow steps similar to those above to predict the direction of the induced current in the inner coil when the switch is moved from "**closed**" to "**open**".

REVIEW CHECKLIST

• Calculate the emf (or current) induced in a circuit when the magnetic flux through the circuit is changing in time. The variation in flux might be due to a change in (a) the area of the circuit, (b) the magnitude of the magnetic field, (c) the direction of the magnetic field, or (d) the orientation/location of the circuit in the magnetic field.

• Calculate the emf induced between the ends of a conducting bar as it moves through a region where there is a constant magnetic field (motional emf). Apply Lenz's law to determine the direction of an induced emf or current. You should also understand that Lenz's law is a consequence of the law of conservation of energy.

• Calculate the inductance of a device of suitable geometry.

• Calculate the magnitude and direction of the self-induced emf in a circuit containing one or more inductive elements when the current changes with time.

• Determine instantaneous values of the current in an RL circuit while the current is either increasing or decreasing with time.

• Calculate the total magnetic energy stored in a magnetic field. You should be able to perform this calculation (1) using Equation 23.20, $U = \frac{1}{2}LI^2$, if you are given the values of the inductance of the device with which the field is associated and the current in the circuit, or (2) using Equation 23.22, $u_B = \frac{B^2}{2\mu_0}$, given the value of the magnitude of the magnetic field throughout the region of space in which the magnetic field exists. In the latter case, you must integrate the expression for the energy density u_B over an appropriate volume.

ANSWERS TO SELECTED OBJECTIVE QUESTIONS

8. If the current in an inductor is doubled, by what factor is the stored energy multiplied? (a) 4 (b) 2 (c) 1 (d) 1/2 (e) 1/4

Answer (a). The energy stored in an inductor carrying a current I is given by $U = \frac{1}{2}LI^2$. Therefore, doubling the current will quadruple the energy stored in the inductor.

10. The bar in Figure OQ23.10 moves on rails to the right with a velocity $\vec{v}$, and the uniform, constant magnetic field is directed out of the page. Which of the following statements are correct? More than one statement may be correct. (a) The induced current in the loop is zero. (b) The induced current in the loop is clockwise. (c) The induced current in the loop is counterclockwise. (d) An external force is required to keep the bar moving at constant speed. (e) No force is required to keep the bar moving at constant speed.

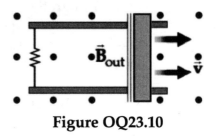

Figure OQ23.10

Answer The correct statements are (b) and (d). The externally-produced magnetic field is out of the paper. As the area A enclosed by the loop increases, the external flux increases according to $\Phi_B = BA\cos\theta = BA$.

As predicted by Lenz's law, due to the increase in flux through the loop, free electrons will produce a current to create a magnetic flux inside the loop to oppose the change in flux.

In this case, the magnetic field due to the current must point into the paper in order to oppose the increasing magnetic flux of the external field coming out of the paper. By the right-hand rule (with your thumb pointing in the direction of the current along each wire), the current must be clockwise, so statement (b) is correct.

Next, the clockwise current is downward in the picture in the bar. The external magnetic field perpendicularly out of the page exerts on this current an $i\vec{L} \times \vec{B}$ magnetic force in the direction (down) × (toward you), which is left in the diagram. A magnetic braking force acts on the bar. The bar would coast to a stop if some outside agent did not exert on it a counterbalancing force to the right to keep the bar's speed constant and to provide energy to maintain the current. Thus answer (d) is also correct.

□   □   □   □

ANSWERS TO SELECTED CONCEPTUAL QUESTIONS

8.   In a hydroelectric dam, how is energy produced that is then transferred out by electrical transmission? That is, how is the energy of motion of the water converted to energy that is transmitted by AC electricity?

Answer   As the water falls, it gains kinetic energy. It is then forced to pass through a turbine (a water wheel), transferring some of its energy to the rotor of a large AC electric generator.

The rotor of the generator is supplied with a small amount of DC current, which powers electromagnets in the rotor. Because the rotor is spinning, the electromagnets then create a magnetic flux that changes with time, according to the equation $\Phi_B = BA \cos \omega t$.

Coils of wire that are placed near the rotor then experience an induced emf described by the equation $\varepsilon = -N d\Phi_B/dt$.

Finally, a small amount of this electricity is used to supply the rotor with its DC current; the rest is sent out over power lines to supply customers with electricity.

In terms of energy, the hydroelectric generating station is a nonisolated system in steady state. It takes in mechanical energy and puts out a precisely equal quantity of energy, nearly all of it by electrical transmission.

□   □   □   □

10.   The current in a circuit containing a coil, a resistor, and a battery has reached a constant value. (a) Does the coil have an inductance? (b) Does the coil affect the value of the current?

Answer   (a) The coil has an inductance regardless of the nature of the current in the circuit. Inductance depends only on the coil geometry and its construction. (b) Since the current is constant, the self-induced emf in the coil is zero, and the coil does not affect the steady-state current. (We assume the resistance of the coil is negligible.)

□   □   □   □

16. After the switch is closed in the *LC* circuit shown in Figure CQ23.16, the charge on the capacitor is sometimes zero, but at such instants the current in the circuit is not zero. How is this behavior possible?

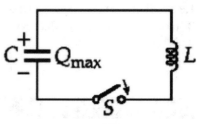

Figure CQ23.16

Answer When the capacitor is fully discharged, the current in the circuit is a maximum. The inductance of the coil is making the charge continue to flow, while the capacitor charge passes through the value zero on its way between positive and negative values. At such an instant, the magnetic field of the coil contains all the energy that was originally stored in the charged capacitor.

□   □   □   □

SOLUTIONS TO SELECTED END-OF-CHAPTER PROBLEMS

5. An aluminum ring of radius $r_1 = 5.00$ cm and resistance 3.00×10^{-4} Ω is placed around one end of a long air-core solenoid with 1 000 turns per meter and radius $r_2 = 3.00$ cm as shown in Figure P23.5. Assume the axial component of the field produced by the solenoid is one-half as strong over the area of the end of the solenoid as at the center of the solenoid. Also assume the solenoid produces negligible field outside its cross-sectional area. The current in the solenoid is increasing at a rate of 270 A/s. (a) What is the induced current in the ring? At the center of the ring, what are (b) the magnitude and (c) the direction of the magnetic field produced by the induced current in the ring?

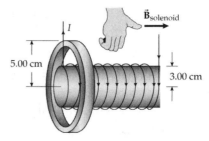

Figure P23.5 (modified)

Solution

Conceptualize: Many turns, a rapid change in the solenoid current, and low resistance in the ring suggest that the current will be many amps. But the ring's field it creates will likely be much less than the solenoid field. If the solenoid current were 50 A at one instant, the solenoid field would be $\mu_0 nI = 63$ mT to the right at the center of the solenoid.

Categorize: We use Faraday's law. In $BA\cos\theta$ the factor that is changing is the magnetic field. We can find its rate of change from the rate of change of the solenoid current. Then the definition of resistance will tell us the current in the ring from the induced emf. At last, the ring current will let us compute its constant field.

Analyze: The symbol for the radius of the ring is r_1, and we use R to represent its resistance. The emf induced in the ring is

$$\mathcal{E} = -\frac{d}{dt}(BA\cos\theta) = -\frac{d}{dt}(0.500\mu_0 nIA \cos 0°) = -0.500\mu_0 nA\frac{dI}{dt}$$

Note that A must be interpreted as the area $A = \pi r_2^2$ of the solenoid, where the field is strong:

$$\mathcal{E} = -0.500(4\pi \times 10^{-7} \text{ T} \cdot \text{m/A})(1\,000 \text{ turns/m})[\pi(0.030\,0 \text{ m})^2](270 \text{ A/s})$$

$$\mathcal{E} = \left(-4.80 \times 10^{-4}\frac{\text{T} \cdot \text{m}^2}{\text{s}}\right)\left(\frac{1 \text{ N} \cdot \text{s}}{\text{C} \cdot \text{m} \cdot \text{T}}\right)\left(\frac{1 \text{ V} \cdot \text{C}}{\text{N} \cdot \text{m}}\right) = -4.80 \times 10^{-4} \text{ V}$$

(a) The negative sign means that the current in the ring is counterclockwise, opposite to the current in the solenoid. Its magnitude is

$$I_{\text{ring}} = \frac{|\mathcal{E}|}{R} = \frac{0.000\,480 \text{ V}}{0.000\,300 \text{ }\Omega} = 1.60 \text{ A}$$ ∎

(b) $$B_{\text{ring}} = \frac{\mu_0 I_{\text{ring}}}{2r_1} = \frac{(4\pi \times 10^{-7}\text{T} \cdot \text{m/A})(1.60 \text{ A})}{2(0.050\,0 \text{ m})} = 2.01 \times 10^{-5} \text{ T}$$ ∎

(c) The solenoid's field points to the right, and is increasing, so B_{ring} points to the left. ∎

Finalize: The ring current is not really very large. Here 270 A/s is not a large rate of change in current, compared to what a switch can cause. The ring field is indeed small compared to an estimate of the solenoid field. Note well that the solenoid current must keep increasing steadily to keep the ring current constant.

10. A long solenoid has $n = 400$ turns per meter and carries a current given by $I = 30.0 \ (1 - e^{-1.60t})$ where I is in amperes and t is in seconds. Inside the solenoid and coaxial with it is a coil that has a radius of 6.00 cm and consists of a total of $N = 250$ turns of fine wire (Fig. P23.10). What emf is induced in the coil by the changing current?

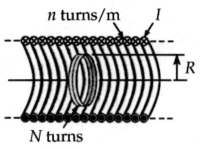

Figure P23.10

Solution

Conceptualize: The solenoid and the flat coil are not part of the same circuit. The transfer of energy between them is mediated by the magnetic field created by the solenoid. The solenoid current starts from zero, increases most rapidly at first, and then asymptotically approaches 30 A as its final value. We think of its rate of change to estimate that the induced voltage will start from its largest value, on the order of one volt, at time zero, and then decay exponentially toward zero.

Categorize: The solenoid current tells us the solenoid field. Faraday's law tells us the induced emf.

Analyze: The solenoid creates a magnetic field

$$B = \mu_0 nI = \left(4\pi \times 10^{-7} \ \text{N/A}^2\right)(400 \ \text{turns/m})(30.0 \ \text{A})\left(1 - e^{-1.60t}\right)$$
$$= \left(1.51 \times 10^{-2} \ \text{N/m} \cdot \text{A}\right)\left(1 - e^{-1.60t}\right)$$

The magnetic flux through one turn of the flat coil is $\Phi_B = \int B \, dA \cos\theta$, but since $BA\cos\theta$ refers to the area perpendicular to the flux, and the magnetic field is uniform over the area A of the flat coil, this integral simplifies to

$$\Phi_B = B \int dA = B\left(\pi R^2\right)$$
$$= \left(1.51 \times 10^{-2} \ \text{N} \cdot \text{m/A}\right)\left(1 - e^{-1.60t}\right)\left[\pi(0.060 \ 0 \ \text{m})\right]$$
$$= \left(1.71 \times 10^{-2} \ \text{N} \cdot \text{m/A}\right)\left(1 - e^{-1.60t}\right)$$

The emf generated in the N-turn coil is $\varepsilon = -N \, d\Phi_B/dt$. Because t has the standard unit of seconds, the factor 1.60 must have the unit s^{-1}.

$$\mathcal{E} = -(250)(1.71 \times 10^{-4}\ \text{N} \cdot \text{m/A})\frac{d(1 - e^{-1.60t})}{dt}$$

$$= -(0.042\ 6\ \text{N} \cdot \text{m/A})(1.60\ \text{s}^{-1})e^{-1.60t}$$

$$\mathcal{E} = -(6.82 \times 10^{-2}\ \text{N} \cdot \text{m/C})e^{-1.60t} = -(6.82 \times 10^{-2})e^{-1.60t} \qquad \blacksquare$$

where $\mathcal{E}$ is in volts and t is in seconds

Finalize: The induced emf shows the predicted exponential decay. The maximum voltage is smaller than we expected. The current in the solenoid is never changing very fast. The minus sign indicates that the emf will produce counterclockwise current in the smaller coil, opposite to the direction of the increasing current in the solenoid.

---

15. Figure P23.15 shows a top view of a bar that can slide on two frictionless rails. The resistor is $R = 6.00\ \Omega$, and a 2.50-T magnetic field is directed perpendicularly downward, into the paper. Let $\ell = 1.20$ m. (a) Calculate the applied force required to move the bar to the right at a constant speed of 2.00 m/s. (b) At what rate is energy delivered to the resistor?

Solution

Conceptualize: The motion of the bar induces an emf in the rectangle. The emf produces a current in the circuit, including the bar. Then the magnetic field exerts a force on the bar, and an outside agent must counterbalance this force to keep the bar moving steadily.

Categorize: We use the rigid body in equilibrium model, together with the definition of resistance and Faraday's law.

Analyze:

(a) At constant speed, the net force on the moving bar equals zero, or

$$\left|\vec{\mathbf{F}}_{\text{app}}\right| = I\left|\vec{\mathbf{L}} \times \vec{\mathbf{B}}\right|$$

where the current in the bar is $\qquad\qquad I = \mathcal{E}/R$

and the motional emf is $\qquad\qquad \mathcal{E} = B\,\ell v$

Therefore,

$$F_{\text{app}} = \left(\frac{B\ell v}{R} \right) \ell B = \frac{B^2 \ell^2 v}{R}$$

$$F_{\text{app}} = \frac{(2.50 \text{ T})^2 (1.20 \text{ m})^2 (2.00 \text{ m/s})}{6.00 \ \Omega} = 3.00 \text{ N to the right} \qquad \blacksquare$$

(b) The outside agent pulling the bar must provide input power

$$P = F_{\text{app}} v = (3.00 \text{ N})(2.00 \text{ m/s}) = 6.00 \text{ W} \qquad \blacksquare$$

Finalize: In terms of energy, the circuit is a nonisolated system in steady state. The energy, six joules every second, delivered to the circuit by work done by the applied force is delivered to the resistor by electrical transmission. The energy can leave the resistor as energy transferred by heat into the surrounding air. When we studied an electric circuit consisting of a resistor connected across a battery, we did not consider the details of the chemical reaction in the battery and the battery's loss of chemical energy. The circuit in this problem is equally simple electrically. The outside agent pulling the bar through the magnetic field takes the place of the battery. This is the first circuit in which we can account completely for the energy transformations. The power could also be computed as $\mathcal{E} I = I^2 / R$.

22. A rectangular coil with resistance R has N turns, each of length ℓ and width w as shown in Figure P23.22. The coil moves into a uniform magnetic field $\vec{\mathbf{B}}$ with constant velocity $\vec{\mathbf{v}}$. What are the magnitude and direction of the total magnetic force on the coil (a) as it enters the magnetic field, (b) as it moves within the field, and (c) as it leaves the field?

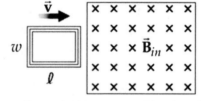

Figure P23.22

Solution

Conceptualize: While it is entering and leaving the field region, the magnetic flux through the coil will be changing, so an emf is induced in it. The emf causes current in the coil. The magnetic field exerts a force on the current. It is reasonable for the force to increase for larger values of N, w, v, and B, and to be inversely proportional to R. We predict the force will be proportional to the square of the magnetic field, because the field causes the emf and acts again to cause the force.

Categorize: We use Faraday's law, the definition of resistance, and the law describing the magnetic force on an electric current. We will need two right-hand rules, for the emf and for the magnetic force.

Analyze:

(a) Call x the distance that the leading edge has penetrated into the strong field region. The flux (Bwx away from you) through the coil increases in time so a voltage of magnitude

$$|\varepsilon| = N\frac{d}{dt}Bwx = NBw\frac{dx}{dt} = NBwv$$

is induced in the coil, tending to produce counterclockwise current so that its own field will be toward you [Figure (a)].

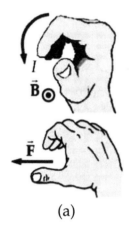

(a)

The current in the coil is $|I| = \dfrac{|\varepsilon|}{R} = \dfrac{NBwv}{R}$.

The current upward in the leading edge experiences a force

$$\vec{F} = N I \vec{L} \times \vec{B} = N\left(\frac{NBwv}{R}\right)w\hat{j} \times B(-\hat{k}) = \left(\frac{N^2B^2w^2v}{R}\right)(-\hat{i}) \qquad \blacksquare$$

(b) The flux through the coil is constant when it is wholly within the high-field region. The induced emf, induced current, and magnetic force are zero. ∎

(c) As the coil leaves the field, the away-from-you flux it encloses decreases. To oppose this change, the coil carries clockwise current to make some away-from-you field of its own [Figure (b)]. Again,

$$|I| = \frac{NBwv}{R}$$

Now the trailing edge carries upward current to experience a force of

$$\vec{F} = \frac{N^2B^2w^2v}{R} \text{ to the left} \qquad \blacksquare$$

(b)

Finalize: The current in the coil is in opposite directions when the coil enters and leaves the field. The magnetic force acts on opposite sides of the coil to exert a backward or retarding force in the same direction, to the left, in both cases. We have here a model for eddy-current damping. The proportionalities of the force to the square of N and the square of w, as well as to the square of B, are interesting. The force is proportional only to the first power of v and the minus-first power of R, as predicted. We can say it is proportional to the zeroth power of ℓ.

───────

28. A magnetic field directed into the page changes with time according to $B = 0.030\ 0t^2 + 1.40$, where B is in teslas and t is in seconds. The field has a circular cross section of radius $R = 2.50$ cm (see Fig. P23.28). When $t = 3.00$ s and $r_2 = 0.020\ 0$ m, what are (a) the magnitude and (b) the direction of the electric field at point P_2?

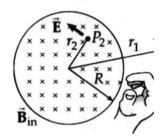

Figure P23.28

Solution

Conceptualize: The electric field is not created by an electric charge, but by a changing magnetic field. There need not be a test charge or a conducing loop at point P_2, but if there were it would respond to the electric field created . . .

Categorize: . . . according to the general form of Faraday's law,

$$\oint \vec{\mathbf{E}} \cdot d\vec{\mathbf{s}} = -\frac{d\Phi_B}{dt}$$

Analyze: Consider a circular integration path of radius r_2. Due to symmetry, Faraday's law becomes

$$E(2\pi r_2) = -\frac{d}{dt}(BA) = -A\left(\frac{dB}{dt}\right)$$

$$|E| = \frac{A}{2\pi r_2}\frac{d}{dt}(0.030\ 0t^2 + 1.40) = \frac{\pi r_2^2}{2\pi r_2}(0.060\ 0t) = \frac{r_2}{2}(0.060\ 0t)$$

At $t = 3.00$ s, the field is

(a) $E = \dfrac{1}{2}(0.020\ 0\ \text{m})(0.060\ 0\ \text{T/sec})(3.00\ \text{sec}) = 1.80 \times 10^{-3}$ N/C ∎

(b) If there were a circle of wire of radius r_2, it would enclose increasing magnetic flux due to a magnetic field away from you. It would carry counterclockwise current to make its own magnetic field toward you, to oppose the change. Even without the wire and current, the counterclockwise electric field that would cause the current is lurking. At point P_2, it is upward and to the left, perpendicular to r_2. ■

Finalize: This is a very weak electric field. We do not have large area, multiple turns in a coil, or sudden change working in our advantage to create it. The large 1.4-T field is constant in time, so it does not contribute to making the electric field large.

---

33. A 10.0-mH inductor carries a current $I = I_{max} \sin \omega t$, with $I_{max} = 5.00$ A and $f = \omega/2\pi = 60.0$ Hz. What is the self-induced emf as a function of time?

Solution

Conceptualize: The AC current keeps changing, so new positive and negative pulses of AC voltage are induced in the coil. The amplitude may be a few volts.

Categorize: We use the definition of self-inductance.

Analyze: $\varepsilon_L = -L\dfrac{dI}{dt} = -L\dfrac{d}{dt}\left(I_{max} \sin \omega t\right)$

$$\varepsilon_L = -L\omega I_{max} \cos \omega t = -(0.010\ 0\ \text{H})\left(120\pi\ \text{s}^{-1}\right)(5.00\text{A}) \cos(120\pi\ t)$$

$$\varepsilon_L = -(18.8) \cos(377\ t), \quad \text{where } \varepsilon \text{ is in volts and } t \text{ is in seconds} \quad ■$$

Finalize: The answer is not a number but an infinity of numbers making up a function, which is the voltage as it depends on time.

---

36. An inductor in the form of a solenoid contains 420 turns and is 16.0 cm in length. A uniform rate of decrease of current through the inductor of 0.421 A/s induces an emf of 175 μV. What is the radius of the solenoid?

Solution

Conceptualize: This is a laboratory tabletop situation. The rate of change in current can be thought of as 0.421 milliamps per millisecond. We expect a radius on the order of a centimeter.

Categorize: We represent the inductance of the solenoid symbolically in terms of its dimensions. Then the definition of inductance will relate the induced voltage to how fast the current is changing. We will be able to solve for the radius as the single unknown.

Analyze: The inductance is $L = \dfrac{\mu_0 N^2 A}{\ell}$ with $A = \pi r^2$. The induced emf as a

function of time is $\varepsilon_L = -L\dfrac{dI}{dt}$. By substitution we have

$$\varepsilon_L = -L\frac{dI}{dt} = -\frac{\mu_0 N^2 \pi r^2}{\ell}\frac{dI}{dt} \quad \text{and} \quad r = \left(\frac{-\varepsilon_L \ell}{\mu_0 N^2 \pi \, dI/dt}\right)^{1/2}$$

Then $\qquad r = \left(\dfrac{-(175 \times 10^{-6}\ \text{V})(0.160\ \text{m})}{(4\pi \times 10^{-7}\ \text{N/A}^2)(420)^2 \pi(-0.421\ \text{A/s})}\right)^{1/2} = 9.77\ \text{mm.}$ ■

Finalize: We did not write down the inductance of the coil, but it is about 416 μH. We were right about the order of magnitude of the radius. The value of the rate of change of current is negative, to represent the current as decreasing.

---

37. A 12.0-V battery is connected into a series circuit containing a 10.0-Ω resistor and a 2.00-H inductor. In what time interval will the current reach (a) 50.0% and (b) 90.0% of its final value?

Solution

Conceptualize: The time constant for this circuit is $\tau = L/R = 0.2$ s. This means that in 0.2 s, the current will reach $1 - 1/e = 63\%$ of its final value, as shown in the graph. We can see from this graph that the time interval to reach 50% of I_{max} should be slightly less than the time constant, perhaps about 0.15 s, and the time interval to reach $0.9I_{max}$ should be about $2.5\tau = 0.5$ s.

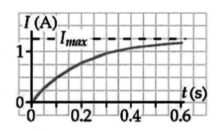

Categorize: The precise time intervals can be found from the equation that describes the rising current in the graph shown and gives the current as a function of time for a known emf, resistance, and time constant.

Analyze: At time t after connecting the circuit, $I(t) = \dfrac{\varepsilon\left(1 - e^{-t/\tau}\right)}{R}$, where, after a

long time, the current is $I_{max} = \dfrac{\varepsilon\left(1 - e^{-\infty}\right)}{R} = \dfrac{\varepsilon}{R}$.

(a) At 50% of this maximum value, $I(t) = .500 I_{max} = I_{max}\left(1 - e^{-t/\tau}\right)$.

This then yields $0.500 = 1 - e^{-t/\tau}$ or $e^{-t/\tau} = 0.500$ or $e^{+t/\tau} = 2.00$.

The natural logarithm is the inverse of the exponential function, so taking the natural logarthm of both sides, we obtain

$$t / \tau = \ln 2 \rightarrow t = \tau \ln 2 = \left(\frac{L}{R}\right) \ln 2 = \left(\frac{2.00 \text{ H}}{10.0 \ \Omega}\right) \ln 2 = 0.139 \text{ s} \qquad \blacksquare$$

(b) Similarly, to reach 90% of I_{max}, we have

$$0.900 = 1 - e^{-t/\tau} \quad \text{or} \quad e^{-t/\tau} = 0.100 \quad \text{or} \quad e^{+t/\tau} = 10.0$$

so $\quad t = \tau \ln 10 = (0.200 \text{ s}) \ln 10 = 0.461 \text{ s}.$ $\qquad \blacksquare$

Finalize: The calculated time intervals agree reasonably well with our predictions. We must be careful to avoid confusing the equation for the rising current with the similar equation for the falling current in a circuit without a battery. Checking our answers against predictions is a good way to prevent such mistakes.

═══════════

47. A 140-mH inductor and a 4.90-Ω resistor are connected with a switch to a 6.00-V battery as shown in Figure P23.47. (a) After the switch is first thrown to a (connecting the battery), what time interval elapses before the current reaches 220 mA? (b) What is the current in the inductor 10.0 s after the switch is closed? (c) Now the switch is quickly thrown from a to b. What time interval elapses before the current in the inductor falls to 160 mA?

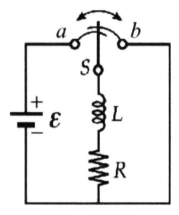

Figure P23.47

Solution

Conceptualize: The resistor limits the *amount* of current. The inductor limits the *rate of increase* of current after it is switched on, and the *rate of decrease* after it is switched off. Noting that the time constant is L/R = 0.14 H/4.9 Ω = 28.6 ms, we estimate some milliseconds as the answers to (a) and (c). The time interval is so long in part (b) that the current will be more than one amp, nearly at its final value of 6V/4.9 Ω = 1.22 A.

Categorize: We use the equations derived in the chapter for the increasing current in an *LR* circuit when the circuit is switched on and for the exponentially decaying current when it is switched off.

Analyze:

(a) The equation for current buildup is obtained by combining Equations 23.14 and 23.15:

$$I = \frac{\mathcal{E}\left(1 - e^{-Rt/L}\right)}{R}$$

We proceed step-by-step to solve for t in terms of the other quantities, all of which are given:

$$IR/\mathcal{E} = 1 - e^{-Rt/L} \quad \text{so} \quad e^{-Rt/L} = 1 - IR/\mathcal{E} \quad \text{and} \quad -Rt/L = \ln(1 - IR/\mathcal{E})$$

$$t = -(L/R)\ln(1 - IR/\mathcal{E})$$

$$t = -(0.140\text{ H}/4.90\ \Omega)\ln[1 - (0.220\text{ A})(4.90\ \Omega)/6.00\text{ V}]$$

$$= -(0.028\ 6\text{ s})\ln(0.820)$$

$$t = -(0.028\ 6\text{ s})(-0.198) = 5.66\text{ ms} \qquad\blacksquare$$

(b) We now make the general equation refer to a different instant. The current after ten seconds is

$$I = \left(\frac{6.00\text{ V}}{4.90\ \Omega}\right)\left(1 - e^{(-35.0\text{ s}^{-1})(10.0\text{ s})}\right) = (1.22\text{ A})(1 - e^{-350}) = 1.22\text{ A} \qquad\blacksquare$$

(c) The equation for current decrease after the battery is removed is $I = \dfrac{\mathcal{E}}{R}e^{-Rt/L}$.

We solve for t: $\dfrac{IR}{\mathcal{E}} = e^{-Rt/L}$ or $\dfrac{\mathcal{E}}{IR} = e^{+Rt/L}$

Then, $\ln(\mathcal{E}/IR) = Rt/L$ and $t = (L/R)\ln(\mathcal{E}/IR)$

Substituting, $t = (0.140\text{ H}/4.90\ \Omega)\ln[6.00\text{ V}/(0.160\text{ A}\cdot 4.90\ \Omega)]$

$$= (0.0286\text{ s})(\ln 7.65) = 58.1\text{ ms} \qquad\blacksquare$$

Finalize: Part (a) is about an instant early in the current buildup. Part (b) is very late in the current buildup, reminding us that we can think of the current as essentially constant after many time constants have passed. Part (c) is about an instant fairly late in the current decay process.

51. On a clear day at a certain location, a 100-V/m vertical electric field exists near the Earth's surface. At the same place, the Earth's magnetic field has a magnitude of 0.500×10^{-4} T. Compute the energy densities of (a) the electric field and (b) the magnetic field.

Solution

Conceptualize: These are typical values for fields we routinely encounter. The fields contain energy, but no energy input is required just to maintain the fields. From its small size in the standard SI unit, we might guess that the magnetic field has lower energy density.

Categorize: We substitute into standard equations for energy per volume in electric and magnetic fields.

Analyze:

(a) The energy density in the electric field is

$$u_E = \frac{\epsilon_0 E^2}{2} = \frac{\left(8.85 \times 10^{-12} \ \text{C}^2/\text{N} \cdot \text{m}^2\right)(100 \ \text{N/C})^2 (1 \ \text{J/N} \cdot \text{m})}{2}$$

$$= 44.3 \ \text{nJ/m}^3 \qquad \blacksquare$$

(b) The energy density in the magnetic field is

$$u_B = \frac{B^2}{2\mu_0} = \frac{\left(5.00 \times 10^{-5} \ \text{T}\right)^2}{2\left(4\pi \times 10^{-7} \ \text{T} \cdot \text{m/A}\right)} = 995 \times 10^{-6} \ \text{T} \cdot \text{A/m}$$

We display explicitly how the units work out:

$$u_B = (995 \times 10^{-6} \ \text{T} \cdot \text{A/m})(1 \ \text{N} \cdot \text{s/T} \cdot \text{C} \cdot \text{m})(1 \ \text{J/N} \cdot \text{m}) = 995 \ \mu\text{J/m}^3 \qquad \blacksquare$$

Finalize: Contrary to our guess, the magnetic energy density is 22 500 times greater than that in the electric field. Unknown in detail, electro-thermal-mechanical currents in dense matter deep within the Earth create its magnetic field. The electric field in this problem is incidental to weather, notably distant storms.

60. At $t = 0$, the open switch in Figure P23.60 is thrown closed. We wish to find a symbolic expression for the current in the inductor for time $t = 0$. Let this current be called I and choose it to be downward in the inductor in Figure P23.60. Identify I_1 as the current to the right through R_1 and I_2 as the current downward through R_2. (a)

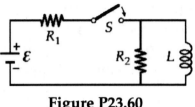

Figure P23.60

Use Kirchhoff's junction rule to find a relation among the three currents. (b) Use Kirchhoff's loop rule around the left loop to find another relationship. (c) Use Kirchhoff's loop rule around the outer loop to find a third relationship. (d) Eliminate I_1 and I_2 among the three equations to find an equation involving only the current I. (e) Compare the equation in part (d) with Equation 23.13 in the text. Use this comparison to rewrite Equation 23.14 in the text for the situation in this problem and show that

$$I(t) = \frac{\mathcal{E}}{R_1}\left[1 - e^{-(R'/L)t}\right]$$

where $R' = R_1R_2/(R_1 + R_2)$.

Solution

Conceptualize: The current in the inductor will start from zero at $t = 0$ and increase to asymptotically approach a final value. The junctions make this circuit more complicated than the LR circuit considered in the chapter text. To find the time-dependent current we must...

Categorize: ...use first principles, in the form of Kirchhoff's rules, and then compare with the solution of the simpler circuit in the chapter text. The equation we are to prove is reasonable in that it starts from $I = 0$ at $t = 0$. Long after the switch is closed, the current will be constant, the voltage across the inductor will be zero, and no current will go through R_2. Then the current will approach the limiting value $\mathcal{E}/R_1$, and the equation we are to prove gets this right.

Analyze:

(a) With I the downward current through the inductor and I_2 the downward current through R_2, Kirchhoff's junction rule says that $I_1 = I + I_2$ is the current in R_1. ∎

(b) Left-hand loop: $\mathcal{E} - (I + I_2)R_1 - I_2R_2 = 0$ ∎

(c) Outside loop: $\mathcal{E} - (I + I_2)R_1 - L\dfrac{dI}{dt} = 0$ ∎

(d) Eliminate I_2, obtaining $I\underbrace{\dfrac{R_1 R_2}{R_1 + R_2}}_{R'} + L\dfrac{dI}{dt} = \underbrace{\dfrac{R_2}{R_1 + R_2}\mathcal{E}}_{\mathcal{E}'}.$ ∎

(e) Thus, the equation has the form $\mathcal{E}' - IR' - L\dfrac{dI}{dt} = 0.$

This is of the same form as Equation 23.13, so the reasoning following that equation in the text shows that the solution is the same form as Equation 23.14,

$$I = \frac{\mathcal{E}'}{R'}\left[1 - e^{-R't/L}\right] \quad \text{with} \quad \frac{\mathcal{E}'}{R'} = \frac{\mathcal{E}R_2/(R_1 + R_2)}{R_1 R_2/(R_1 + R_2)} = \frac{\mathcal{E}}{R_1},$$

and we have $I(t) = \dfrac{\mathcal{E}}{R_1}\left[1 - e^{-(R't/L)}\right].$ ∎

Finalize: The current grows in a pattern like that in a circuit with an inductor and a single resistor. The time constant is L/R' with R' being equal to the equivalent resistance of R_1 and R_2 in parallel. We would not guess this from the original circuit, but the Kirchhoff equations demonstrate it.

━━━━━━━━━━━━━━━━

61. The magnetic flux through a metal ring varies with time t according to $\Phi_B = at^3 - bt^2$, where Φ_B is in webers, $a = 6.00 \text{ s}^{-3}$, $b = 18.0 \text{ s}^{-2}$, and t is in seconds. The resistance of the ring is 3.00 Ω. For the interval from $t = 0$ to $t = 2.00$ s, determine the maximum current induced in the ring.

Solution

Conceptualize: The given equation says that the flux is zero at time zero and goes through negative values immediately thereafter. But what really counts is its rate of change…

Categorize: …which we will find by doing the derivative in Faraday's law. To identify the extreme value for emf we will differentiate again and use the calculus idea that the derivative of a continuous function is zero when the function is a maximum. Knowing the time when the emf is a maximum, we will substitute to evaluate that maximum and at last use the definition of resistance to find the current.

Analyze: Substituting the given values, $\Phi_B = 6.00\,t^3 - 18.0\,t^2$.

Therefore, the emf induced is $\mathcal{E} = -\dfrac{d\Phi_B}{dt} = -18.0t^2 + 36.0t$.

The maximum $\mathcal{E}$ occurs when

$$\frac{d\mathcal{E}}{dt} = -36.0t + 36.0 = 0$$

which gives $t = 1.00$ s. The maximum current at $t = 1.00$ s is

$$I_{max} = \frac{\mathcal{E}}{R} = \frac{(-18.0 + 36.0)\ \text{V}}{3.00\ \Omega} = 6.00\ \text{A} \qquad\blacksquare$$

Finalize: In many problems we can think of the answer as a number. But Faraday's law wants a function, magnetic flux as it depends on time, as input data, and returns a function, emf as it depends on time, as the answer. The function $\mathcal{E}(t)$ is much richer than a single number, but numbers can be pulled from it.

---

65. A long, straight wire carries a current given by $I = I_{max} \sin(\omega t + \phi)$. The wire lies in the plane of a rectangular coil of N turns of wire, as shown in Figure P23.65. The quantities I_{max}, ω, and ϕ are all constants. Assume $I_{max} = 50.0$ A, $\omega = 200\pi\ \text{s}^{-1}$, $N = 100$, $h = w = 5.00$ cm, and $L = 20.0$ cm. Determine the emf induced in the coil by the magnetic field created by the current in the straight wire.

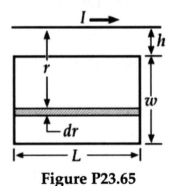

Figure P23.65
(Modified)

Solution

Conceptualize: If the current were constant in time, it would create a magnetic field at the coil but induce no voltage in the coil. The time-varying current acts across empty space to create changing flux through the rectangle, and so to induce an emf in it, perhaps with an amplitude of a few volts.

Categorize: The magnetic field varies in time and also in space between the top and the bottom of the rectangle. We will identify the flux through a ribbon of width dr running horizontally across the rectangle, and do an integral to find the total flux. Then a time derivative in Faraday's law will tell us the emf.

Analyze: The coil is the boundary of a rectangular area. The magnetic field produced by the current in the straight wire is perpendicular to the plane of the area at all points. The magnitude of the field is

$$B = \frac{\mu_0 I}{2\pi r}$$

Thus the flux through the rectangle is

$$\Phi_B = \frac{\mu_0 L}{2\pi} I \int_h^{h+w} \frac{dr}{r} = \frac{\mu_0 L}{2\pi} I_{max} \ln\left(\frac{h+w}{h}\right)\sin(\omega t + \phi)$$

Finally, the induced emf is

$$\mathcal{E} = -N\frac{d\Phi_B}{dt} = -\frac{\mu_0 NL}{2\pi} I_{max}\omega \ln\left(\frac{h+w}{h}\right)\cos(\omega t + \phi)$$

$$\mathcal{E} = -\left(4\pi \times 10^{-7}\,\frac{T \cdot m}{A}\right)\frac{(100)(0.200\ m)}{2\pi}(50.0\ A)$$

$$\times\left(200\pi\ s^{-1}\right)\ln\left(\frac{10.0\ cm}{5.00\ cm}\right)\cos(\omega t + \phi)$$

$$= -(87.1\ mV)\cos(200\pi t\ rad/s + \phi)$$

$$= -87.1\cos(200\pi t + \phi) \qquad\qquad \blacksquare$$

where $\mathcal{E}$ is in millivolts and t is in seconds.

Finalize: With these numbers the sinusoidally varying emf has an amplitude only on the order of a tenth of a volt. Practical transformers have many turns in the primary as well as the secondary.

Related Comment: The factor $\sin(\omega t + \phi)$ in the expression for the current in the straight wire does not change appreciably when ωt changes by 0.1 rad or less. Thus, the current does not change appreciably during a time interval

$$\Delta t < \frac{0.100}{200\pi\ s^{-1}} = 1.59 \times 10^{-4}\,s$$

We define a critical length,

$$c\Delta t = (3.00 \times 10^8\ m/s)(1.59 \times 10^{-4}\ s) = 4.77 \times 10^4\ m$$

equal to the distance to which field changes could be propagated during an interval of 1.59×10^{-4} s. This length is so much larger than any dimension of the loop or its distance from the wire that, although we consider the straight wire to be infinitely long, we can also safely ignore the field propagation effects in the vicinity of the loop. Moreover, the phase angle can be considered to be constant along the wire in the vicinity of the loop. If the angular frequency ω were much larger, say $200\pi \times 10^5$ s^{-1}, the corresponding critical length would be only 48 cm. In this situation, propagation effects would be important and the above expression of $\mathcal{E}$ would require modification. As a rough rule, we can consider field propagation effects for circuits of laboratory size to be negligible for frequencies, $f = \omega/2\pi$, that are less than about 10^6 Hz.

---

72. A bar of mass m and resistance R slides without friction in a horizontal plane, moving on parallel rails as shown in Figure P23.72. The rails are separated by a distance d. A battery that maintains a constant emf $\mathcal{E}$ is connected between the rails, and a constant magnetic field $\vec{B}$ is directed perpendicularly out of the page. Assuming the bar starts from rest at time $t = 0$, show that at time t it moves with a speed

$$v = \frac{\mathcal{E}}{Bd}\left(1 - e^{-B^2 d^2 t/mR}\right)$$

Solution

Conceptualize: The battery establishes a counterclockwise current in the circuit, a current that in the bar is downward toward the bottom of the page. Then the external magnetic field exerts a force on the bar in the direction ☞ (down) × (toward you) = left. The bar accelerates toward the left. Now the magnetic flux through the rectangle changes, so an extra generated or induced emf appears, to oppose the battery. Its presence will make the current somewhat smaller, to make the magnetic force smaller, to reduce the acceleration of the bar. But the leftward velocity of the bar keeps increasing gradually, until the induced emf approaches balancing the battery voltage. The current and the magnetic force on the bar approach zero. The bar approaches a terminal speed.

That's a lot of causally linked simultaneous physical processes, and…

Categorize: …we must use the laws describing all of them. Mathematics can follow the whole process of v increasing, first rapidly and then approaching a limit. We use Kirchhoff's loop rule, the definition describing how a magnetic field exerts a force, Newton's second law for the bar as a particle under a net force as it approaches equilibrium, Faraday's law, and the definition of acceleration.

Analyze: Mass m, distance d, battery voltage $\mathcal{E}$, and magnetic field B are all constants. Think of them as known. Time t is a variable, upon which depend the unknown functions I, $\mathcal{E}_{induced}$, magnetic force F, speed v, and acceleration dv/dt.

From Kirchhoff's loop rule, the current is $I = \dfrac{\mathcal{E} + \mathcal{E}_{\text{induced}}}{R}$,

where the induced emf is $\mathcal{E}_{\text{induced}} = -\dfrac{d}{dt}(BA) = -Bvd$.

(Because this equation contains a minus sign, we were careful to show a plus sign in the equation for current—don't count a minus sign twice.)

The only force on the bar is the magnetic force, causing its speed to change according to Newton's second law, $F = m\dfrac{dv}{dt} = IBd$.

We rearrange and substitute to eliminate unknowns:

$$\frac{dv}{dt} = \frac{IBd}{m} = \frac{Bd}{mR}\left(\mathcal{E} + \mathcal{E}_{\text{induced}}\right)$$

$$\frac{dv}{dt} = \frac{Bd}{mR}\left(\mathcal{E} - Bvd\right)$$

We have used all of the physical principles to express the unknown function v in terms of the known quantities. Now we proceed with mathematics instead of physics. The equation involves dv/dt as well as v. It is a differential equation. To solve it, we define a new unknown function $u = \mathcal{E} - Bvd$. Then $\dfrac{du}{dt} = -Bd\dfrac{dv}{dt}$ and the differential equation becomes $-\dfrac{1}{Bd}\dfrac{du}{dt} = \dfrac{Bd}{mR}u$.

Now we "separate variables," getting the unknown u on one side in

$$\int_{u_0}^{u} \frac{du}{u} = -\int_{0}^{t} \frac{(Bd)^2}{mR}\,dt$$

We have chosen to integrate from the release point $t = 0$ to any particular later time $t = t$. You know how to integrate to get

$$\ln\frac{u}{u_0} = -\frac{(Bd)^2}{mR}t \quad \text{or} \quad \frac{u}{u_0} = e^{-B^2 d^2 t/mR}$$

Now we put the answer in terms of v instead of u.

Since $v = 0$ when $t = 0$, we have $u_0 = \mathcal{E}$,

and our definition was $u = \varepsilon - Bvd$.

Substitution gives $\varepsilon - Bvd = \varepsilon e^{-B^2 d^2 t / mR}$.

So we can solve for the speed: $v = \dfrac{\varepsilon}{Bd}\left(1 - e^{-B^2 d^2 t / mR}\right)$ ■

Finalize: The equation we have proved says that at $t = 0$, the speed is $v = (\varepsilon / Bd)(1 - e^0) = (\varepsilon / Bd)(1 - 1) = 0$. This is a true statement, a "boundary condition" that we included in the solution. As t increases without limit, the speed keeps increasing, approaching asymptotically a definite limit, $v_{max} = (\varepsilon / Bd)(1 - e^{-\infty}) = (\varepsilon / Bd)(1 - 0) = \varepsilon / Bd$. Check that this equation is dimensionally correct:

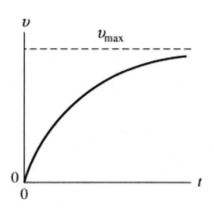

$$V/T \cdot m = (N \cdot m/C)/(N \cdot s/C \cdot m)\, m = m/s$$

In between, the speed shows a pattern of rapid increase followed by slower and slower increase, as sketched in the graph.

You have seen this shape twice before: For a falling object feeling a force of air resistance proportional to its speed in Section 5.4, and for the charge on a capacitor charging up in an *RC* circuit with a battery, in Section 21.9. You will see this shape again, for the current in a coil after it is connected to a battery to form an *RL* circuit, in Section 23.6.

You can profitably study this example repeatedly. First work to understand all of the physical processes described in the Conceptualize step. Then identify how each is expressed in an equation. After that, find a pattern in the mathematics of eliminating unknown functions by substitution and at last solving the differential equation. In a math class you may solve differential equations by doing indefinite integrals and dealing with "constants of integration." We prefer to show definite integrals with physically identifiable limits.

If Lenz's law were not true, if the equation to be solved were $\dfrac{dv}{dt} = \dfrac{Bd}{mR}(\varepsilon + Bvd)$ with a plus sign instead of a minus sign, then the speed would increase exponentially, like a population of bacteria with unlimited resources. This pattern of speed increase could not sustain itself physically and would violate conservation of energy for the circuit-field system.

75. The plane of a square loop of wire with edge length a = 0.200 m is oriented vertically and along an east-west axis. The Earth's magnetic field at this point is of magnitude B = 35.0 μT and directed northward at 35.0° below the horizontal. The total resistance of the loop and the wires connecting it to a sensitive ammeter is 0.500 Ω. If the loop is suddenly collapsed by horizontal forces as shown in Figure P23.75, what total charge enters one terminal of the ammeter?

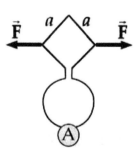

Figure P23.75

Solution

Conceptualize: For the situation described, the maximum current is probably less than 1 mA. So if the loop is closed in 0.1 s, then the total charge would be less than

$$Q = I\Delta t \sim (1 \text{ mA})(0.1 \text{ s}) = 100 \ \mu\text{C}$$

Categorize: We do not know how quickly the loop is collapsed, but we can find the total charge by integrating the change in magnetic flux due to the change in area of the loop (from a^2 to 0).

Analyze: The normal to the loop is horizontally north, at 35.0° to the magnetic field. We assume that 0.500 Ω is the total resistance around the circuit, including the ammeter.

$$Q = \int I\,dt = \int \frac{\mathcal{E}\,dt}{R} = \frac{1}{R}\int -\left(\frac{d\Phi_B}{dt}\right)dt = -\frac{1}{R}\int d\Phi_B = -\frac{1}{R}\int d(BA\cos\theta)$$

$$= -\frac{B\cos\theta}{R}\int_{A_1=a^2}^{A_2=0} dA$$

$$Q = -\left[\frac{B\cos\theta}{R}A\right]_{A_1=a^2}^{A_2=0} = \frac{B\cos\theta a^2}{R}$$

$$= \frac{(35.0\times10^{-6}\text{ T})(\cos 35.0°)(0.200\text{ m})^2}{0.500\ \Omega}$$

$$= 2.29\times10^{-6}\text{ C} \qquad\blacksquare$$

Finalize: The total charge is less than the maximum charge we predicted, so the answer seems reasonable. It is interesting that this charge can be calculated without knowing either the current or the time to collapse the loop.

Chapter 24
Electromagnetic Waves

NOTES FROM SELECTED CHAPTER SECTIONS

Section 24.2   Maxwell's Equations and Hertz's Discoveries

Electromagnetic waves are generated by accelerating electric charges. The radiated waves consist of oscillating electric and magnetic fields.

The fundamental laws describing the behavior of electric and magnetic fields are Maxwell's equations. In this unified theory of electromagnetism Maxwell showed that electromagnetic waves are a natural consequence of these fundamental laws.

Maxwell's theory of electromagnetic radiation is based on the following:

- A charge creates an electric field. Electric field lines originate on positive charges and terminate on negative charges. The relationship between charges and the fields they produce is described by **Gauss's law.**

- Magnetic field lines always form closed loops; they do not begin or end anywhere.

- A varying magnetic field induces an emf and hence an electric field. This is a statement of **Faraday's law.**

- A moving charge (constituting a current) creates a magnetic field as summarized in **Ampère's law.**

- A varying electric field creates a magnetic field. This is **Maxwell's addition to Ampère's law.**

Section 24.3 Electromagnetic Waves

Following is a summary of the properties of electromagnetic waves:

- The solutions of Maxwell's third and fourth equations are wavelike, where both $\vec{E}$ and $\vec{B}$ satisfy the same wave equation.

- Electromagnetic waves travel through empty space with the speed of light:

$$c = \frac{1}{\sqrt{\epsilon_0\, \mu_0}}$$

- The electric and magnetic field components of plane electromagnetic waves are perpendicular to each other and also perpendicular to the direction of wave propagation; electromagnetic waves are transverse waves.

- The magnitudes of $\vec{E}$ and $\vec{B}$ in empty space are related by $\frac{E}{B} = c$.

- Electromagnetic waves obey the principle of superposition.

Section 24.4 Energy Carried by Electromagnetic Waves

The magnitude of the **Poynting vector** represents the rate at which energy flows through a unit surface area perpendicular to the flow.

For an electromagnetic wave the instantaneous energy density associated with the magnetic field equals the instantaneous energy density associated with the electric field. Therefore, in a given volume the energy is equally shared by the two fields.

Section 24.5 Momentum and Radiation Pressure

Electromagnetic waves have momentum and exert pressure on surfaces on which they are incident. *The pressure exerted by a normally incident wave on a totally reflecting surface is double that exerted on a surface that completely absorbs the incident wave.*

Section 24.6 The Spectrum of Electromagnetic Waves

All **electromagnetic waves are produced by accelerating charges**. Types of electromagnetic waves can be characterized by "typical" ranges of wavelength.

- **Radio waves** $\left(\sim 10^4 \text{ m} > \lambda > \sim 0.1 \text{ m}\right)$ are the result of electric charges accelerating through a conducting wire (antenna).

- **Microwaves** $\left(\sim 0.3 \text{ m} > \lambda > \sim 10^{-4} \text{ m}\right)$ are generated by electronic devices.

- **Infrared waves** $\left(\sim 10^{-3} \text{ m} > \lambda > \sim 7 \times 10^{-7} \text{ m}\right)$ are produced by high temperature objects and molecules.

- **Visible light** $\left(\sim 7 \times 10^{-7} \text{ m} > \lambda > \sim 4 \times 10^{-7} \text{ m}\right)$ is produced by the rearrangement of electrons in atoms and molecules.

- **Ultraviolet (UV) light** $\left(\sim 4 \times 10^{-7} \text{ m} > \lambda > \sim 6 \times 10^{-10} \text{ m}\right)$ is an important component of radiation from the Sun.

- **X-rays** $\left(\sim 10^{-8} \text{ m} > \lambda > \sim 10^{-12} \text{ m}\right)$ are produced when high-energy electrons bombard a metal target.

- **Gamma rays** $\left(\sim 10^{-10} \text{ m} > \lambda > \sim 10^{-14} \text{ m}\right)$ are electromagnetic waves emitted by radioactive nuclei.

Remember, the wavelength ranges stated above are approximate; the adjacent regions of the electromagnetic spectrum overlap in wavelength. On the long wavelength end radio waves can be arbitrarily long; and on the short wavelength end gamma rays can be arbitrarily short.

Section 24.7   Polarization of Light Waves

A linearly polarized (or plane polarized) light beam is one for which the electric field vector of all waves vibrates in the same plane (the plane of polarization) at all times. If an unpolarized light beam is incident on an ideal polarizer the transmitted beam will be plane polarized with the electric field vectors vibrating parallel to the transmission axis of the polarizer.

EQUATIONS AND CONCEPTS

Maxwell's equations are the fundamental laws governing the behavior of electric and magnetic fields. Electromagnetic waves are a natural consequence of these laws.

Gauss's law: The total electric flux through any closed surface equals the net charge enclosed by the surface divided by ϵ_0.

$$\oint \vec{E} \cdot d\vec{A} = \frac{q}{\epsilon_0} \qquad (24.4)$$

Gauss's law for magnetism: The net magnet flux through a closed surface is zero.

$$\oint \vec{B} \cdot d\vec{A} = 0 \qquad (24.5)$$

Faraday's law of induction: The line integral of the electric field around any closed path equals the rate of change of magnetic flux through any surface area bounded by the path.

$$\oint \vec{E} \cdot d\vec{s} = -\frac{d\Phi_B}{dt} \qquad (24.6)$$

$$\Phi_B = \vec{B} \cdot \vec{A}$$

The Ampère-Maxwell law: The line integral of the magnetic field around any closed path is determined by the net current and the rate of change of electric flux through any surface bounded by the path.

$$\oint \vec{B} \cdot d\vec{s} = \mu_0 I + \epsilon_0 \mu_0 \frac{d\Phi_E}{dt} \qquad (24.7)$$

$$\Phi_E = \vec{E} \cdot \vec{A}$$

The **wave equations for electromagnetic waves in free space** are differential equations. Equations 24.16 and 24.17 represent linearly polarized waves traveling with a speed c.

$$\frac{\partial^2 E}{\partial x^2} = \mu_0 \epsilon_0 \frac{\partial^2 E}{\partial t^2} \qquad (24.16)$$

$$\frac{\partial^2 B}{\partial x^2} = \mu_0 \epsilon_0 \frac{\partial^2 B}{\partial t^2} \qquad (24.17)$$

The **speed of electromagnetic waves in vacuum** is the same as the speed of light in vacuum.

$$c = \frac{1}{\sqrt{\epsilon_0 \mu_0}} \qquad (24.18)$$

The **electric and magnetic fields** propagate as sinusoidal transverse waves varying in position and time. *Their planes of vibration are perpendicular to each other and perpendicular to the direction of propagation.*

$$E = E_{max} \cos(kx - \omega t) \qquad (24.19)$$

$$B = B_{max} \cos(kx - \omega t) \qquad (24.20)$$

$$k = 2\pi/\lambda \qquad \omega = 2\pi f$$

The **ratio of the magnitudes of the electric magnetic fields** is constant and equal to the speed of light.

$$\frac{E_{max}}{B_{max}} = \frac{E}{B} = c \qquad (24.22)$$

The **Poynting vector** $\vec{S}$ describes the energy flow associated with an electromagnetic wave. *The direction of $\vec{S}$ is along the direction of propagation and the magnitude of $\vec{S}$ is the rate at which electromagnetic energy crosses a unit surface area perpendicular to the direction of $\vec{S}$.*

$$\vec{S} \equiv \frac{1}{\mu_0} \vec{E} \times \vec{B} \qquad (24.24)$$

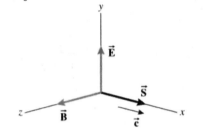

The Pointing vector has units of W/m^2.

The **wave intensity** is the time average of the magnitude of the Poynting vector. E_{max} and B_{max} are the maximum values of the field magnitudes. The SI units of I are W/m^2.

$$I = S_{av} = \frac{E_{max}B_{max}}{2\mu_0} = \frac{E_{max}^2}{2\mu_0 c} = \frac{cB_{max}^2}{2\mu_0}$$
$$(24.26)$$

The **instantaneous energy densities** of the electric and magnetic fields are equal.

$$u_E = \tfrac{1}{2}\epsilon_0 E^2 \qquad (24.27)$$

$$u_B = \frac{B^2}{2\mu_0} \qquad (24.28)$$

$$u_B = u_E$$

The **total instantaneous energy density** u is proportional to E^2 and to B^2.

$$u = \epsilon_0 E^2 = \frac{B^2}{\mu_0}$$

The **total average energy density** is proportional to E_{max}^2 and to B_{max}^2. The average energy density is also proportional to the wave intensity.

$$u_{avg} = \tfrac{1}{2}\epsilon_0 E_{max}^2 = \frac{B_{max}^2}{2\mu_o} \qquad (24.29)$$

$$I = S_{avg} = c u_{avg} \qquad (24.30)$$

The **linear momentum** (p) **and the radiation pressure** (P) **delivered to an absorbing surface** by an electromagnetic wave at normal incidence depends on the intensity of the wave and the fraction of the total energy absorbed. T_{ER} equals the total energy transmitted by a wave during a time interval Δt.

$$p = T_{ER}/c \quad \text{(complete absorption)} \tag{24.31}$$

$$p = 2T_{ER}/c \quad \text{(complete reflection)}$$

$$P = S/c \quad \text{(complete absorption)} \tag{24.32}$$

$$P = 2S/c \quad \text{(complete reflection)}$$

Malus's law states that the fraction of polarized light (from a polarizer) that will be transmitted by a second sheet of polarizing material (the analyzer) depends on the square of the cosine of the angle between the transmission axes of the two materials.

$$I = I_{max}\cos^2\theta \tag{24.36}$$

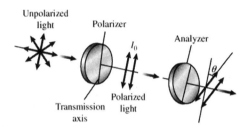

Two polarizing disks whose transmission axes make an angle θ with each other.

SUGGESTIONS, SKILLS, AND STRATEGIES

You should notice that:

- $\oint \vec{E} \cdot d\vec{A} = \dfrac{Q}{\epsilon_0}$ and $\oint \vec{B} \cdot d\vec{A} = 0$ are surface integrals.

  The normal components of electric and magnetic fields are integrated over a closed surface.

- $\oint \vec{E} \cdot d\vec{s} = -\dfrac{d\Phi_B}{dt}$ and $\oint \vec{B} \cdot d\vec{s} = \mu_0 I + \epsilon_0 \mu_0 \dfrac{d\Phi_E}{dt}$ are line integrals.

  The tangential components of electric and magnetic fields are integrated around a closed path.

REVIEW CHECKLIST

- Describe the essential features of the apparatus and procedure used by Hertz in his experiments leading to the discovery and understanding of the source and nature of electromagnetic waves.

- For a properly described plane electromagnetic wave, calculate the values for the Poynting vector (magnitude), wave intensity, and instantaneous and average energy densities.

- Calculate the radiation pressure on a surface and the linear momentum delivered to a surface by an electromagnetic wave.

- Understand the production of electromagnetic waves and radiation of energy by an oscillating dipole. Use a diagram to show the relative directions for $\vec{E}$, $\vec{B}$, and $\vec{S}$.

ANSWER TO AN OBJECTIVE QUESTION

5. Assume you charge a comb by running it through your hair and then hold the comb next to a bar magnet. Do the electric and magnetic fields produced constitute an electromagnetic wave? (a) Yes they do, necessarily. (b) Yes they do because charged particles are moving inside the bar magnet. (c) They can, but only if the electric field of the comb and the magnetic field of the magnet are perpendicular. (d) They can, but only if both the comb and the magnet are moving. (e) They can, if either the comb or the magnet or both are accelerating.

Answer   (e). Charge on the comb creates an electric field and the bar magnet sets up a magnetic field. If these fields are constant or are steadily changing, they will not start to recreate each other and move as a wave. Only nonzero acceleration of the source charge or current radiates a wave.

ANSWERS TO SELECTED CONCEPTUAL QUESTIONS

5. Radio stations often advertise "instant news." If that means you can hear the news the instant the radio announcer speaks it, is the claim true? What approximate time interval is required for a message to travel from Maine to California by radio waves? (Assume the waves can be detected at this range.)

Answer   The claim is untrue, but the transit time delay is usually too short to notice. Radio waves move at the speed of light. They can travel around the curved surface of the Earth, bouncing between the ground and the ionosphere, which has an altitude that is small when compared to the radius of the Earth. The distance across the lower forty-eight states is approximately 5 000 km, requiring a time interval of $(5 \times 10^6 \, \text{m})/(3 \times 10^8 \, \text{m/s}) \sim 10^{-2} \, \text{s}$. To go halfway around the Earth takes only 0.07 s. In other words, a speech can be heard on the other side of the world before it is heard at the back of a large room.

□   □   □   □

10. What does a radio wave do to the charges in the receiving antenna to provide a signal for your car radio?

Answer Consider a typical metal rod antenna for a car radio. The rod detects the electric field portion of the carrier wave. Variations in the amplitude of the carrier wave cause the electrons in the rod to vibrate with amplitudes emulating those of the carrier wave. Likewise, for frequency modulation, the variations of the frequency of the carrier wave cause constant-amplitude vibrations of the electrons in the rod but at frequencies that imitate those of the carrier.

□   □   □   □

12. Suppose a creature from another planet has eyes that are sensitive to infrared radiation. Describe what the alien would see if it looked around your library. In particular, what would appear bright and what would appear dim?

Answer There is a lot of uniform dark gray, for the walls, the wooden tables, the metal shelves, the books, and the illustrated magazines and art books on display. The incandescent bulbs in the reading lamps are very bright, much brighter than the fluorescent tubes in the ceiling. It is the evening of a hot day, and a window air conditioner is making a large puddle of black on the floor in front of it. With a similar shape, a small puddle of brightness has spilled onto the tabletop from the side vent in a student's notebook computer. The computer screen shows more of the uniform dark gray. People of all races have skin with the same light gray color, but there are subtle variations for the woman checking her email. Her face gets a little darker when she is obligated to read something boring, and her face lights up a bit as soon as she opens a message from a good friend. She pulls a black cafeteria apple from an insulated lunch bag. Clothing is dark as a rule, but the seat of another student's pants glows like a baboon's bottom when he gets up from his chair, and he leaves behind a patch of the same blush on the chair. When he comes back from the restroom, you can tell that he has washed his hands with hot water. His face appears lit from within, like a jack-o-lantern. His nostrils and the openings of his ear canals are bright; brighter still are just the pupils of his eyes.

□   □   □   □

SOLUTIONS TO SELECTED END-OF-CHAPTER PROBLEMS

3.   A 0.100-A current is charging a capacitor that has square plates 5.00 cm on each side. The plate separation is 4.00 mm. Find (a) the time rate of change of electric flux between the plates and (b) the displacement current between the plates.

Solution

Conceptualize: The charge on the capacitor plates is changing, so the electric field between the plates is changing. This situation is described by a displacement current, …

Categorize: … which we can compute from its definition.

Analyze: The electric field in the space between the plates is $E = \dfrac{\sigma}{\epsilon_0} = \dfrac{Q}{\epsilon_0 A}$.

The flux of this field is $\Phi_E = \vec{E} \cdot \vec{A} = \left(\dfrac{Q}{\epsilon_0 A} \right) A \cos 0° = \dfrac{Q}{\epsilon_0}$.

(a)   The rate of change of flux is $\dfrac{d\Phi_E}{dt} = \dfrac{d}{dt} \dfrac{Q}{\epsilon_0} = \dfrac{1}{\epsilon_0} \dfrac{dQ}{dt} = \dfrac{I}{\epsilon_0}$.

$$\dfrac{d\Phi_E}{dt} = \left(\dfrac{0.100 \text{ A}}{8.85 \times 10^{-12} \text{ C}^2/\text{N} \cdot \text{m}^2} \right)(1 \text{ C/A} \cdot \text{s}) = 1.13 \times 10^{10} \text{ N} \cdot \text{m}^2/\text{C} \cdot \text{s} \quad ■$$

(b)   The displacement current is defined as

$$I_d = \epsilon_0 \dfrac{d\Phi_E}{dt} = (8.85 \times 10^{-12} \text{ C}^2/\text{N} \cdot \text{m}^2)(1.13 \times 10^{10} \text{ N} \cdot \text{m}^2/\text{C} \cdot \text{s})$$

$$= 0.100 \text{ A} \quad ■$$

Finalize: The changing electric field would create magnetic field in the surrounding space in the same way that 100 mA of electric current would. No electric current exists in the space between the capacitor plates, but Maxwell called his measure of the changing flux of an electric field a displacement current because it has the units of current, creates magnetic field as a current does, and can even be included in a generalization of Kirchhoff's junction rule.

4. A 1.05-μH inductor is connected in series with a variable capacitor in the tuning section of a short wave radio set. What capacitance tunes the circuit to the signal from a transmitter broadcasting at 6.30 MHz?

Solution

Conceptualize: It is difficult to predict a value for the capacitance without doing the calculations, but we might expect a typical value in the μF or pF range.

Categorize: We want the resonance frequency of the circuit to match the broadcasting frequency, and for a simple LC circuit, the resonance frequency only depends on the magnitudes of the inductance and capacitance.

Analyze: The resonance frequency is $f_0 = \dfrac{1}{2\pi\sqrt{LC}}$.

Thus, $C = \dfrac{1}{\left(2\pi f_0\right)^2 L} = \dfrac{1}{\left[2\pi\left(6.30\times10^6 \text{ Hz}\right)\right]^2\left(1.05\times10^{-6}\text{ H}\right)} = 608 \text{ pF}.$ ∎

Finalize: This is indeed a typical capacitance, so our calculation appears reasonable. You probably would not hear any familiar music on this broadcast frequency. The frequency range for FM radio broadcasting is 88.0 MHz to 108.0 MHz, and AM radio is 535 kHz to 1 605 kHz. The 6.30-MHz frequency falls in the Maritime Mobile SSB Radiotelephone range, so you might hear a ship captain instead of Top 40 tunes! This and other information about the radio frequency spectrum can be found on the National Telecommunications and Information Administration (NTIA) website, which at the time of this writing is at http://www.ntia.doc.gov/files/ntia/publications/2003-allochrt.pdf.

---

5. A proton moves through a region containing a uniform electric field given by $\vec{\mathbf{E}} = 50.0\hat{\mathbf{j}}$ V/m and a uniform magnetic field $\vec{\mathbf{B}} = (0.200\hat{\mathbf{i}} + 0.300\hat{\mathbf{j}} + 0.400\hat{\mathbf{k}})$ T. Determine the acceleration of the proton when it has a velocity $\vec{\mathbf{v}} = 200\hat{\mathbf{i}}$ m / s.

Solution

Conceptualize: The electric field is weak, the magnetic field strong, and the velocity small. Both fields will contribute to the net force on the proton, which will be a tiny fraction of a newton but will cause some huge acceleration. Gravity will be a negligible influence by comparison.

Categorize: We use the Lorentz force equation. It amounts to adding the electric force and the magnetic force as vectors. Then we use Newton's second law, or the particle under net force model.

Analyze: The net force on the proton is the Lorentz force, as described by

$$\sum \vec{F} = m\vec{a} = q\vec{E} + q\vec{v} \times \vec{B} \quad \text{so that} \quad \vec{a} = \frac{e}{m}\left[\vec{E} + \vec{v} \times \vec{B}\right].$$

Taking the cross product of $\vec{v}$ and $\vec{B}$,

$$\vec{v} \times \vec{B} = \begin{vmatrix} \hat{i} & \hat{j} & \hat{k} \\ 200 & 0 & 0 \\ 0.200 & 0.300 & 0.400 \end{vmatrix} = -200(0.400)\hat{j} + 200(0.300)\hat{k}$$

$$\vec{a} = \left(\frac{1.60 \times 10^{-19}}{1.67 \times 10^{-27}}\right)\left[50.0\,\hat{j} - 80.0\,\hat{j} + 60.0\,\hat{k}\right] \text{ m/s}^2$$

$$= \left(-2.87 \times 10^9\hat{j} + 5.75 \times 10^9\hat{k}\right) \text{ m/s}^2 \qquad \blacksquare$$

Finalize: Don't forget that a proton has charge $+e$, with the same magnitude as the electron charge. It may seem obvious that a particle feels force $q\vec{E} + q\vec{v} \times \vec{B}$ when it is in both an electric and a magnetic field, but the elucidation of this part of the relationship between electricity and magnetism was a real step that came comparatively late.

---

13. Figure P24.13 shows a plane electromagnetic sinusoidal wave propagating in the x direction. Suppose the wavelength is 50.0 m and the electric field vibrates in the xy plane with an amplitude of 22.0 V/m. Calculate (a) the frequency of the wave and (b) the magnetic field $\vec{B}$ when the electric field has its maximum value in the negative y direction. (c) Write an expression for $\vec{B}$ with the correct unit vector, with numerical values for B_{max}, k, and ω, and with its magnitude in the form $B = B_{max}\cos(kx - \omega t)$.

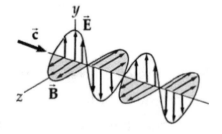

Figure P24.13

Solution

Conceptualize: With a 50-m wavelength, this is a radio wave with a frequency in megahertz that an electronic timer could count. The electric field amplitude is weak, so the magnetic field amplitude should be a tiny fraction of a tesla. The wave function will contain x and t as variables, so that it can represent the moving graph at all points in space and time.

Categorize: We are reviewing the description of "a wave" from Chapter 13. Amplitude (here two vector amplitudes related by $E = Bc$), wavelength, frequency, speed, wave number, and angular frequency are all constants of the motion.

Analyze:

(a) $c = f\lambda$ gives the frequency as

$$f = \frac{c}{\lambda} = \frac{3.00 \times 10^8 \text{ m/s}}{50.0 \text{ m}} = 6.00 \times 10^6 \text{ Hz} \qquad \blacksquare$$

(b) $c = E/B$ gives the magnetic field amplitude as

$$B = \frac{E}{c} = \frac{22.0 \text{ V/m}}{3.00 \times 10^8 \text{ m/s}} = 7.33 \times 10^{-8} \text{ T} = 73.3 \text{ nT} \qquad \blacksquare$$

$\vec{\mathbf{B}}$ must be directed along **negative z direction** when $\vec{\mathbf{E}}$ is in the negative y direction, so that $\vec{\mathbf{S}} = \vec{\mathbf{E}} \times \vec{\mathbf{B}}/\mu_0$ will propagate in the direction $\left(-\hat{\mathbf{j}}\right) \times \left(-\hat{\mathbf{k}}\right) = +\hat{\mathbf{i}}.$ $\blacksquare$

(c) In $B = B_{max} \cos(kx - \omega t)$, we require $k = \dfrac{2\pi}{\lambda} = \dfrac{2\pi}{50.0 \text{ m}} = 0.126 \text{ m}^{-1}$

and $\omega = 2\pi f = (2\pi \text{ rad})(6.00 \times 10^6 \text{ Hz}) = 3.77 \times 10^7 \text{ rad/s}$

Thus $\vec{\mathbf{B}} = 73.3\cos\left[0.126x - 3.77 \times 10^7 t\right]\left(-\hat{\mathbf{k}}\right),$ $\blacksquare$

where B is in nanoteslas, x is in meters, and t is in seconds

Finalize: In Figure P24.13, hold your right hand so that the curving fingers show the rotation from the electric field direction into the magnetic field direction in the diagram. Check that your extended thumb points in the direction of the wave's motion. Figure out where $c = f\lambda$ came from. It will be essential for the rest of the book.

16. In SI units, the electric field in an electromagnetic wave is described by

$$E_y = 100 \sin\left(1.00 \times 10^7 x - \omega t\right)$$

Find (a) the amplitude of the corresponding magnetic field oscillations, (b) the wavelength λ, and (c) the frequency f.

Solution

Conceptualize: The rules about electromagnetic wave propagation are so strict that two numbers describing a continuous wave, here the electric-field amplitude and the wave number, determine the required values for several other constants characterizing the wave.

Categorize: We identify the meaning of 100 N/C and 1.00×10^7 by comparison to a general sinusoidal wave function. Then $E = Bc$ and $c = f\lambda$ tell us the rest. We assume the wave is moving in vacuum.

Analyze:

(a) 100 V/m is the amplitude of the electric field, so the amplitude of the magnetic field is

$$B_{max} = \frac{E_{max}}{c} = \frac{100 \text{ V/m}}{3.00 \times 10^8 \text{ m/s}} = 3.33 \times 10^{-7} \text{ T} \qquad \blacksquare$$

(b) We compare the given wave function with $y = A \sin(kx - \omega t)$ to see that the wave number is $k = 1.00 \times 10^7 \text{ m}^{-1}$.

With $k = 2\pi/\lambda$, we then have the wavelength as

$$\lambda = \frac{2\pi}{k} = \frac{2\pi}{1.00 \times 10^7 \text{ m}^{-1}} = 6.28 \times 10^{-7} \text{ m} \qquad \blacksquare$$

(c) The frequency is $\quad f = \dfrac{c}{\lambda} = \dfrac{3.00 \times 10^8 \text{ m/s}}{6.28 \times 10^{-7} \text{ m}} = 4.77 \times 10^{14} \text{ Hz.} \qquad \blacksquare$

Finalize: The magnetic field is very small. The frequency is too high for an electronic counter to keep up with. But this is a familiar wave: orangey-red visible light. The wavelength tells us that.

23. What is the average magnitude of the Poynting vector 5.00 mi from a radio transmitter broadcasting isotropically (equally in all directions) with an average power of 250 kW?

Solution

Conceptualize: As the distance from the source is increased, the power per unit area will decrease, so at a distance of 5 miles from the source, the power per unit area will be a small fraction of the Poynting vector near the source.

Categorize: The Poynting vector is the power per unit area, where A is the surface area of a sphere with a 5-mile radius.

Analyze: The Poynting vector is $\quad S_{\text{avg}} = \dfrac{Power}{A} = \dfrac{Power}{4\pi r^2}.$

In meters, $r = (5.00 \text{ mi})(1\,609 \text{ m/mi}) = 8\,045 \text{ m,}$

and the intensity of the wave is $\quad S = \dfrac{250 \times 10^3 \text{ W}}{4\pi(8\,045 \text{ m})^2} = 3.07 \times 10^{-4} \text{ W/m}^2.$ ■

Finalize: The magnitude of the Poynting vector ten meters from the source is 199 W/m^2, on the order of a million times larger than it is 5 miles away! It is surprising how little power is actually received by a radio. At the 5-mile distance, the signal would only be about 30 nW, assuming a receiving area of about 1 cm^2.

---

29. A community plans to build a facility to convert solar radiation to electrical power. The community requires 1.00 MW of power, and the system to be installed has an efficiency of 30.0% (that is, 30.0% of the solar energy incident on the surface is converted to useful energy that can power the community). If sunlight has a constant intensity of 1 000 W/m^2, what must be the effective area of a perfectly absorbing surface used in such an installation?

Solution

Conceptualize: We are not worrying about nights, cloudy days, orienting the collector to face the Sun, or storing the energy. Then just some hundred-meter by hundred-meter area should be adequate, like the roof of a shopping mall.

Categorize: We use the definition of intensity S_{avg} as power transferred by the wave per perpendicular area.

Analyze: At 30.0% efficiency, the collected power is *Power* $= 0.300\, SA.$

Then the area is $A = \dfrac{Power}{0.300\ S} = \dfrac{1.00 \times 10^6\ \text{W}}{0.300\left(1\ 000\ \text{W/m}^2\right)} = 3\ 330\ \text{m}^2 \approx 0.8\ \text{acre}.$ ■

Finalize: Bad news: Storing the energy can be a serious concern. It does not make sense to use solar-voltaic electricity to run an electric laundry dryer—it is much more efficient to hang the clothes out on a line. Good news: A million watts can keep a lot of computers and telephones running. With efficient light-emitting diode battery lanterns, a lot of students can do their homework.

37. What are the wavelengths of electromagnetic waves in free space that have frequencies of (a) 5.00×10^{19} Hz and (b) 4.00×10^9 Hz?

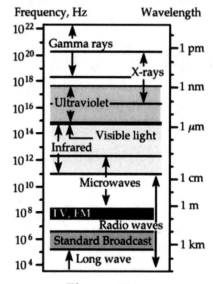

Figure 24.11

Solution

Conceptualize: Both frequencies are very high compared to frequencies for sound or mechanical waves we have studied. The spectrum chart shows that ~10^{20} Hz characterizes an x-ray or gamma ray with a picometer wavelength. And ~10^{10} Hz characterizes a very different microwave or radio wave with an ordinary-size wavelength in centimeters.

Categorize: We use just the $v = f\lambda$ relationship that applies to any continuous wave. For an electromagnetic wave in vacuum, the speed is c.

Analyze:

(a) $\lambda = \dfrac{c}{f} = \dfrac{3.00 \times 10^8\ \text{m/s}}{5.00 \times 10^{19}\ \text{s}^{-1}} = 6.00\ \text{pm}$ ■

(b) $\lambda = \dfrac{c}{f} = \dfrac{3.00 \times 10^8\ \text{m/s}}{4.00 \times 10^9\ \text{s}^{-1}} = 7.50\ \text{cm}$ ■

Finalize: The wave in (a) would be called an x-ray if it were emitted when an inner electron in an atom or an electron in a vacuum tube loses energy. It would be called a gamma ray if it were radiated by an atomic nucleus. By Figure 24.11, the finger-length wave in part (b) is called a radio wave or a microwave.

45. Plane-polarized light is incident on a single polarizing disk with the direction of $\vec{E}_0$ parallel to the direction of the transmission axis. Through what angle should the disk be rotated so that the intensity in the transmitted beam is reduced by a factor of (a) 3.00, (b) 5.00, and (c) 10.0?

Solution

Conceptualize: The disk should be called an analyzer. Maximum intensity is transmitted when its axis is at $0°$ to the incident light's polarization direction. Turning the disk more and more, approaching $90°$, will cut down the transmitted intensity by larger and larger factors.

Categorize: We use Malus's law.

Analyze: We define the initial angle, at which all the light is transmitted, to be $\theta = 0$. Turning the disk to another angle will then reduce the transmitted light by an intensity factor as described by $I = I_{max} \cos^2\theta$.

Then $\theta = \cos^{-1}(I/I_{max})^{1/2}$.

(a)   For $I = I_{max}/3.00$, $\theta = \cos^{-1}\left(\dfrac{I}{I_{max}}\right)^{1/2} = \cos^{-1}\dfrac{1}{\sqrt{3.00}} = \cos^{-1} 0.577 = 54.7°$.   ■

(b)   Now $\theta = \cos^{-1}\left(\dfrac{I}{I_{max}}\right)^{1/2} = \cos^{-1}\dfrac{1}{\sqrt{5.00}} = \cos^{-1} 0.447 = 63.4°$.   ■

(c)   The largest factor of intensity reduction requires the largest crossing angle,

$$\theta = \cos^{-1}\left(\dfrac{I}{I_{max}}\right)^{1/2} = \cos^{-1}\dfrac{1}{\sqrt{10.0}} = \cos^{-1} 0.316 = 71.6°$$ ■

Finalize: You can think of the cosine in Malus's law as taking the component of the incident electric field along the direction of the polarizing filter's transmission axis. It is squared because the energy of any oscillator is proportional to the square of its amplitude and the intensity of any traveling wave is proportional to the square of its amplitude.

---

57. A ruby laser delivers a 10.0-ns pulse of 1.00 MW average power. If the photons have a wavelength of 694.3 nm, how many are contained in the pulse?

Solution

Conceptualize: Lasers generally produce concentrated beams that are bright (except for IR or UV lasers that produce invisible beams). Since our eyes can detect light levels as low as a few photons, there are probably at least 1 000 photons in each pulse.

Categorize: From the pulse width and average power, we can find the energy delivered by each pulse. The number of photons can then be found by dividing the pulse energy by the energy of each photon, which is determined from the photon wavelength.

Analyze: The energy in each pulse is

$$E = P\Delta t = \left(1.00 \times 10^6 \text{ W}\right)\left(1.00 \times 10^{-8} \text{ s}\right) = 1.00 \times 10^{-2} \text{ J}$$

The energy of each photon is

$$E_\gamma = hf = \frac{hc}{\lambda} = \frac{\left(6.626 \times 10^{-34} \text{ J} \cdot \text{s}\right)\left(3.00 \times 10^8 \text{ m/s}\right)}{694.3 \times 10^{-9} \text{ m}} = 2.86 \times 10^{-19} \text{ J}$$

So the number of photons in the pulse is

$$N = \frac{E}{E_\gamma} = \frac{1.00 \times 10^{-2} \text{ J}}{2.86 \times 10^{-19} \text{ J/photon}} = 3.49 \times 10^{16} \text{ photons} \qquad \blacksquare$$

Finalize: With 10^{16} photons/pulse, this laser beam should produce a bright red spot when the light reflects from a surface, even though the time between pulses is generally much longer than the width of each pulse. For comparison, this laser produces more photons in a single ten-nanosecond pulse than a typical 5 mW helium-neon laser produces over a full second (about 1.6×10^{16} photons/second). Human eyes require many more than 1 000 photons for something to appear bright.

---

62. Review. In the absence of cable input or a satellite dish, a television set can use a dipole-receiving antenna for VHF channels and a loop antenna for UHF channels. In Figure CQ24.1, the "rabbit ears" form the VHF antenna and the smaller loop of wire is the UHF antenna. The UHF antenna produces an emf from the changing magnetic flux through the loop. The television station broadcasts a signal with a frequency f, and the signal has an electric-field amplitude E_{max} and a magnetic-field amplitude B_{max} at the location of the receiving antenna. (a) Using Faraday's law, derive an expression for the amplitude of the emf that appears in a single-turn, circular loop antenna with a radius r that is small compared with the wavelength of the wave. (b) If the electric field in the signal points vertically, what orientation of the loop gives the best reception?

Solution

Conceptualize: The amplitude should be proportional to the amplitude of the magnetic field, the area of the loop, and the frequency.

Figure CQ24.1

Categorize: The magnetic flux through the loop changes as the wave moves past, so Faraday's law is appropriate for finding the induced voltage.

Analyze: We can approximate the magnetic field as uniform over the area of the loop while it oscillates in time as $B = B_{max} \cos \omega t$. The induced voltage is

$$\mathcal{E} = -\frac{d\Phi_B}{dt} = -\frac{d}{dt}(BA \cos\theta) = -A\frac{d}{dt}\left(B_{max} \cos\omega t \cos\theta\right)$$

$$\mathcal{E} = AB_{max} \,\omega \,(\sin \,\omega t \cos\theta)$$

(a) Since the angular frequency is $\omega = 2\pi f$, the amplitude of this emf is $\mathcal{E}_{max} = 2\pi^2 r^2 f B_{max} \cos\theta$, where θ is the angle between the magnetic field and the normal to the loop. ∎

(b) If $\vec{E}$ is vertical, then $\vec{B}$ is horizontal, so the plane of the loop should be vertical, and the plane should contain the line of sight to the transmitter. This will make $\theta = 0°$, so $\cos\theta$ takes on its maximum value. ∎

Finalize: The derived equation shows the predicted proportionalities to r^2, to B_{max}, and to f. We see also proportionality to $\cos\theta$, and that the proportionality constant is $2\pi^2$. The moral is that the laws set up to describe static or local or individual electric and magnetic fields also give a complete theory of electromagnetic waves. Maxwell was right when he said his theory was "great guns!"

63. A dish antenna having a diameter of 20.0 m receives (at normal incidence) a radio signal from a distant source as shown in Figure P24.63. The radio signal is a continuous sinusoidal wave with amplitude E_{max} = 0.200 μV/m. Assume the antenna absorbs all the radiation that falls on the dish. (a) What is the amplitude of the magnetic field in this wave? (b) What is the intensity of the radiation received by this antenna? (c) What is the power received by the antenna? (d) What force is exerted by the radio waves on the antenna?

Figure P24.63

Solution

Conceptualize: Radio waves can be detected with remarkably low amplitudes. The force especially will be very small.

Categorize: Increasing any one of the electric-field amplitude, magnetic-field amplitude, intensity, or radiation pressure would make the others all increase in a predictable way. These quantities characterize the wave in itself. The energy received and the force are proportional to the area of the antenna as an absorbing surface intercepting the wave.

Analyze:

(a) The magnetic-field amplitude is

$$B_{max} = E_{max}/c = 6.67 \times 10^{-16} \text{ T} \qquad \blacksquare$$

(b) The intensity is the Poynting vector averaged over one or more cycles, given by

$$S_{avg} = E^2_{max}/2\mu_0 c = 5.31 \times 10^{-17} \text{ W/m}^2 \qquad \blacksquare$$

(c) The power tells how fast the antenna receives energy. It is

$$P_{avg} = S_{avg}A = S_{avg}\pi r^2 = 1.67 \times 10^{-14} \text{ W} \qquad \blacksquare$$

(d) The force tells how fast the antenna receives momentum. It is

$$F = PA = (S_{avg}/c)A = 5.56 \times 10^{-23} \text{ N} \qquad \blacksquare$$

Finalize: Do not confuse the power $P = S_{avg}A$ with the pressure $P = S_{avg}/c$, which we used to find the force. The force is immeasurably small, like the weight of a few thousand hydrogen atoms, but the radio wave can be detected and so carries measurable power.

---

68. In 1965, Arno Penzias and Robert Wilson discovered the cosmic microwave radiation left over from the big bang expansion of the Universe. Suppose the energy density of this background radiation is 4.00×10^{-14} J/m^3. Determine the corresponding electric field amplitude.

Solution

Conceptualize: The primordial background radiation is moving past all points in all directions. Penzias and Wilson observed it as arriving in New Jersey from a sphere around us, about 14 billion light-years in radius. The problem does not quote an intensity, but we can still solve it…

Categorize: …by using the relationship between energy density and electric field magnitude that we studied previously for a static field.

Analyze: The energy density can be written as $u = \dfrac{1}{2} \epsilon_0 E_{max}^2$,

so

$$E_{max} = \sqrt{\frac{2u}{\epsilon_0}} = \sqrt{\frac{2\left(4.00 \times 10^{-14} \text{ N} \cdot \text{m}^2\right)}{8.85 \times 10^{-12} \text{ C}^2/\text{N} \cdot \text{m}^2}} = 95.1 \text{ mV/m}.$$ ∎

Finalize: This is a weak field, but an antenna built for microwave relaying of telephone calls detected the radiation as a hum, isotropic and constant in time.

---

73. A linearly polarized microwave of wavelength 1.50 cm is directed along the positive x axis. The electric field vector has a maximum value of 175 V/m and vibrates in the xy plane. Assuming the magnetic field component of the wave can be written in the form $B = B_{max} \sin(kx - \omega t)$, give values for (a) B_{max}, (b) k, and (c) ω. (d) Determine in which plane the magnetic field vector vibrates. (e) Calculate the average value of the Poynting vector for this wave. (f) If this wave were directed at normal incidence onto a perfectly reflecting sheet, what radiation pressure would it exert? (g) What acceleration would be imparted to a 500-g sheet (perfectly reflecting and at normal incidence) with dimensions of 1.00 m × 0.750 m?

Solution

Conceptualize: This is a microwave that could be used for communication. It will have an intensity of some watts per square meter. But the other answers will be a small magnetic field, a small fraction of a pascal for the pressure, and a small acceleration for the sail.

Categorize: The given wavelength will tell us the values of wave number and angular frequency. The given electric-field amplitude will tell us the magnetic-field amplitude, intensity, and radiation pressure. The particle under net force model will tell us the acceleration of the sail.

Analyze:

(a) The magnetic field has amplitude

$$B_{max} = \frac{E_{max}}{c} = \frac{175 \text{ V/m}}{3.00 \times 10^8 \text{ m/s}} = 5.83 \times 10^{-7} \text{ T} \qquad \blacksquare$$

(b) The wave number is $k = \dfrac{2\pi}{\lambda} = \dfrac{2\pi}{0.015 \text{ m}} = 419 \text{ m}^{-1}.$ ■

(c) The angular frequency is

$$\omega = kc = (419 \text{ m}^{-1})(3.00 \times 10^8 \text{ m/s}) = 1.26 \times 10^{11} \text{ rad/s} \qquad \blacksquare$$

(d) The magnetic field must be in the z direction so that $\vec{\mathbf{S}} = \vec{\mathbf{E}} \times \vec{\mathbf{B}}/\mu_0$ can be in the $\hat{\mathbf{j}} \times \hat{\mathbf{k}} = \hat{\mathbf{i}}$ direction. The magnetic field vibrates in the xz plane. ■

(e) The magnitude of the average Poynting vector is the wave intensity

$$S_{avg} = \frac{E_{max}B_{max}}{2\mu_0} = \frac{(175 \text{ V/m})(5.83 \times 10^{-7} \text{ T})}{2(4\pi \times 10^{-7} \text{ N/A}^2)} = 40.6 \text{ W/m}^2$$

The Poynting vector itself points in the direction of energy transport:

$$\vec{\mathbf{S}}_{avg} = (40.6 \text{ W/m}^2)\,\hat{\mathbf{i}} \qquad \blacksquare$$

(f) For perfect reflection, the pressure is

$$P_r = \frac{2S}{c} = \frac{2(40.6 \text{ W/m}^2)}{3.00 \times 10^8 \text{ m/s}} = 2.71 \times 10^{-7} \text{ N/m}^2 \qquad \blacksquare$$

(g) We use Newton's second law. With magnitude

$$a = \frac{F}{m} = \frac{P_r A}{m} = \frac{\left(2.71 \times 10^{-7} \text{ N/m}^2\right)\left(0.750 \text{ m}^2\right)}{0.500 \text{ kg}} = 4.06 \times 10^{-7} \text{ m/s}^2$$

the acceleration is $\vec{a} = (406 \text{ nm/s}^2) \, \hat{i}.$ ■

Finalize: We have not forgotten that acceleration is a vector, like the magnetic field and the Poynting vector. Pressure and angular frequency are scalars. Wave number could be defined as a vector in the direction of wave velocity, but we treat it as a scalar.

━━━━━━━━━━━━━━━

75. An astronaut, stranded in space 10.0 m from her spacecraft and at rest relative to it, has a mass (including equipment) of 110 kg. Because she has a 100-W flashlight that forms a directed beam, she considers using the beam as a photon rocket to propel herself continuously toward the spacecraft. (a) Calculate the time interval required for her to reach the spacecraft by this method. (b) **What If?** Suppose she throws the 3.00-kg flashlight in the direction away from the spacecraft instead. After being thrown, the flashlight moves at 12.0 m/s relative to the recoiling astronaut. After what time interval will the astronaut reach the spacecraft?

Solution

Conceptualize: Based on our everyday experience, the force exerted by photons is too small to feel, so it may take a very long time (maybe days!) for the astronaut to travel 10 m with her "photon rocket." Using the momentum of the thrown lamp seems like a better solution, but it will still take a while (maybe a few minutes) for the astronaut to reach the spacecraft because her mass is so much larger than the mass of the flashlight.

Categorize: In part (a), the radiation pressure can be used to find the force that accelerates the astronaut toward the spacecraft. Then we will use the particle under a net force and particle under constant acceleration models to find the travel time interval. In part (b), the principle of conservation of momentum can be applied to the flashlight-astronaut system to find the time required to travel the 10 m.

Analyze:

(a) Light exerts on the astronaut a pressure $P = F/A = S/c$, and a force of

$$F = \frac{SA}{c} = \frac{Power}{c} = \frac{100 \text{ J/s}}{3.00 \times 10^8 \text{ m/s}} = 3.33 \times 10^{-7} \text{ N}$$

By Newton's second law,

$$a = \frac{F}{m} = \frac{3.33 \times 10^{-7} \text{ N}}{110 \text{ kg}} = 3.03 \times 10^{-9} \text{ m/s}^2$$

This acceleration is constant, so the distance traveled is $x = \frac{1}{2}at^2$, and the amount of time she travels is

$$t = \sqrt{\frac{2x}{a}} = \sqrt{\frac{2(10.0 \text{ m})}{3.03 \times 10^{-9} \text{ m/s}^2}} = 8.12 \times 10^4 \text{ s} = 22.6 \text{ h} \qquad \blacksquare$$

(b) Because there are no external forces, the momentum of the astronaut-flashlight system before throwing the lamp is the same as afterwards, when the now 107-kg astronaut is moving at speed v towards the spacecraft and the flashlight is moving away from the spacecraft at (12.0 m/s – v).

Thus, $\sum \vec{\mathbf{p}}_i = \sum \vec{\mathbf{p}}_f$ gives $0 = (107 \text{ kg}) v - (3.00 \text{ kg})(12.0 \text{ m/s} - v)$:

$$0 = (107 \text{ kg}) v - (36.0 \text{ kg} \cdot \text{m/s}) + (3.00 \text{ kg})v$$

$$v = \frac{36.0 \text{ kg} \cdot \text{m/s}}{110 \text{ kg}} = 0.327 \text{ m/s}$$

Then the time interval for the astronaut's trip is

$$t = \frac{x}{v} = \frac{10.0 \text{ m}}{0.327 \text{ m/s}} = 30.6 \text{ s} \qquad \blacksquare$$

Finalize: Throwing the light source away is certainly a quicker way to reach the spacecraft, but if you are a little off with the direction, you will miss the ship. You will not be able to retrieve the flashlight and try again unless it has a very long cord. How long would the cord need to be, and does its length depend on how hard the astronaut throws the lamp? (You can verify that the minimum cord length is 367 m, independent of the speed that the lamp is thrown.)

Chapter 25
Reflection and
Refraction of Light

Section 25.2   The Ray Model in Geometric Optics

In the simplification model called the ray model or ray approximation, it is assumed that a wave travels through a medium along a straight line in the direction of its rays. A ray for a given wave is a straight line perpendicular to the wave front; geometric models are based on these straight lines. This approximation also neglects diffraction effects.

Section 25.3   Analysis Model: Wave Under Reflection

Specular reflection occurs when the reflecting surface is smooth; reflection from a rough surface is called **diffuse reflection**. In the case of reflection from a smooth surface, the angle of incidence equals the angle of reflection; the angles are measured between the normal to the surface and the respective rays. *The path of a light ray is reversible, an important property in geometric optics.*

Section 25.4   Analysis Model: Wave Under Refraction

Refraction refers to the change in direction of a light ray upon striking obliquely the interface between two transparent media. The angle of refraction (between the refracted ray and the normal) depends on the properties of the two media and the angle of incidence as expressed by Snell's law (Equation 25.7, $n_1 \sin \theta_1 = n_2 \sin \theta_2$). The incident ray, reflected ray, and refracted (transmitted) ray lie in the same plane. *As a ray travels from one medium to another, the speed of the wave changes but the frequency does not change.*

Section 25.5   Dispersion and Prisms

For a given material the index of refraction is a function of the wavelength of the light passing through the material. This effect is called **dispersion**. In particular, when light passes through a prism, a given ray is refracted at two surfaces and emerges bent away from its original direction by an angle of deviation (δ). Due to the dependence of the index of refraction on wavelength, δ is different for different wavelengths.

Section 25.6   Huygens's Principle

Every point on a given wave front can be considered as a point source for a secondary wavelet. At some later time, the new position of the wave front is determined by the surface tangent to the set of secondary wavelets.

Section 25.7   Total Internal Reflection

Total internal reflection is possible only when light rays traveling in one medium are incident on the boundary between the first medium and a second medium of lesser index of refraction. The angle θ_c shown in the figure is called the **critical angle**.

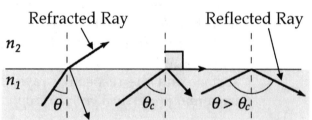

Total internal reflection of light occurs for angles of incidence equal to or greater than the critical angle (θ_c) and when $n_1 > n_2$.

EQUATIONS AND CONCEPTS

Consider the situation shown in the figure. A light ray is incident obliquely on a smooth, planar surface that forms the boundary between two transparent media with different indices of refraction. A portion of the ray will be reflected back into the original medium, while the remaining fraction will be transmitted into the second medium (the refracted ray).

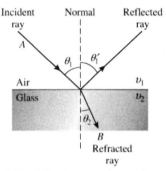

Light incident on the boundary is partially reflected and partially refracted. The three rays and the normal are in the same plane.

The **law of reflection** states that the angle of incidence equals the angle of reflection. This condition is valid when the reflecting surface is smooth (specular reflection). The angle of incidence and the angle of reflection are measured between the normal to the surface and the respective rays.

$$\theta_1' = \theta_1 \qquad (25.1)$$

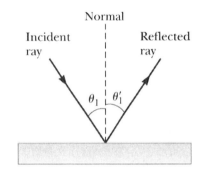

The incident ray, reflected ray and normal lie in the same plane.

Snell's law of refraction can be expressed in terms of the speed of light in the media on either side of the refracting surface (Equation 25.2) or in terms of the indices of refraction of the two media (Equation 25.7).

$$\frac{\sin\theta_2}{\sin\theta_1} = \frac{v_2}{v_1} \qquad (25.2)$$

The **most practical form of Snell's law** is given in terms of a parameter called the index of refraction (*n*), the value of which is characteristic of a particular medium.

$$n_1 \sin \theta_1 = n_2 \sin \theta_2 \qquad (25.7)$$

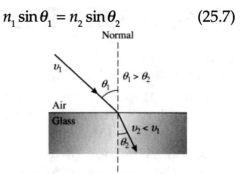

A light ray is refracted as it crosses the boundary between two media with different indices of refraction.

The **index of refraction** of a transparent medium is defined as the ratio of the speed of light in vacuum (*λ*) to the speed of light in the medium (*λ_n*). It can also be stated as a ratio of wavelengths.

$$n \equiv \frac{\text{Speed of light in vacuum}}{\text{Speed of light in a medium}} = \frac{c}{v}$$

$$n = \frac{\lambda}{\lambda_n} \qquad (25.6)$$

The **frequency of a wave is characteristic of the source.** *Therefore, as light travels from one medium into another of different index of refraction, the frequency remains constant but the wavelength changes.*

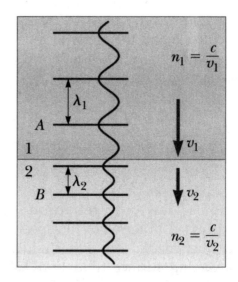

The light beam travels with lower speed in medium 2 where $n_2 > n_1$.

The **critical angle for total internal reflection** is the minimum angle of incidence for which total internal reflection can occur. *Total internal reflection is possible only when a light ray is directed from a medium of higher index of refraction into a medium of lower index of refraction.*

$$\sin \theta_c = \frac{n_2}{n_1} \quad \left(\text{for } n_1 > n_2 \right) \quad (25.8)$$

θ_c = critical angle

A light ray is internally reflected when the angle of incidence is $\geq \theta_c$ and $n_1 > n_2$.

REVIEW CHECKLIST

- Understand Huygens's principle and the use of this technique to construct the subsequent position and shape of a given wave front.

- Determine the directions of the reflected and refracted rays when a light ray is incident obliquely on the interface between two optical media.

- Make calculations using Snell's law and the law of reflection.

- Understand the basis of dispersion by a prism and the formation of rainbows.

- Understand the conditions under which total internal reflection can occur in a medium and determine the critical angle for a given pair of adjacent media.

ANSWER TO AN OBJECTIVE QUESTION

8. Suppose you find experimentally that two colors of light, A and B, originally traveling in the same direction in air, are sent through a glass prism, and A changes direction more than B. Which travels more slowly in the prism, A or B? Alternatively, is there insufficient information to determine which moves more slowly?

Answer As the light slows down upon entering the prism, a light ray that is bent more suffers a greater loss of speed as it enters the new medium. Therefore, light ray A travels more slowly.

ANSWER TO A CONCEPTUAL QUESTION

6. Explain why a diamond sparkles more than a glass crystal of the same shape and size.

Answer Diamond has a larger index of refraction than glass, and consequently has a smaller critical angle for internal reflection. A brilliant-cut diamond is shaped to admit light from above, reflect it totally at the converging facets on the underside of the jewel, and let the light escape only at the top. Glass will have less light internally reflected.

SOLUTIONS TO SELECTED END-OF-CHAPTER PROBLEMS

1. A prism that has an apex angle of 50.0° is made of cubic zirconia. What is its minimum angle of deviation?

Solution

Conceptualize: Shining a beam of laser light through a D-shaped piece of glass or plastic is a good way to study Snell's law and the law of reflection too. After doing that, obtain any prism and shine the narrow beam through it. Turn the prism slowly clockwise about an axis perpendicular to the direction of the light. You will see the exiting beam swinging counterclockwise, slowing down, stopping, and reversing. At the point of stopping, the net change in direction of the light is a minimum. In experiments it is not hard to set up minimum deviation and it is easy to relate the measured deviation angle to the index of refraction of the material.

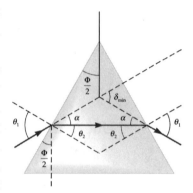

Categorize: We use several geometric results shown in the figure, and the refractive index of the material from the table in the text.

Analyze: Consider the geometry in the figure, where we have used symmetry to label several angles. The reproduction of the angle $\Phi/2$ at the location of the incoming light ray shows that $\theta_2 = \Phi/2$. The theorem that an exterior angle of any triangle equals the sum of the two opposite interior angles shows that $\delta_{min} = 2\alpha$. The geometry also shows that $\theta_1 = \theta_2 + \alpha$.

Combine these three geometric results:

$$\theta_1 = \theta_2 + \alpha = \frac{\Phi}{2} + \frac{\delta_{min}}{2} = \frac{\Phi + \delta_{min}}{2}$$

Apply the wave under refraction model at the left surface and solve for n:

$$(1.00)\sin\theta_1 = n\sin\theta_2 \quad \rightarrow \quad n = \frac{\sin\theta_1}{\sin\theta_2}$$

Substitute for the incident and refracted angles:

$$n = \frac{\sin\left[(\Phi + \delta_{min})/2\right]}{\sin(\Phi/2)}$$

Solving for δ_{min},

$$\delta_{min} = 2\sin^{-1}\left(n\sin\frac{\Phi}{2}\right) - \Phi = 2\sin^{-1}\left[(2.20)\sin(25.0°)\right] - 50.0°$$

$$= 86.8° \qquad\qquad ■$$

Finalize: The prism need not be a triangular block. Φ refers to the angle between the surface of entry and the surface of exit, whatever the shape of the refracting object. So here is a prettier and perhaps more convenient demonstration of minimum deviation: Obtain a single crystal from a chandelier. Hang it by a string from the top of a sunny window and set it spinning. On the wall or floor opposite, spots of light will sweep toward the shadow of the crystal, stop at a certain radial distance, and sweep away again. The index of refraction depends on the wavelength, so each of these spots of light will show a full spectrum.

---

8. An underwater scuba diver sees the Sun at an apparent angle of 45.0° above the horizontal. What is the actual elevation angle of the Sun above the horizontal?

Solution

Conceptualize: The sunlight refracts as it enters the water from the air. Because the water has a higher index of refraction, the light slows down and bends toward the vertical line that is normal to the interface. Therefore, the elevation angle of the Sun above the water will be less than 45°, as shown in the diagram, even though it appears to the diver that the Sun is 45° above the horizon.

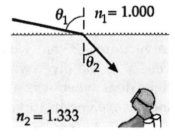

Categorize: We can use the wave under refraction model to find the precise angle of incidence.

Analyze: In Snell's law $n_1 \sin\theta_1 = n_2 \sin\theta_2$, we know n_1, n_2, and θ_2, so we find

$$\theta_1 = \sin^{-1}\left(\frac{n_2 \sin\theta_2}{n_1}\right) = \sin^{-1}\left(\frac{1.333 \sin 45.0°}{1.000}\right) = \sin^{-1} 0.943 = 70.5°$$

The sunlight impinges at $\theta_1 = 70.5°$ to the vertical, so the Sun is 19.5° above the horizon. ∎

Finalize: The calculated result agrees with our prediction. When applying Snell's law, it is easy to mix up the index values and to confuse angles-with-the-normal and angles-with-the-surface. Making a sketch and a prediction as we did here helps avoid careless mistakes.

Bonus problem: The following extension will remind you that we are studying wave motion. Consider yellow light from the Sun with wavelength 580 nm in air. (The diver's face mask might block out other colors.) Find its frequency and speed in air and in water, and its wavelength in water.

Solution: We answer the questions in a convenient order. The speed of light in air is $v = c/n = (3.00 \times 10^8 \text{ m/s})/1.000 = 300 \text{ Mm/s}$.

The frequency of this yellow light is $f = v/\lambda$. We take the speed in air and the wavelength in air to find $f = v/\lambda = (3.00 \times 10^8 \text{ m/s})/5.80 \times 10^{-7} \text{ m} = 517 \text{ THz}$. The frequency in water has the same value, 517 THz, because every wave crest coming down to a point on the water surface immediately propagates as one wave crest into the water. The same number of wavefronts in air and in water pass a point on the interface in any particular time interval, notably in one second.

In water the speed is $v = c/n = (3.00 \times 10^8 \text{ m/s})/1.333 = 225 \text{ Mm/s}$.

The wavelength in water is $\lambda = v/f = (2.25 \times 10^8 \text{ m/s})/(5.17 \times 10^{14}/\text{s}) = 435 \text{ nm}$. This wavelength could also be computed as the vacuum wavelength divided by water's index of refraction: 580 nm/1.333 = 435 nm.

In vacuum, 435 nm would be the wavelength of blue light. The light still looks yellow to the diver when she is under water. The frequency controls what the light does when it is absorbed. It is conventional to classify different colors by their wavelengths in vacuum because wavelength can be measured (easily) and the $\sim 10^{15}$ Hz frequency cannot be measured directly. But it is the frequency that is set by the source and intrinsically carried by the light ray.

16. A ray of light strikes a flat block of glass ($n = 1.50$) of thickness 2.00 cm at an angle of 30.0° with the normal. Trace the light beam through the glass and find the angles of incidence and refraction at each surface.

Solution

Conceptualize: The ray changes direction sharply at the point of incidence on the block. Then it travels straight through the block to the opposite face and switches direction there again. Bending toward the normal at entry, it will bend away from the normal at exit.

Categorize: We use Snell's law about entry and separately about exit.

Analyze: At entry, the wave under refraction model, expressed as $n_1 \sin\theta_1 = n_2 \sin\theta_2$, gives

$$\theta_2 = \sin^{-1}\left(\frac{n_1 \sin\theta_1}{n_2}\right) = \sin^{-1}\left(\frac{1.000 \sin 30.0°}{1.50}\right) = \sin^{-1}\frac{0.500}{1.50} = 19.5° \quad \blacksquare$$

To do geometric optics, you must remember some geometry. The surfaces of entry and exit are parallel so their normals are parallel. Then angle θ_2 of refraction at entry and the angle θ_3 of incidence at exit are alternate interior angles formed by the ray as a transversal cutting parallel lines.

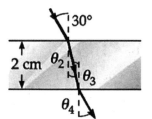

Therefore $\theta_3 = \theta_2 = 19.5°$. $\quad \blacksquare$

At the exit point, $n_2 \sin\theta_3 = n_1 \sin\theta_4$ gives

$$\theta_4 = \sin^{-1}\left(\frac{n_2 \sin\theta_3}{n_1}\right) = \sin^{-1}\left(\frac{1.50 \sin 19.5°}{1.000}\right) = \sin^{-1}0.333 = 30.0° \quad \blacksquare$$

Because θ_1 and θ_4 are equal, the departing ray in air is parallel to the original ray.

Finalize: This is a useful theorem, with practical applications. A car windshield of uniform thickness will not distort, but shows the driver the actual direction to every object outside. In the next chapter we will use the rule that a ray going through the center of a lens is unchanged in direction, because it encounters parallel surfaces of entry and exit.

23. The index of refraction for violet light in silica flint glass is 1.66 and that for red light is 1.62. What is the angular spread of visible light passing through a prism of apex angle 60.0° if the angle of incidence is 50.0°? See Figure P25.23.

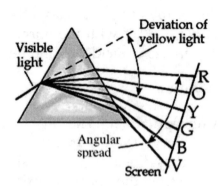

Figure P35.33

Solution

Conceptualize: The prism will produce a "rainbow," separating the colors of a narrow beam of incoming white light. Even if the angular dispersion is small, by placing a white card far away from the prism we can get a clear spectrum.

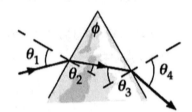

Categorize: We trace rays of violet light and red light through the prism, finding the direction of each as it exits. Then subtraction gives the angular width of the spectrum between them.

Analyze: Call the angles of incidence and refraction, at the surfaces of entry and exit, $\theta_1, \theta_2, \theta_3,$ and θ_4, in the order as shown. The apex angle ($\phi = 60.0°$) is the angle between the surfaces of entry and exit. The ray in the glass forms a triangle with these surfaces, in which the interior angles must add to 180°.

Thus, $\qquad (90° - \theta_2) + \phi + (90° - \theta_3) = 180°$

and $\qquad \theta_3 = \phi - \theta_2$.

This is a general rule for light going through prisms.

For the incoming ray, $\sin\theta_2 = \dfrac{\sin\theta_1}{n}$:

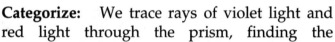

$$\left(\theta_2\right)_{\text{violet}} = \sin^{-1}\left(\frac{\sin 50.0°}{1.66}\right) = 27.48°$$

$$\left(\theta_2\right)_{\text{red}} = \sin^{-1}\left(\frac{\sin 50.0°}{1.62}\right) = 28.22°$$

For the ray exiting the glass, $\theta_3 = 60° - \theta_2$ $\quad$ and $\quad \sin\theta_4 = n\sin\theta_3$:

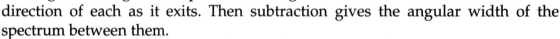

$$\left(\theta_4\right)_{\text{violet}} = \sin^{-1}\left[1.66\sin 32.52°\right] = 63.17°$$

$$(\theta_4)_{\text{red}} = \sin^{-1}[1.62 \sin 31.78°] = 58.56°$$

The dispersion or angular spread is the difference between these two angles:

$$\Delta\theta_4 = 63.17° - 58.56° = 4.61°$$ ■

Finalize: Figure 25.15 in the textbook shows the effect nicely. Reflection cannot separate the colors of white light. Refraction does it, provided that the speeds of different wavelengths are different in the prism material. To help you understand lenses in the next chapter, you should learn that all colors of the light bend around the thicker part of the prism, away from the apex angle.

29. A triangular glass prism with apex angle $\Phi = 60.0°$ has an index of refraction $n = 1.50$ (Fig. P25.29). What is the smallest angle of incidence θ_1 for which a light ray can emerge from the other side?

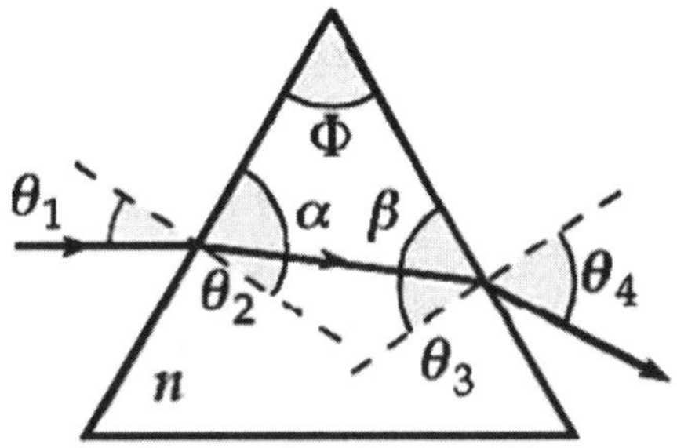

Solution

Conceptualize: The alternative to emerging is undergoing total internal reflection. Figure 25.23 in the textbook

Figure P25.29 (modified)

shows the effect nicely. Going down through the incident rays in that figure, if θ_1 is too small, θ_2 will be too small and θ_3 will be too large for refraction to take place at the second surface of the prism. Then the ray will be totally internally reflected there instead of emerging.

Categorize: We must trace a generic ray through the prism, finding how θ_2 and then θ_3 depend on θ_1. Then we apply the threshold condition for total internal reflection to θ_3 and evaluate the corresponding θ_1.

Analyze: Call the angles of incidence and refraction, at the surfaces of entry and exit, θ_1, θ_2, θ_3, and θ_4, in order as shown. The apex angle Φ is the angle between the surface of entry and the second surface. The ray in the glass forms a triangle with these surfaces, in which the interior angles must add to 180°. Thus, with $\Phi = 60.0°$,

we have $(90° - \theta_2) + 60° + (90° - \theta_3) = 180°$,

so $\theta_2 + \theta_3 = 60.0°$,   [1] which exemplifies a general rule for light going through prisms. At the first refraction, Snell's law gives

$$\sin \theta_1 = 1.50 \sin \theta_2 \qquad\qquad [2]$$

At the second boundary, we want to almost reach the condition for total internal reflection:

$$1.50 \sin \theta_3 = 1.00 \sin 90° = 1.00$$

or $\qquad \theta_3 = \sin^{-1}(1.00/1.50) = 41.8°.$

Now by equation [1] above, $\qquad\qquad \theta_2 = 60.0° - 41.8° = 18.2°,$

while by Equation [2], we find that $\quad \theta_1 = \sin^{-1}(1.50 \sin 18.2°),$

so $\qquad \theta_1 = 27.9°.$ ∎

Finalize: The solution of this problem is significantly less complicated than the solution of its symbolic version, Problem 30. Whether a given problem is more conveniently solved analytically or numerically depends on the complexity of the problem.

---

31. Consider a common mirage formed by superheated air immediately above a roadway. A truck driver whose eyes are 2.00 m above the road, where $n = 1.000\ 293$, looks forward. She perceives the illusion of a patch of water ahead on the road. The road appears wet only beyond a point on the road at which her line of sight makes an angle of 1.20° below the horizontal. Find the index of refraction of the air immediately above the road surface.

Solution

Conceptualize: A wet-road mirage is indeed common. Think of the air as in two discrete layers, the first medium being cooler air on top with $n_1 = 1.000\ 293$ and the second medium being hot air with a lower index, which reflects 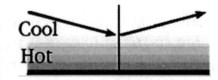 light from the sky by total internal reflection. Then reflection from the cool-air-hot-air interface will mimic reflection from a smooth water surface, at all points where the angle of the trucker's sightline is less than 1.20 degrees with the horizontal, which means at all points more than a certain distance away from the driver.

Categorize: We use the condition for total internal reflection, which is …

Analyze: … $n_1 \sin \theta_1 \geq n_2 \sin 90°$.

At the edge of the mirage of wetness $1.000\ 293 \sin 88.8° = n_2$,

so $n_2 = 1.000\ 073\ 6$.   ∎

Finalize: Perhaps a more realistic model is that a nearly vertical wavefront continuously distorts as it moves through the nonuniform medium, with its lower portions moving faster than its upper portions. The direction of propagation (the ray) bends upward continuously as the wave first enters and then leaves sequentially hotter, lower-density layers closer to the road surface.

Do you think that the answer contains too many significant digits? O thou of little faith! A technique using a Michelson interferometer, which was introduced in Chapter 9, makes it possible to measure just the amount by which the index of refraction of a gas differs from the exact value 1. Let this difference be represented by $x = n - 1$. Snell's law of refraction can be put in terms of angles between rays and the interface (so-called grazing angles) instead of angles that rays make with the normal. If in this problem situation we define $\alpha_1 = 1.20°$ and $\alpha_2 = 0$ by $\alpha_1 = 90° - \theta_1$ and $\alpha_2 = 90° - \theta_2$, the condition for total internal reflection becomes

$$(1 + x_1) \cos \alpha_1 \geq (1 + x_2)1$$

We have next $(1 + x_1)(1 - \alpha_1^2/2 + \alpha_1^4/4 + ...) \geq 1 + x_2$ for α in radians.

For a small angle this gives simply $x_1 - \alpha_1^2/2 \geq x_2$ and the answer as stated above.

---

34. *Why is the following situation impossible?* A laser beam strikes one end of a slab of material of length $L = 42.0$ cm and thickness $t = 3.10$ mm as shown in Figure P25.34 (not to scale). It enters the material at the center of the left end, striking it at an angle of incidence of $\theta = 50.0°$. The index of refraction of the slab is $n = 1.48$. The light makes 85 internal reflections from the top and bottom of the slab before exiting at the other end.

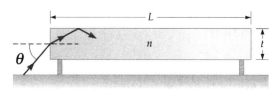

Figure P25.34

Solution

Conceptualize: There are two possible reasons why this situation is impossible. First, it is possible that the laser beam doesn't undergo total internal reflection at the slab-air boundary shown in Figure P25.34, but this is unlikely given the index of refraction of the slab. Second, the number of internal reflections may not be 85. This is the more probable answer to this problem.

Categorize: We use the condition for total internal reflection and count the number of bounces along the length L of the slab.

Analyze: The number N of reflections the beam makes before exiting at the other end is equal to the length of the slab divided by the component of the displacement of the beam for each reflection:

$$N = \frac{L}{(t / \tan\theta_2)} = \frac{L\tan\theta_2}{t}$$

where θ_2 is the refracted angle as the beam enters the material. Substitute for this refracted angle in terms of the incident angle by using Snell's law:

$$N = \frac{L}{t}\tan\left[\sin^{-1}\left(\frac{n_1 \sin\theta_1}{n_2}\right)\right]$$

Substitute numerical values:

$$N = \frac{0.420 \text{ m}}{0.003\ 10 \text{ m}}\tan\left[\sin^{-1}\left(\frac{(1)\sin 50.0°}{1.48}\right)\right]$$

$$= 81.96 \rightarrow 81 \text{ reflections}$$

Therefore, the beam will exit after making 81 reflections, so it does not make 85 reflections. ∎

Finalize: Our second guess was the correct one. Note that we can use Snell's law to show that for a slab immersed in air, total internal reflection will always occur at the slab-air boundary for $n > 1.44$.

═══════════

37. A small light fixture on the bottom of a swimming pool is 1.00 m below the surface. The light emerging from the still water forms a circle on the water surface. What is the diameter of this circle?

Solution

Conceptualize: Only the light that is directed upwards and hits the water's surface at less than the critical angle will be transmitted to the air so that someone outside can see it. The light that hits the surface farther from the center at an angle greater than θ_c will be totally reflected within the water, and cannot be seen from the outside. From the diagram, the diameter of this circle of light appears to be about 2 m.

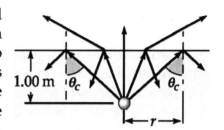

Categorize: We apply Snell's law to find the critical angle, and then find the diameter from geometry.

Analyze: The critical angle is found by imagining the refracted ray just grazing the surface ($\theta_2 = 90°$). The index of refraction of water is $n_1 = 1.333$, and $n_2 = 1.00$ for air, so $n_1 \sin \theta_c = n_2 \sin 90°$ gives $\theta_c = \sin^{-1}(1/1.333) = \sin^{-1}(0.750) = 48.6°$.

The radius then satisfies $\quad \tan \theta_c = \dfrac{r}{1.00 \text{ m}}$.

So the diameter is $\quad d = 2r = 2(1.00 \text{ m})(\tan 48.6°) = 2.27 \text{ m}.$ ∎

Finalize: Only the light rays within a 97.2° cone above the lamp escape the water and can be seen by an outside observer. This angle does not depend on the depth of the light source. The path of a light ray is always reversible, so if a person were located beneath the water, they could see the whole hemisphere above the water surface within this cone; this is a good experiment to try the next time you go swimming!

---

41. A light ray enters the atmosphere of the Earth and descends vertically to the surface a distance $h = 100$ km below. The index of refraction where the light enters the atmosphere is precisely 1, and it increases linearly with distance to have the value $n = 1.000\ 293$ at the Earth's surface. (a) Over what time interval does the light traverse this path? (b) By what percentage is the time interval larger than that required in the absence of the Earth's atmosphere?

Solution

Conceptualize: The electromagnetic wave has its speed drop to lower and lower values. It gets bogged down by having to cause oscillation of charges in molecules of denser and denser air. If the air were absent, the time interval to traverse 100 km would be $h/c = 100 \text{ km}/(300 \text{ Mm/s}) = 333 \text{ } \mu\text{s}$.

Because the index of refraction of even the densest air is close to 1, the time for the light to come down through the atmosphere will be just a little longer, by a fraction of one percent.

Categorize: We used the particle under constant velocity model for our estimate in the Conceptualize step, but that will not work for the calculation itself, and neither will the particle under constant acceleration model, because the speed of light changes linearly with distance and not with time. We must drop back to the basic identification of speed as $v = dx/dt$ …

Analyze: … which lets us say that the incremental distance traveled in an incremental time interval is

$$dx = v\,dt = (c/n)dt$$

(a) Then the time elapsing as the light moves down by distance dx is

$$dt = n\,dx/c \qquad\qquad \textbf{[1]}$$

Here n is a function of x while c is of course constant. We choose to let x represent the distance traversed downward from the top of the atmosphere. From the description in the problem we can identify the functional form of n as

$$n = 1.000\,000 + (0.000\,293/100\text{ km})x = 1 + (0.000\,293/h)x$$

just because this is the linear function that fits the boundary conditions that n is 1 exactly at $x = 0$ and $n = 1.000\,293$ at $x = 100$ km.

Now the whole time for the light to pass vertically through the atmosphere is given by integrating both sides of our equation [1] from the entry of the light at the top to its arrival at the bottom:

$$\Delta t = \int_0^t dt = \int_0^h n\,dx/c = \frac{1}{c}\int_0^{100\text{ km}}\left(1 + \frac{0.000\,293}{h}x\right)dx$$

$$\Delta t = \frac{1}{c}\left(x + \frac{0.000\,293}{h}\frac{x^2}{2}\right)\Bigg|_0^h = \frac{h}{c} + \frac{0.000\,293}{ch}\frac{h^2}{2}$$

$$= \frac{1.00\times10^5\text{ m}}{3.00\times10^8\text{ m/s}}\left(1 + \frac{0.000\,293}{2}\right)$$

$$= 3.33\times10^{-4}\text{ s} + 4.88\times10^{-8}\text{ s}$$

So to three significant digits we have 3.33×10^{-4} s   ∎

(b) As noted in the Conceptualize step, the time interval without air would be *h/c*, so the fractional excess time attributable to the presence of the air is

$$\frac{\Delta t - h/c}{h/c} = \frac{\dfrac{h}{c} + \dfrac{0.000\ 293\ h}{2}\dfrac{h}{c} - \dfrac{h}{c}}{\dfrac{h}{c}} = \frac{0.000\ 293}{2} \times 100\% = 0.014\ 6\%$$ ■

Finalize: For the actual atmosphere, the excess travel time might be hard to measure directly. But we will see in Chapter 27, about interference, that remarkably precise measurements can be made of the number of extra wavelengths traveled by light going through a layer of glass or even of air, compared to light going through vacuum.

47. A hiker stands on an isolated mountain peak near sunset and observes a rainbow caused by water droplets in the air at a distance of 8.00 km along her line of sight to the most intense light from the rainbow. The valley is 2.00 km below the mountain peak and entirely flat. What fraction of the complete circular arc of the rainbow is visible to the hiker?

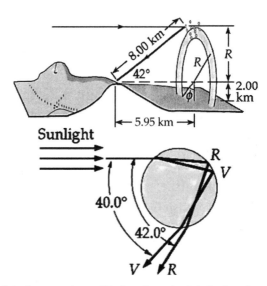

Solution

Conceptualize: The textbook's chapter-opening photograph, its Figures 25.16 and 25.17, and our diagram contain a lot of information. If the Sun is high in the sky you will see only a small arc of rainbow, or none at all. Without the mountain, at sunset you will see just a 180° arc, a half-circle above the level ground.

Categorize: Horizontal light rays from the setting Sun pass above and around the hiker. The light rays are twice refracted and once reflected, as in Figure 25.16. The most intense light reaching the hiker, that which represents the visible rainbow, is located between angles of 40.0° and 42.0° from the hiker's shadow. The hiker sees a greater percentage of the violet inner edge, so we consider the red outer edge. The radius *R* of the circle of droplets is …

Analyze: *R* = (8.00 km)sin 42.0° = 5.35 km.

Then the angle ϕ between the vertical and the radius where the bow touches the ground is given by

$$\cos\phi = \frac{2.00\text{ km}}{R} \quad \text{so} \quad \phi = \cos^{-1}\left(\frac{2.00\text{ km}}{5.35\text{ km}}\right) = \cos^{-1}0.374 = 68.1°.$$

The angle filled by the visible bow is 360° − 2(68.1°) = 224°, so the visible bow is

$$\frac{224°}{360°} = 62.2\% \text{ of a circle.} \qquad\qquad\blacksquare$$

Finalize: This striking view motivated Charles Wilson's 1906 invention of the cloud chamber, a standard tool of nuclear physics. The effect is mentioned in the Bible, Ezekiel 1:28. Look for a full circle of color around your shadow when you fly in an airplane. With a stepladder over a lawn sprinkler you can show children a full-circle rainbow when the summer Sun is high in the sky. Do not let the wet children fall from the ladder.

---

53. The light beam in Figure P25.53 strikes surface 2 at the critical angle. Determine the angle of incidence θ_1.

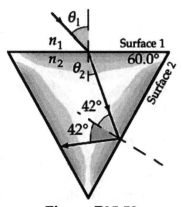

Figure P25.53

Solution

Conceptualize: From the diagram it appears that the angle of incidence is about 40°.

Categorize: We can find θ_1 by applying Snell's law at the first interface where the light is refracted. At surface 2, knowing that the 42.0° angle of reflection is the critical angle, we can work backwards to find θ_1.

Analyze: Define n_1 to be the index of refraction of the surrounding medium and n_2 to be that for the prism material. We can use the critical angle of 42.0° to find the ratio n_2/n_1:

$$n_2 \sin 42.0° = n_1 \sin 90.0°$$

So, $\quad \dfrac{n_2}{n_1} = \dfrac{1}{\sin 42.0°} = 1.49.$

(If we had started with the assumption $n_1 = 1.000$, we would have $n_2 = 1.49$, and the remainder of the solution would proceed with no real change, to give the same answer that we will obtain.)

Call the angle of refraction θ_2 at the surface 1. The ray inside the prism forms a triangle with surfaces 1 and 2, so the sum of the interior angles of this triangle must be 180°.

Thus, $(90.0° - \theta_2) + 60.0° + (90.0° - 42.0°) = 180°$.

Therefore, $\theta_2 = 18.0°$.

Applying Snell's law at surface 1, $n_1 \sin\theta_1 = n_2 \sin\theta_2$:

$$\theta_1 = \sin^{-1}\left(\frac{n_2 \sin\theta_2}{n_1}\right) = \sin^{-1}(1.49\sin 18.0°) = 27.5° \qquad \blacksquare$$

Finalize: The result is a bit less than the 40.0° we expected, but this is probably because the figure is not drawn to scale. This problem was a bit tricky because it required four key concepts (refraction, reflection, critical angle, and geometry) in order to find the solution. One practical extension of this problem is to consider what would happen to the exiting light if the angle of incidence were varied slightly. Would all the light still be reflected off surface 2, or would some light be refracted and pass through this second surface? See Figure 35.23 in the textbook.

---

54. A light ray of wavelength 589 nm is incident at an angle θ on the top surface of a block of polystyrene as shown in Figure P25.54. (a) Find the maximum value of θ for which the refracted ray undergoes total internal reflection at the point P located at the left vertical face of the block. **What If?** Repeat the calculation for the case in which the polystyrene block is immersed in (b) water and (c) carbon disulfide. Explain your answers.

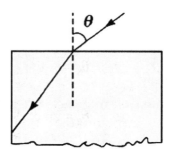

Figure P25.54

Solution

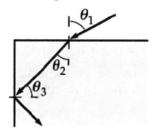

Conceptualize: If θ_1 is too large, θ_2 will be too large and θ_3 too small for total internal reflection to happen at the vertical surface. We might estimate an angle not far from 45°.

Categorize: We can find θ_3 from the condition for total internal reflection. Then we work backwards to find θ_2 and θ_1.

Analyze: We look up the index of refraction (for 589-nm light) for each material as listed here:

Air   1.00

Water   1.33

Polystyrene   1.49

Carbon disulfide   1.63

(a) For polystyrene **surrounded by air**, the critical angle for total internal reflection is

$$\theta_3 = \sin^{-1}(1/1.49) = 42.2°$$

and then from geometry, $\theta_2 = 90.0° - \theta_3 = 47.8°$.

From Snell's law, $\sin\theta_1 = 1.49\sin 47.8° = 1.10$.

This has no solution; thus, the real maximum value for θ_1 is 90.0°.

Total internal reflection always occurs.   ■

(b) For polystyrene **surrounded by water**, we have

$$\theta_3 = \sin^{-1}\left(\frac{1.33}{1.49}\right) = 63.2°$$

and $\theta_2 = 26.8°$.

From Snell's law, $n_1\sin\theta_1 = n_2\sin\theta_2$: $1.33\sin\theta_1 = 1.49\sin 26.8°$

and $\theta_1 = 30.3°$.   ■

(c) Total internal reflection at the second surface is not possible because the beam is initially traveling in polystyrene, a medium of lower index of refraction. When the block is surrounded by carbon disulfide, total internal reflection never happens.   ■

Finalize: Note that $\sin\theta = 1.10$ does not describe an angle larger than 90°. It describes no angle whatsoever. We had to think to figure out the answer to part (a). It was not what we guessed. Thinking is a good habit in any case.

Chapter 26
Image Formation by Mirrors and Lenses

Section 26.1　　Images Formed by Flat Mirrors

An image formed by a flat mirror has the following properties:

- The image is as far behind the mirror as the object is in front.

- The image is actual-size, virtual, and upright. (Upright is used to mean that if the object is an arrow that points upward, so does the image arrow.)

- The image has an apparent right-left reversal.

Section 26.2　　Images Formed by Spherical Mirrors
Section 26.4　　Images Formed by Thin Lenses

A **spherical mirror** is a reflecting surface that has the shape of a segment of a sphere. A **concave mirror** is one that reflects light from the inner concave surface; a convex mirror reflects light from the outer convex surface. When using the mirror equation (Equation 26.6, $\frac{1}{p} + \frac{1}{q} = \frac{1}{f}$) carefully adhere to the sign conventions stated in the Suggestions, Skills, and Strategies section.

Real images are formed at a point when reflected light actually passes through the image point. **Virtual images** are formed at a point when light rays appear to diverge from the image point.

A **ray diagram** can be constructed to locate the image position formed by a mirror or lens if the location of the object, focal point and center of curvature are known. *A ray diagram must be carefully drawn to scale.* **Study the examples on the following pages.**

For a concave mirror ($p > f$):

Ray 1 from the top of the object parallel to the principal axis is reflected back through the focal point, F.

Ray 2 from the top of the object through the focal point is reflected parallel to the axis.

Ray 3 from the top of the object through the center of curvature, C is reflected back on itself.

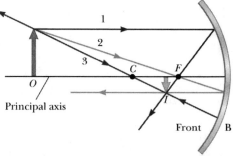

Ray diagram for concave mirror when the object distance is greater than two focal lengths. The image is real, inverted and smaller than the object.

For a concave mirror ($p < f$):

Ray 1 from the top of the object parallel to the principal axis is reflected back through the focal point, F.

Ray 2 from the top of the object as if coming from the focal point (dashed line) is reflected parallel to the axis.

Ray 3 from the top of the object as coming from the center of curvature is reflected back on itself.

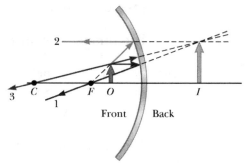

Ray diagram for a concave mirror when the object distance is less than the focal length. The image is virtual, upright and larger than the object.

For a convex mirror (any value of p):

Ray 1 from the top of the object parallel to the principal axis is reflected *away* from the focal point, F.

Ray 2 from the top of the object toward the focal point *on the back side of the mirror* is reflected parallel to the axis.

Ray 3 from the top of the object toward the center of curvature C *on the back side* of the *mirror* is reflected back on itself.

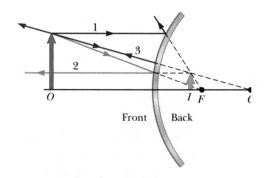

Ray diagram for convex mirror. The image will be virtual, upright and smaller for all object distances.

For a convex lens ($p > f$):

Ray 1 from the top of the object parallel to the principal axis and after being refracted by the lens passes through the focal point *on the back side of the lens.*

Ray 2 from the top of the object through the focal point on the front side of the lens emerges from the lens parallel to the principal axis.

Ray 3 from the top of the object toward the center of the lens continues in a straight line.

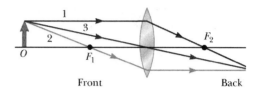

Ray diagram for convex lens when the object distance is greater than two focal lengths. The image is real, inverted and smaller than the object.

For a convex lens ($p < f$):

Ray 1 from the top of the object parallel to the principal axis and after being refracted by the lens passes through the focal point *on the back side of the lens.*

Ray 2 from the top of the object as if coming from the focal point emerges from the lens parallel to the principal axis.

Ray 3 from the top of the object toward the center of the lens continues in a straight line.

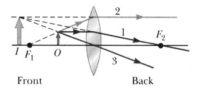

Ray diagram for convex lens when the object distance is less than the focal length. The image is virtual, upright and larger than the object. The image is located at the intersection of the backward extensions of the three refracted rays.

For a concave (diverging) lens:

Ray 1 is drawn parallel to the principal axis, and after refraction emerges directed away from the focal point on the front side of the lens.

Ray 2 is drawn in the direction toward the focal point *on the back side of the lens* and emerges from the lens parallel to the principle axis.

Ray 3 is drawn through the center of the lens and continues in a straight line.

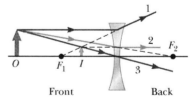

Ray diagram for a concave (diverging) lens. The image will be virtual, upright and smaller for *all object distances.*

EQUATIONS AND CONCEPTS

The **sign conventions** appropriate for the equations stated here are summarized in SUGGESTIONS, SKILLS AND STRATEGIES.

The **mirror equation** is used to locate the position of an image formed by reflection of paraxial rays. The focal point of a spherical mirror is located midway between the center of curvature and the center of the mirror.

$$\frac{1}{p} + \frac{1}{q} = \frac{1}{f} \tag{26.6}$$

$$f = \frac{R}{2} \tag{26.5}$$

The **lateral magnification of a spherical mirror** can be stated either as a ratio of image size to object size or in terms of the ratio of image distance to object distance.

$$M = \frac{h'}{h} = -\frac{q}{p} \tag{26.2}$$

A magnified image of an object can be formed by a **single spherical refracting surface** of radius R, which separates two media whose indices of refraction are n_1 and n_2. *If the calculated value of M is negative, the image will be inverted.* A special case is that of the virtual image formed by a **planar refracting surface**.

$$\frac{n_1}{p} + \frac{n_2}{q} = \frac{n_2 - n_1}{R} \tag{26.8}$$

$$M = -\frac{n_1 q}{n_2 p} \tag{26.9}$$

$$q = -\frac{n_2}{n_1} p \tag{26.10}$$

The **focal length of a thin lens** is determined by the characteristic properties of the lens (index of refraction n and radii of curvature R_1 and R_2).

$$\frac{1}{f} = (n - 1)\left(\frac{1}{R_1} - \frac{1}{R_2}\right) \tag{26.13}$$

The **lateral magnification of a lens** (M) will be negative when the image is inverted.

$$M = \frac{h'}{h} = -\frac{q}{p} \tag{26.11}$$

The **thin lens equation** is identical to the mirror equation and can be used with both converging (positive f) and diverging (negative f) lenses.

$$\frac{1}{p} + \frac{1}{q} = \frac{1}{f} \tag{26.12}$$

SUGGESTIONS, SKILLS, AND STRATEGIES

Sign conventions for mirrors, refracting surfaces, and thin lenses are described on the following pages.

Sign Conventions for Mirrors

Equations:
$$\frac{1}{p}+\frac{1}{q}=\frac{1}{f}=\frac{2}{R} \qquad M=\frac{h'}{h}=-\frac{q}{p}$$

The front side of the mirror is the region on which light rays are incident and reflected.

p is + if the object is in front of the mirror (real object).
p is − if the object is in back of the mirror (virtual object).

q is + if the image is in front of the mirror (real image).
q is − if the image is in back of the mirror (virtual image).

Both f and R are + if the center of curvature is in front (concave mirror).
Both f and R are − if the center of curvature is in back (convex mirror).

If M is positive, the image is upright.
If M is negative, the image is inverted.

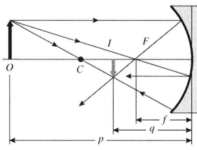

(a) Concave Mirror ($p > 2f$)
     Image: real, inverted, diminished

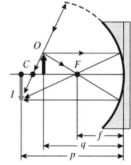

(b) Concave Mirror ($2f > p > f$)
     Image: real, inverted, enlarged

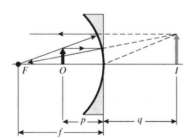

(c) Concave Mirror ($p < f$)
     Image: virtual, upright, enlarged

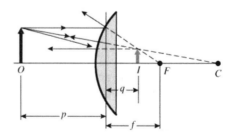

(d) Convex Mirror (any value of p)
     Image: virtual, upright, diminished

Figures describing sign conventions for mirrors.

Sign Conventions for Refracting Surfaces

Equations:
$$\frac{n_1}{p} + \frac{n_2}{q} = \frac{n_2 - n_1}{R} \qquad M = \frac{h'}{h} = -\frac{n_1 q}{n_2 p}$$

In the following list, the front side of the surface is the side from which the light is incident.

> p is + if the object is in front of the surface (real object).
> p is − if the object is in back of the surface (virtual object).

> q is + if the image is in back of the surface (real image).
> q is − if the image is in front of the surface (virtual image).

> R is + if the center of curvature is in back of the surface.
> R is − if the center of curvature is in front of the surface.

> n_1 refers to the index of refraction of the first medium (before refraction).

> n_2 is the index of refraction of the second medium (after refraction).

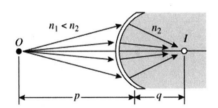

(a) Convex Refracting Surface
(object outside surface)
Image: real, inside surface

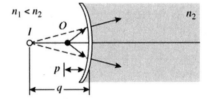

(b) Concave Refracting Surface
(object outside surface)
Image: virtual, outside surface

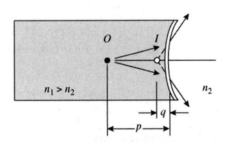

(c) Concave Refracting Surface
(object inside surface)
Image: virtual, inside surface

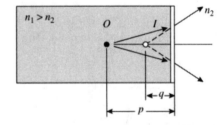

(d) Flat Refracting Surface
(object inside surface)
Image: virtual, inside surface

Figures describing sign conventions for refracting surfaces.

Sign Conventions for Thin Lenses

Equations:
$$\frac{1}{p}+\frac{1}{q}=\frac{1}{f}=(n-1)\left(\frac{1}{R_1}-\frac{1}{R_2}\right) \qquad M=\frac{h'}{h}=-\frac{q}{p}$$

In the following list, the front of the lens is the side from which the light is incident.

p is + if the object is in front of the lens.
p is − if the object is in back of the lens.

q is + if the image is in back of the lens.
q is − if the image is in front of the lens.

f is + if the lens is thickest at the center.
f is − if the lens is thickest at the edges.

R_1 and R_2 are + if the center of curvature is in back of the lens.
R_1 and R_2 are − if the center of curvature is in front of the lens.

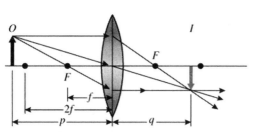

(a) Converging Lens ($p > 2f$)
Image: real, inverted, diminished

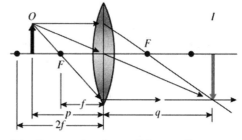

(b) Converging Lens ($2f > p > f$)
Image: real, inverted, enlarged

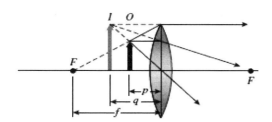

(c) Converging Lens ($p < f$)
Image: virtual, upright, enlarged

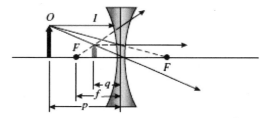

(d) Diverging Lens (any value of p)
Image: virtual, upright, diminished

Figures describing sign conventions for various thin lenses.

REVIEW CHECKLIST

- Identify the following properties that characterize an image formed by a lens or mirror system with respect to an object: position, magnification, orientation (i.e., inverted or upright) and whether real or virtual.

- Understand the relationship of the algebraic signs associated with calculated quantities to the nature of the image and object: real or virtual, upright or inverted.

- Determine the location of the image of a specified object as formed by a plane mirror, spherical mirror, plane refracting surface, spherical refracting surface, thin lens, or a combination of two or more of these devices. Determine the magnification and character of the image in each case.

- Construct ray diagrams to determine the location and nature of the image of a given object when the geometrical characteristics of the optical device (lens or mirror) are known.

ANSWER TO AN OBJECTIVE QUESTION

9. Lulu looks at her image in a makeup mirror. It is enlarged when she is close to the mirror. As she backs away, the image becomes larger, then impossible to identify when she is 30.0 cm from the mirror, then upside down when she is beyond 30.0 cm, and finally small, clear, and upside down when she is much farther from the mirror. **(i)** Is the mirror (a) convex, (b) plane, or (c) concave? **(ii)** Is the magnitude of its focal length (a) 0, (b) 15.0 cm, (c) 30.0 cm, (d) 60.0 cm, or (e) ∞ ?

Answer **(i)** (c) A plane mirror always produces a right side up, actual-size image. A convex mirror always produces a right side up image of a real object, diminished in size. A concave mirror has the property of producing an enlarged, right side up image of an object closer than its focal point, and producing an inverted image of an object beyond its focal point. **(ii)** (c) For an object at the focal point, the image is at infinity, or the reflected rays originating from one point on the object are parallel, or no image is formed.

ANSWERS TO SELECTED CONCEPTUAL QUESTIONS

6. Lenses used in eyeglasses, whether converging or diverging, are always designed so that the middle of the lens curves away from the eye like the center lenses of Figures 26.21a and 26.21b. Why?

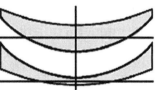

Answer With this so-called meniscus design, when you direct your gaze near the outer circumference of the lens you receive a ray that has passed through glass with more nearly parallel surfaces of entry and exit. Thus, the lens minimally distorts the direction to the object you are looking at. If you wear glasses, you can demonstrate this by turning them around and looking through them the wrong way, maximizing the distortion.

8. Explain why a fish in a spherical goldfish bowl appears larger than it really is.

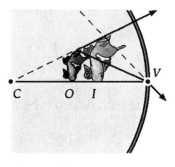

Answer As in the diagram, let the center of curvature C of the fishbowl and the bottom point of the fish define the optical axis, intersecting the fishbowl at vertex V. A ray from the top point of the fish that reaches the bowl along a radial line through C has angle of incidence zero and angle of refraction zero. This ray exits the bowl unchanged in direction. A ray from the top of the fish to V is refracted to bend away from the normal. Its extension back inside the fishbowl determines the location of the image and the characteristics of the image. It is upright, virtual, and enlarged.

SOLUTIONS TO SELECTED END-OF-CHAPTER PROBLEMS

1. Determine the minimum height of a vertical flat mirror in which a person 178 cm tall can see his or her full image. *Suggestion:* Drawing a ray diagram would be helpful.

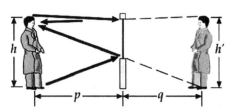

Solution

Conceptualize: Almost anyone looking for a full-length mirror would look for one around 2 meters tall. Many people think that their image lies on the mirror surface.

Categorize: We will use general techniques based on the mirror-lens equation, the magnification equation, and a diagram.

Analyze: The flatness of the mirror is described by $R = \infty$, $f = \infty$, and $1/f = 0$. By our general mirror and lens equation,

$$\frac{1}{p} + \frac{1}{q} = \frac{1}{f} \quad \text{we then have} \quad q = -p.$$

Thus, the image is as far behind the mirror as the person is in front. The magnification is then

$$M = \frac{-q}{p} = 1 = \frac{h'}{h} \quad \text{so} \quad h' = h' = 178 \text{ cm.}$$

The required height of the mirror is defined by the triangle from the person's eyes to the top and bottom of the image, as shown. From the geometry of the triangle, we see that the mirror height must be:

$$h'\left(\frac{p}{p-q}\right) = h'\left(\frac{p}{2p}\right) = \frac{h'}{2}$$

Thus, the mirror must be at least 89.0 cm high.   ■

Finalize: On the bedroom or hallway wall you would really want a mirror a bit taller than 90 cm, to accommodate people whose eyes are different distances from the floor. But the mirror only has to be half as wide as your shoulders. Knowing some physics saves you some money.

Look back at the first line of the Analyze section. The flatness of the mirror is perfectly real, and the equation $R = \infty$ is a perfectly real and meaningful mathematical description of it. The variables p, q, h, h', R, f, and M range over an "augmented" number line, incorporating the usual real numbers and a single extra infinite value. An object at $p = \infty$, for example, is the source of incoming light rays that can be modeled as parallel. You do not need an ∞ key on your calculator, but recognize that $1/\infty = 0$.

———————————————

7.   A convex spherical mirror has a radius of curvature of magnitude 40.0 cm. Determine the position of the virtual image and the magnification for object distances of (a) 30.0 cm and (b) 60.0 cm. (c) Are the images in parts (a) and (b) upright or inverted?

Solution

Conceptualize:　Think of your reflection in a backyard gazing globe or, if it is more familiar, the back of a shiny spoon. We expect right side up images, diminished in size, behind the reflecting surface. As the object moves farther away from (a) to (b), will the image move closer to the reflecting surface or farther away?

Categorize:　This problem is in standard cause-and-effect, locate-and-describe format. The lens-mirror equation will tell us the image location and $M = -q/p$ will tell us the magnification. A ray diagram is the best way to determine and understand the image description.

Analyze:　The convex mirror is described by

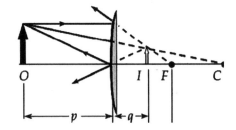

$$f = \frac{R}{2} = \frac{-40.0 \text{ cm}}{2} = -20.0 \text{ cm}$$

(a)　Then $\dfrac{1}{p} + \dfrac{1}{q} = \dfrac{1}{f}$ gives

$$q = \frac{1}{1/f - 1/p} = \frac{1}{1/(-20.0 \text{ cm}) - 1/(30.0 \text{ cm})} = -12.0 \text{ cm} \qquad ∎$$

The magnification factor is $M = -\dfrac{q}{p} = -\left(\dfrac{-12.0 \text{ cm}}{30.0 \text{ cm}}\right) = +0.400.$ ∎

(c)　As shown by the figure above, the image is behind the mirror, upright, virtual, and diminished. ∎

(b)　Following the same steps,

$$q = \frac{1}{1/f - 1/p} = \frac{1}{1/(-20.0 \text{ cm}) - 1/(60.0 \text{ cm})} = -15.0 \text{ cm} \qquad ∎$$

and　$M = \dfrac{-q}{p} = -\left(\dfrac{-15.0 \text{ cm}}{60.0 \text{ cm}}\right) = +0.250.$ ∎

(c)　The principal ray diagram is an essential complement to the numerical description of the image. Add rays on this diagram for a 60-cm object distance.

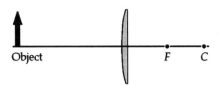

Use your diagram to confirm that the image is behind the mirror, upright, virtual, and diminished. ∎

Finalize: A negative answer for q indicates that the image is behind the mirror and virtual. A positive answer for M indicates that the image is upright. A value of M less than 1 in absolute value indicates that the image is diminished. Drawing the diagram really is easier and clearer than trying to remember those different rules.

11. A concave spherical mirror has a radius of curvature of magnitude 20.0 cm. (a) Find the location of the image for object distances of **(i)** 40.0 cm, **(ii)** 20.0 cm, and **(iii)** 10.0 cm. For each case state whether the image is (b) real or virtual and (c) upright or inverted. (d) Find the magnification in each case.

Solution

Conceptualize: It is a good idea to draw a principal-ray diagram as early as possible in any optics problem. This gives a qualitative sense of how the image appears relative to the object. From the ray diagrams below, we see that when the object is 40 cm from the mirror, the image is real, inverted, diminished, and located a bit more than 10 cm from the mirror. When the object is at 20 cm, the image is real, inverted, actual-size, and also at 20 cm. When the object is at 10 cm, the reflected rays do not intersect, so we can say that no image is formed.

Categorize: The mirror equation gives a precise value for the image distance. Then $M = -q/p$ gives a precise value for the magnification. Together they complement the ray diagrams.

Analyze: We apply the mirror equation using the sign conventions listed in *Suggestions, Skills, and Strategies*.

(i) (a) $1/p + 1/q = 2/R$ becomes

$$q = \frac{1}{2/R - 1/p} = \frac{1}{2/(20.0 \text{ cm}) - 1/(40.0 \text{ cm})} = 13.3 \text{ cm}$$ ∎

(b) The positive value for q indicates that the image is real. ∎

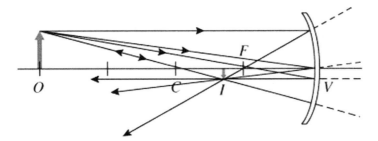

The ray diagram shows this identification more clearly, and (c) that the image is inverted. ∎

(d) The magnification is $M = -\dfrac{q}{p} = -\dfrac{13.3 \text{ cm}}{40.0 \text{ cm}} = -0.333$. ∎

Its value indicates that the image is inverted and one-third the height of the object.

(ii) (a) The object is now at the center of curvature. Following the same steps gives

$$q = \frac{1}{2/R - 1/p} = \frac{1}{2/(20.0 \text{ cm}) - 1/(20.0 \text{ cm})} = 20.0 \text{ cm}$$ ∎

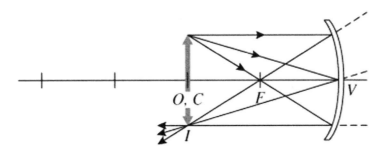

(b) The image is real, as shown by the ray diagram and by the positive value for q. ∎

(c) The ray diagram shows that the image is inverted. ∎

(d) The magnification is $M = -\dfrac{q}{p} = -\dfrac{20.0 \text{ cm}}{20.0 \text{ cm}} = -1.00$. ∎

Its value indicates that the image is inverted and the same height as the object in this special case.

(iii) (a) The object is now at the focal point of the mirror. Following the same steps gives

$$q = \frac{1}{2/R - 1/p} = \frac{1}{2/(20.0 \text{ cm}) - 1/(10.0 \text{ cm})} = \frac{1}{0} = \infty$$

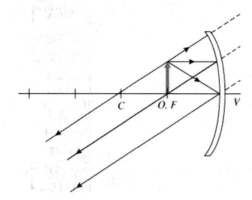

We can say that no image is formed, or that the image is at an infinite distance. ∎

(b) In this special case the reflected rays do not intersect. We cannot classify the image as real or virtual. ∎

(c) We cannot classify the image as upright or inverted. ∎

A screen placed at a large distance in front of the mirror can intercept the reflected light energy, showing the appearance of an upside-down real image, but it is not sharp for any finite distance. You can look into the mirror to view the image as a right side up virtual image, with your eye focused on infinity. ∎

(d) The magnification is $M = -\dfrac{q}{p} = -\dfrac{\infty}{20.0 \text{ cm}} = \infty$. ∎

In this special case, if we say no image is formed at a finite distance, it has no finite magnification. If we say the image is at infinity, then its height and its magnification are also infinite. There is no physical difference between $+\infty$ and $-\infty$.

Finalize: The calculations of image characteristics agree well with the conclusions from our ray diagrams. Especially in trial (iii), the ray diagram helps to explain the meaning of the calculated results. It is easy to miss a minus sign or to make a computational mistake in using the mirror-lens equation, so the characteristics and approximate values obtained from the ray diagrams are useful as a check on the calculated values.

Each of the three situations has a practical application. Situation (i) can be considered a model for a parabolic microphone, used for concentrating the sound of a bird on an electronic transducer. If the object distance were much greater than the radius of curvature, it would be clearly a model for a satellite dish, a reflecting telescope, and a solar-energy stove. Situation (ii) is used for hallway displays in physics buildings. The image of a lightbulb looks just like a lightbulb when your eyes intercept the reflected rays as they diverge from the image after converging to it; but you can move your finger through the image. Situation (iii) models a searchlight.

In the ray diagram for situation (ii), with the object at the center of curvature, it is not possible to draw a meaningful ray from O to C. In situation (iii) we could not draw a ray from O to F. So in all of the diagrams we supplement the rays mentioned in the textbook with a ray from the top of the object to point V, the mirror vertex or origin of coordinates. This ray reflects, making an equal angle on the other side of the axis.

---

19. A spherical mirror is to be used to form an image 5.00 times the size of an object on a screen located 5.00 m from the object. (a) Is the mirror required concave or convex? (b) What is the required radius of curvature of the mirror? (c) Where should the mirror be positioned relative to the object?

Solution

Conceptualize: A mirror is not often used as a slide projector, but it can work well. The image must be real to be on a screen.

Categorize: We are careful not to assume that the image is right side up. The standard equations, supplemented by the diagram, will give us the answer.

Analyze: The object's distance from the mirror is p and the image's distance from the mirror is $q = p + 5.00$ m.

Further, from $M = -\dfrac{q}{p}$ we have $|M| = 5.00$.

Since the image must be real, q must be positive, so M must be negative:

$$M = -5.00 \quad \text{and} \quad q = 5.00\,p$$

(c) Solving first for the distance of the mirror from the object,

$$p + 5.00 \text{ m} = 5.00p \quad \text{and} \quad p = 1.25 \text{ m}$$

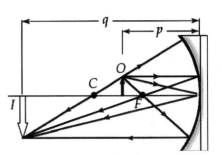

■

Applying the lens-mirror equation

$$\frac{1}{f} = \frac{1}{p} + \frac{1}{q}$$

gives $\quad \dfrac{1}{f} = \dfrac{1}{1.25 \text{ m}} + \dfrac{1}{6.25 \text{ m}},$

so that the focal length of the mirror is $f = 1.04$ m.

(a, b) Noting that the image is real, inverted, and enlarged, we can say that the mirror must be concave, and must have a radius of curvature of $R = 2f = 2.08$ m. ∎

Finalize: With a shaving mirror or telescope mirror, a darkened room, and a small light bulb you can set up this situation to scale. Putting the light bulb precisely at the focal point makes a model of a searchlight. It is very instructive to put the light bulb at different locations in front of the mirror and find and observe the images.

---

24. A cubical block of ice 50.0 cm on a side is placed over a speck of dust on a level floor. Find the location of the image of the speck as viewed from above. The index of refraction of ice is 1.309.

Solution

Conceptualize: Light rays diverging from the dust are changed in direction when they leave the top surface of the ice. They can be traced back to a point from which they apparently diverge, and that is the image, somewhere (we think) inside the block.

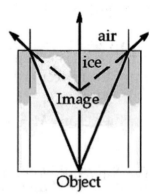

Categorize: The upper surface of the block is a single refracting surface with zero curvature, and with infinite radius of curvature. We use the image-formation-by-a-single-refracting-surface equation.

Analyze:

The equation is $\dfrac{n_1}{p} + \dfrac{n_2}{q} = \dfrac{n_2 - n_1}{R}$, or

$$q = \frac{n_2}{(n_2 - n_1)/R - n_1/p} = \frac{1.00}{(1.00 - 1.309)/\infty - 1.309/(50.0 \text{ cm})}$$

$$= \frac{-50.0 \text{ cm}}{1.309} = -38.2 \text{ cm}$$

The dust speck appears to be 38.2 cm below the upper surface of the ice. ∎

Finalize: The image is virtual, upright, and actual size. A virtual image is a perfectly visible thing at a perfectly definite location. It is the intersection point of exiting rays. In optics an object is not necessarily a chunk of material. An object is an intersection point for rays entering a mirror, lens, or refracting surface.

---

27. A glass sphere ($n = 1.50$) with a radius of 15.0 cm has a tiny air bubble 5.00 cm above its center. The sphere is viewed looking down along the extended radius containing the bubble. What is the apparent depth of the bubble below the surface of the sphere?

Solution

Conceptualize: This problem is just like Problem 24, immediately above, except for different numbers and a curved rather than a flat refracting surface. We guess that the image will be quite close below the upper surface of the glass.

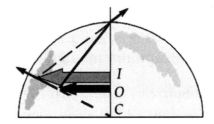

Categorize: The same methods as in Problem 24 should work.

Analyze: From the equation about image formation by a single refracting surface,

$$\frac{n_1}{p} + \frac{n_2}{q} = \frac{n_2 - n_1}{R}$$

We solve for q to find $\quad q = \dfrac{n_2 Rp}{p(n_2 - n_1) - n_1 R}.$

In this case, $n_1 = 1.50$, $n_2 = 1.00$, $p = 10.0$ cm, and $R = -15.0$ cm.

So the image location is

$$q = \frac{(1.00)(-15.0 \text{ cm})(10.0 \text{ cm})}{(10.0 \text{ cm})(1.00 - 1.50) - (1.50)(-15.0 \text{ cm})} = -8.57 \text{ cm}$$

The depth of the image is 8.57 cm. ■

Finalize: Oops! If the surface had been flat rather than curved, the image position would have been –10 cm/1.5 = –6.67 cm. This can be seen from our symbolic equation $q = \dfrac{n_2 Rp}{p(n_2 - n_1) - n_1 R}$ by taking $R = \infty$. The curvature of the surface makes the image distance larger, not smaller as we guessed. To sketch a ray diagram we use the wave under refraction model, as shown. The image is virtual, upright, and enlarged.

---

30. An object is located 20.0 cm to the left of a diverging lens having a focal length $f = -32.0$ cm. Determine (a) the location and (b) the magnification of the image. (c) Construct a ray diagram for this arrangement.

Solution

Conceptualize: If you are nearsighted, your eyeglass lenses are diverging, likely with longer focal length that the lens considered here. We expect the image distance to be negative and smaller in absolute value than 20 cm. The magnification should be positive and less than 1. That is, the image should be virtual, upright, and diminished.

Categorize: The mirror-and-lens equation and the magnification equation should do the job for precise answers.

Analyze: The mirror-and-lens equation, $\dfrac{1}{p} + \dfrac{1}{q} = \dfrac{1}{f}$,

becomes here $\quad q = \dfrac{1}{1/f - 1/p}$.

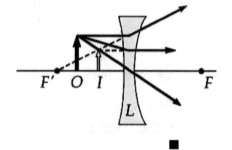

(a) So $\quad q = -\left(\dfrac{1}{20.0 \text{ cm}} + \dfrac{1}{32.0 \text{ cm}}\right)^{-1} = -12.3$ cm.

(b) The magnification is

$$M = -\dfrac{q}{p} = -\dfrac{(-12.3 \text{ cm})}{20.0 \text{ cm}} = 0.615$$

(c) The image is virtual, upright, and diminished; the ray diagram is shown. ■

Finalize: Another practical application of a diverging lens is a wide-angle lens in a peephole in a door.

In drawing the ray diagram, the focal points might give you trouble. Review how light goes through a prism, in the textbook's Figure 25.14 or the drawing accompanying Problem 25.1 in this manual. In the diagram for this problem, draw first the ray from the top of the object through the center of the lens, continuing on straight. Draw next the ray from the top of the object parallel to the axis. It bends "around the thicker part of the prism," up in the diagram, to go off as if it were coming from F'. Then F has not yet been used. A ray from the top of the object toward F bends around the thicker part of the "prism" it encounters to go off parallel to the axis. Extending the outgoing rays backwards determines the image by their intersection.

––––––––––

33. The left face of a biconvex lens has a radius of curvature of magnitude 12.0 cm, and the right face has a radius of curvature of magnitude 18.0 cm. The index of refraction of the glass is 1.44. (a) Calculate the focal length of the lens for light incident from the left. (b) **What If?** After the lens is turned around to interchange the radii of curvature of the two faces, calculate the focal length of the lens for light incident from the left.

Solution

Conceptualize: Since this is a biconvex lens, the center is thicker than the edges, and the lens will tend to converge incident light rays. Therefore, it has a positive focal length. Exchanging the radii of curvature amounts to turning the lens around so the light enters the opposite side first. However, this does not change the fact that the center of the lens is thicker than the edges, so we should not expect the focal length of the lens to be different.

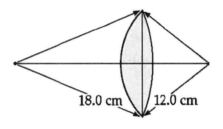
18.0 cm   12.0 cm

Categorize: The lens makers' equation can be used to find the focal length of this lens.

Analyze: The centers of curvature of the lens surfaces are on opposite sides, so the second surface has a negative radius:

(a) $\dfrac{1}{f} = (n-1)\left(\dfrac{1}{R_1} - \dfrac{1}{R_2}\right)$ gives

$$f = \frac{1}{(n-1)(1/R_1 - 1/R_2)}$$

$$= \frac{1}{(1.44-1)(1/(12.0 \text{ cm}) - 1/(-18.0 \text{ cm}))} = 16.4 \text{ cm}$$ ∎

(b) Similarly,

$$f = \frac{1}{(n-1)(1/R_1 - 1/R_2)}$$

$$= \frac{1}{(0.440)(1/(18.0 \text{ cm}) - 1/(-12.0 \text{ cm}))} = 16.4 \text{ cm}$$ ∎

Finalize: As expected, reversing the orientation of the lens does not change what it does to the light, as long as the lens is relatively thin (variations may be noticed with a thick lens). If you can find a converging lens that is flat on one side, set it up to make an image of a distant object on a card, and ask your lab partner or sibling what will happen when you flip the lens so that the other surface is facing the object. They may think that the image will turn to be right side up. An explanation based on how the lens alters an incoming flat wavefront may be best for them.

═══════════

44. A nearsighted person cannot see objects clearly beyond 25.0 cm (her far point). If she has no astigmatism and contact lenses are prescribed for her, what (a) power and (b) type of lens are required to correct her vision?

Solution

Conceptualize: We expect a strong (short-focal-length) diverging lens.

Categorize: We use the mirror-lens equation for a very distant object. If the patient can see this object clearly as an image at her far point and has normal accommodation, she will be able to see closer objects clearly as well.

Analyze: The lens should take parallel light rays from a very distant object ($p = \infty$) and make them diverge from a virtual image at the woman's far point, which is 25.0 cm beyond the lens, at $q = -25.0$ cm.

Thus, $\quad \dfrac{1}{f} = \dfrac{1}{p} + \dfrac{1}{q} = \dfrac{1}{\infty} + \dfrac{1}{-25.0 \text{ cm}}$.

(a) Hence, the power of the lens is $P = \dfrac{1}{f} = -\dfrac{1}{0.250 \text{ m}} = -4.00$ diopters.   ■

(b) With negative focal length, this is a diverging lens.   ■

Finalize: A goal of vision correction is always to provide a clear view of objects at infinity. A car down the highway and a player across the football field are so far away, compared to the size of the eye, that you receive essentially parallel rays from them. Thus they are well described as at infinity.

46. A patient has a near point of 45.0 cm and far point of 85.0 cm. (a) Can a single lens correct the patient's vision? Explain the patient's options. (b) Calculate the power lens needed to correct the near point so that the patient can see objects 25.0 cm away. Neglect the eye–lens distance. (c) Calculate the power lens needed to correct the patient's farpoint, again neglecting the eye–lens distance.

Solution

Conceptualize: The patient needs corrective action in both the near vision (to allow clear viewing of objects between 45.0 cm and the normal near point of 25.0 cm) and the distant vision (to allow clear viewing of objects more than 85.0 cm away).

Categorize: We use the mirror-lens equation for the patient's near point and for a very distant object.

Analyze:

(a) Yes, a single lens can correct the patient's vision. A single lens solution is for the patient to wear a bifocal or progressive lens. Alternately, the patient must purchase two pairs of glasses, one for reading, and one for distant vision.   ■

(b) To correct the near vision, the lens must form an upright, virtual image at the patient's near point ($q = -45.0$ cm) when a real object is at the normal near point ($p = +25.0$ cm). The thin-lens equation gives the needed focal length as

$$f = \frac{pq}{p+q} = \frac{(25.0 \text{ cm})(-45.0 \text{ cm})}{25.0 \text{ cm} - 45.0 \text{ cm}} = +56.3 \text{ cm}$$

so the required power in diopters is

$$P = \frac{1}{f_{\text{in meters}}} = \frac{1}{+0.563 \text{ m}} = +1.78 \text{ diopters} \qquad \blacksquare$$

(c) To correct the distant vision, the lens must form an upright, virtual image at the patient's far point ($q = -85.0$ cm) for the most distant objects ($p \rightarrow \infty$). The thin-lens equation gives the needed focal length as $f = q = -85.0$ cm, so the needed power is

$$P = \frac{1}{f_{\text{in meters}}} = \frac{1}{-0.850 \text{ m}} = -1.18 \text{ diopters} \qquad \blacksquare$$

Finalize: Until the invention of bifocal lenses by Benjamin Franklin, it would not have been possible for a patient to correct nearsightedness and farsightedness with a single lens.

---

49. A person is to be fitted with bifocals. She can see clearly when the object is between 30 cm and 1.5 m from the eye. (a) The upper portions of the bifocals (Fig. P26.49) should be designed to enable her to see distant objects clearly. What power should they have? (b) The lower portions of the bifocals should enable her to see objects located 25 cm in front of the eye. What power should they have?

Far vision

Near vision

Figure P26.49

Solution

Conceptualize: The upper portion of the lens should form an upright, virtual image of very distant objects $\left(p \approx \infty \right)$ at the far point of the eye ($q = -1.50$ m), and the lower part of the lens should form an upright, virtual image at the near point of the eye ($q = -30.0$ cm) when the object distance is $p = 25.0$ cm.

Categorize: We use the mirror-lens equation for a very distant object and for the patient's near point.

Analyze:

(a) With $p \approx \infty$ at the far point of the eye ($q = -1.50$ m), the thin-lens equation gives $f = q = -1.50$ m, so the needed power is

$$P = \frac{1}{f_{\text{in meters}}} = \frac{1}{-1.50 \text{ m}} = -0.667 \text{ diopters} \qquad \blacksquare$$

(b) From the thin-lens equation, with $p = 25.0$ cm and $q = -30.0$ cm,

$$f = \frac{pq}{p+q} = \frac{(25.0 \text{ cm})(-30.0 \text{ cm})}{25.0 \text{ cm} - 30.0 \text{ cm}} = +1.50 \times 10^2 \text{ cm} = +1.50 \text{ m}$$

Therefore, the power is $P = \dfrac{1}{f} = \dfrac{1}{+1.50 \text{ m}} = +0.667$ diopters. ∎

Finalize: Bifocals are slowly being replaced by new progressive spectacle lenses. In these lenses, lens power increases gradually from the top portion of the lens, which corrects for distant objects, to the bottom, which corrects for the near point.

---

54. In a darkened room, a burning candle is placed 1.50 m from a white wall. A lens is placed between the candle and the wall at a location that causes a larger, inverted image to form on the wall. When the lens is in this position, the object distance is p_1. When the lens is moved 90.0 cm toward the wall, another image of the candle is formed on the wall. From this information, we wish to find p_1 and the focal length of the lens. (a) From the lens equation for the first position of the lens, write an equation relating the focal length f of the lens to the object distance p_1, with no other variables in the equation. (b) From the lens equation for the second position of the lens, write another equation relating the focal length f of the lens to the object distance p_1. (c) Solve the equations in (a) and (b) simultaneously to find p_1. (d) Use the value in (c) to find the focal length f of the lens.

Solution

Conceptualize: Light rays are always reversible. We guess that the object distance for the second image is just the same as the image distance for the first and the image distance the same as the first object distance. This symmetry implies that when the lens has moved 45 of the 90 cm it must be halfway between candle and wall. The pair of distances must be 75 cm − 45 cm and 75 cm + 45 cm.

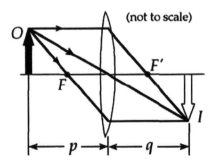

Categorize: The mirror-lens equation will be useful, together with the recognition that $p + q = 1.5$ m. We note that p_1 and q_1 must both be positive because the image must be real to be formed on a wall.

Analyze: The ray diagram is not to scale, but it shows correctly that originally and finally

$$p_1 + q_1 = 1.50 \text{ m} = p_2 + q_2$$

In the final situation, $p_2 = p_1 + 0.900$ m,

and $q_2 = q_1 - 0.900 \text{ m} = (1.50 \text{ m} - p_1) - 0.900 \text{ m} = 0.600 \text{ m} - p_1.$

Our lens equation is $\dfrac{1}{p_1} + \dfrac{1}{q_1} = \dfrac{1}{f} = \dfrac{1}{p_2} + \dfrac{1}{q_2}.$

(a) We read the first answer from the first equality sign,

$$\frac{1}{p_1} + \frac{1}{1.50 \text{ m} - p_1} = \frac{1}{f} \qquad\blacksquare$$

(b) and the second answer from the second,

$$\frac{1}{p_1 + 0.900 \text{ m}} + \frac{1}{0.600 \text{ m} - p_1} = \frac{1}{f} \qquad\blacksquare$$

(c) The two expressions for the lens power $1/f$ must be equal, and so we have an equation with only one unknown:

$$\frac{1}{p_1} + \frac{1}{1.50 \text{ m} - p_1} = \frac{1}{p_1 + 0.900} + \frac{1}{0.600 - p_1}$$

Adding the fractions, $\dfrac{1.50 \text{ m} - p_1 + p_1}{p_1(1.50 \text{ m} - p_1)} = \dfrac{0.600 - p_1 + p_1 + 0.900}{(p_1 + 0.900)(0.600 - p_1)}.$

Simplified, this becomes $p_1(1.50 \text{ m} - p_1) = (p_1 + 0.900)(0.600 - p_1).$

Thus, $p_1 = \dfrac{0.540}{1.80} \text{ m} = 0.300 \text{ m} \qquad\blacksquare$

and $p_2 = p_1 + 0.900 \text{ m} = 1.20 \text{ m}.$

(d) $\dfrac{1}{f} = \dfrac{1}{0.300\ \text{m}} + \dfrac{1}{1.50\ \text{m} - 0.300\ \text{m}}$

and $\qquad f = 0.240\ \text{m}.$ ∎

The second image is real, inverted, and diminished, with

$$M = -\dfrac{q_2}{p_2} = -\dfrac{0.600\ \text{m} - 0.300\ \text{m}}{1.20\ \text{m}} = -0.250 \qquad ∎$$

Finalize: Do this as a demonstration. Obtain any convenient converging lens. Measure its focal length by making an image of a distant object. Set up the light source in a darkened room, distant from the wall by more than four times the focal length. Form the first image on the wall. This arrangement is a model for a movie projector. Now move the lens toward the wall until you get the second image. It is like the first in being real and inverted, but different in being diminished. This arrangement models a camera or your eye.

In effect, the problem required solving the four simultaneous equations

$$p_1 + q_1 = 1.50\ \text{m}, \qquad p_2 + q_2 = 1.50\ \text{m}, \qquad p_2 = p_1 + 0.900\ \text{m}.$$

and $\quad \dfrac{1}{p_1} + \dfrac{1}{q_1} = \dfrac{1}{p_2} + \dfrac{1}{q_2} \quad$ for p_1. We did it by substitution. If someone tries to tell you to solve by typing it all into a calculator, or by determinants or by matrix inversion, do not believe them. If you wish you could believe them, try it and find out what is inconvenient and what doesn't work at all.

———

61. An object is placed 12.0 cm to the left of a diverging lens of focal length –6.00 cm. A converging lens of focal length 12.0 cm is placed a distance d to the right of the diverging lens. Find the distance d so that the final image is infinitely far away to the right.

Solution

Conceptualize: We must think separately about how the first lens affects the light and how the second lens then again changes how rays intersect. We expect the first lens to produce a virtual image less than 6 cm to the left of it. This should be at the focal point of the second lens so that the second lens can render the rays parallel. So we estimate 9 cm for distance d.

Categorize: We use the mirror-lens equation to locate the image formed by the first lens. This virtual image will be a real object for the second lens at a distance from it that we can represent in an expression involving d. Then the mirror-lens equation applied to the second lens will let us evaluate the object distance and so d.

Analyze: From the mirror-and-lens equation $1/p + 1/q = 1/f$

we have
$$q_1 = \frac{f_1 p_1}{p_1 - f_1} = \frac{(-6.00 \text{ cm})(12.0 \text{ cm})}{12.0 \text{ cm} - (-6.00 \text{ cm})} = -4.00 \text{ cm}.$$

The first lens forms an image 4.00 cm to its left. The rays between the lenses diverge from this image, so the second lens receives diverging light. It sees a real object at distance $p_2 = d - (-4.00 \text{ cm}) = d + 4.00 \text{ cm}$.

For the second lens, when we require that $q_2 \rightarrow \infty$,

The mirror-lens equation becomes $p_2 = f_2 = 12.0 \text{ cm}$.

Since the object for the converging lens must be 12.0 cm to its left, and since this object is the image for the diverging lens, which is 4.00 cm to **its** left, the two lenses must be separated by 8.00 cm.

Mathematically, $f_2 = 12.0 \text{ cm} = p_2 = d + 4.00 \text{ cm}$ and $d = 8.00 \text{ cm}$, ∎

We could draw separate ray diagrams for the two lenses, but we choose to combine them here.

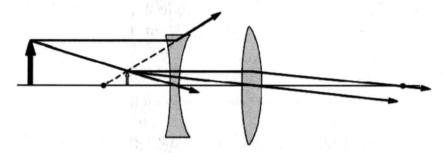

Finalize: We could say that no final image is formed, because the outgoing parallel rays do not intersect. Using the idea of an augmented number line including a point at infinity, we could say that the final image is at infinity on the right, real, enlarged, and inverted. The arrangement could be a useful invention in that it takes light from a fairly large source, concentrates it, and turns it into a fairly narrow searchlight beam. Looking into the outgoing rays, an observer will see a virtual image at infinity to the left, upright and magnified—this is an equally accurate description.

67. The disk of the Sun subtends an angle of 0.533° at the Earth. What are (a) the position and (b) the diameter of the solar image formed by a concave spherical mirror with a radius of curvature of magnitude 3.00 m?

Solution

Conceptualize: The Sun is so far away that the image will be very nearly at the focal point, 1.5 m in front of the collector. For the diameter we guess something small, perhaps a millimeter.

Categorize: We can take the distance to our star as known. The mirror-lens equation and the magnification equation $h'/h = -q/p$ will give us the answers, but we may need to watch our approximations.

Analyze: For the mirror, the focal length is $f = R/2 = +1.50$ m. In the mirror equation, because the distance to the Sun is so much larger than f and q, we can take $p = \infty$. Alternatively, we could look up the distance to the Sun and substitute it in. In either case, the mirror equation, $\dfrac{1}{p} + \dfrac{1}{q} = \dfrac{1}{f}$ then gives $q = f = 1.50$ m.   ∎

Now, in $M = -q/p = h'/h$, the magnification is nearly zero, but we can be more precise: the definition of radian measure means that h/p is the angular diameter of the object. Thus the image size is

$$h' = -\frac{hq}{p} = (-0.533°)\left(\frac{\pi}{180}\,\text{rad / deg}\right)(1.50\ \text{m}) = -0.014\ 0\ \text{m} = -1.40\ \text{cm}$$

The negative sign refers to the image being inverted. Anyone looking at the image would say its diameter is 1.40 cm.   ∎

Finalize: We could take $p = \infty$ in one equation but not in the other. In the magnification equation, we could have substituted in the diameter of the Sun in kilometers and its distance away, to obtain the same result. This method would in effect verify that the angular diameter of the Sun is 0.533°. The image diameter is bigger than we guessed. It is quite suitable for toasting a marshmallow, and this could be done quite quickly if the face diameter of the mirror is large. Especially for large mirror areas on sunny days, the arrangement described here is dangerous. It might set things on fire, produce burns, and very quickly produce eye damage. It is the face diameter of the mirror, its "objective diameter," that controls how much solar power is collected. Both the objective diameter and the radius of curvature affect the temperature attained at the image.

69. A parallel beam of light enters a glass hemisphere perpendicular to the flat face as shown in Figure P26.69. The magnitude of the radius of the hemisphere is $R = 6.00$ cm, and its index of refraction is $n = 1.560$. Assuming paraxial rays, determine the point at which the beam is focused.

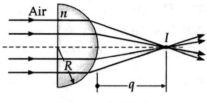

Figure P26.69

Solution

Conceptualize: "Focused" is the right word because the object is at infinity. We must calculate the image distance.

Categorize: A hemisphere is too thick to be described as a thin lens. The light is undeviated on entry into the flat face. Having noted this, we consider the light's exit from the second surface, for which $R = -6.00$ cm. We use the equation about image formation by a single refracting surface.

Analyze: The incident rays are parallel, so $p = \infty$. Then

$$\frac{n_1}{p} + \frac{n_2}{q} = \frac{n_2 - n_1}{R}$$

becomes $q = \dfrac{n_2}{(n_2 - n_1)/R - n_1/p} = \dfrac{1.00}{(1.00 - 1.560)/(-6.00 \text{ cm}) - 0} = 10.7 \text{ cm.}$ ∎

Finalize: The image is real, diminished, and inverted. You can tell it is inverted by noticing that the image point would move down in the picture if the object rays came in from the upper right. What if you did try to use the thin-lens model? You would compute a focal length for the hemisphere from

$$\frac{1}{f} = (n-1)\left(\frac{1}{R_1} - \frac{1}{R_2}\right)$$

giving $f = \dfrac{1}{(n-1)\left(1/R_1 - 1/R_2\right)} = \dfrac{1}{(1.560 - 1)\left(1/\infty - 1/(-6.00 \text{ cm})\right)}$

$$= \frac{1}{0.093 \ 3/\text{cm}} = 10.7 \text{ cm}$$

You would likely interpret this as a distance from the center of mass of the hemisphere, whereas really the focal point is 16.7 cm from the center of the flat face and 10.7 cm from the point where the axis passes through the curved face. Further, if the object were at a finite distance, the thin-lens approximation would give an unambiguously wrong answer.

Chapter 27
Wave Optics

Section 27.1   Conditions for Interference

In order to observe **sustained interference** in light waves, the following conditions must be met:

- The sources must be **coherent**; they must maintain a **constant phase** with respect to each other.

- The sources must be **monochromatic**—of a **single wavelength.**

Section 27.2   Young's Double-Slit Experiment

A schematic diagram illustrating the geometry used in Young's double-slit experiment is shown in the figure below. The two slits S_1 and S_2 serve as coherent monochromatic sources. When L (the distance from source plane to viewing screen) is much greater than d (the distance between sources), the **path difference** is $\delta = r_2 - r_1 = d \sin\theta$.

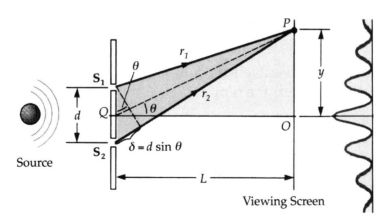

Young's double slit experiment. The relative intensity of light on the screen is shown on the right.

Section 27.4   Change of Phase Due to Reflection

Consider a light wave traveling in a medium with an index of refraction n_1. When partial reflection occurs at the surface of a medium with an index of refraction n_2:

- If $n_1 < n_2$, (top figure) the reflected ray experiences a phase change of 180°.

- If $n_1 > n_2$, (bottom figure) there is no phase change in the reflected ray.

- There is no phase change in the transmitted ray regardless of the relative values of n_1 and n_2.

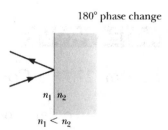

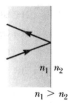

Section 27.5 Interference in Thin Films

In order to predict constructive or destructive interference in thin films you must consider:

- the difference in path-length traveled by the two interfering waves;

- any expected phase changes due to reflection;

- the change in wave length of the light as it enters the film.

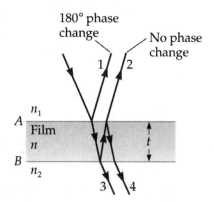

Constructive or destructive interference in a film depends on the thickness of the film t, the wavelength of light incident on the film, and the relative values of the indices of refraction of the film and the media above and below the film.

There are two general cases to consider (see figure on previous page):

(a) **Reflection resulting in a phase change at only one surface of the film.**

n_1 and n_2 both less than n (phase change at the top surface)
n_1 and n_2 both greater than n (phase change at the bottom surface)

Constructive interference will occur under these conditions, when the path difference (which equals $2t$) is an odd number of half wavelengths, $2t = (2m+1)\dfrac{\lambda_n}{2}$. Therefore the film thickness for constructive interference in this case is $t = \left(\dfrac{m+1}{2}\right)\dfrac{\lambda_n}{2}$. *Here,* $\lambda_n = \dfrac{\lambda}{n}$ *is the wavelength as measured in the film (the wavelength in vacuum divided by the index of refraction of the film).*

Destructive interference will occur under these conditions when the path difference, $2t$, equals an integer number of wavelengths so that $t = m\lambda\,/\,2n$.

(b) **Reflection resulting in phase changes at both top and bottom surfaces of the film**
or at neither surface.

$n_1 < n$ and $n_2 > n$ (phase change at both surfaces)
$n_1 > n$ and $n_2 < n$ (no phase change at either surface)

In these cases, two phase changes are offsetting and interference of the reflected rays depends only on the difference in distance traveled by the two reflected rays and the index of refraction of the film.

Constructive interference will occur when the path difference ($2t$) equals an integer number of wavelengths; the film thickness must be an integer number of half wavelengths, $t = \dfrac{m\lambda}{2n}$.

Destructive interference in this case will be observed when the path difference equals an odd number of half wavelengths; that is, when $t = \left(m + \tfrac{1}{2}\right)\dfrac{\lambda}{2n}$.

Section 27.6   Diffraction Patterns

Diffraction occurs when light waves deviate (or spread) from their initial direction of travel when passing through small openings, around obstacles, or by sharp edges.

The diffraction pattern produced by a single narrow slit consists of a broad, intense central band (the central maximum) flanked by a series of narrower and less intense secondary bands (called secondary maxima) alternating with a series of dark bands or minima.

In the case of single-slit diffraction, each portion of the slit acts as a source of waves; light from one portion of the slit can interfere with light from another portion. The resultant intensity on the screen depends on the angle θ. Fraunhofer diffraction occurs when the rays reaching the observing screen are approximately parallel.

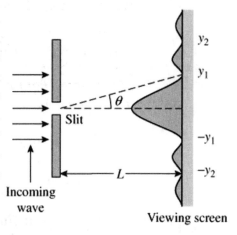

Single-slit diffraction pattern

Section 27.7   Resolution of Single-Slit and Circular Apertures

Rayleigh's criterion states the limiting condition for the resolution of the images due to nearby sources: *Two images are at the limit of resolution when the central maximum of one image falls on the first minimum of the other.* The dashed lines in the figure show the sum of the two diffraction patterns.

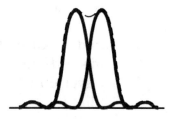

Diffraction patterns at the limit of resolution

Section 27.8   The Diffraction Gratting

A diffraction grating, consisting of many equally spaced parallel slits separated by a distance d, will produce a diffraction pattern. The figure illustrates the case for which the incident light contains a single wavelength component. If the incident light contains a *second wavelength component, a second series of principle maxima will be present in the diffraction pattern.*

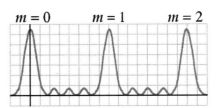

Intensity distribution for a grating. The central maximum and the 1st and 2nd maxima are shown.

EQUATIONS AND CONCEPTS

In **Young's double-slit experiment** two slits, S_1 and S_2 separated by a distance d, serve as monochromatic coherent sources. The light intensity at any point on the screen is the resultant of light reaching the screen from both slits. *Point P on the screen can be identified by the angle θ or by the distance y from the center of the screen.* When using $y = L \tan \theta$, y is measured from the center of the interference pattern and L is the distance from the double slit to the screen. Light from the two slits reaching any point on the screen (except the center) travels unequal path lengths. This difference in length of path (δ) is called the **path difference**.

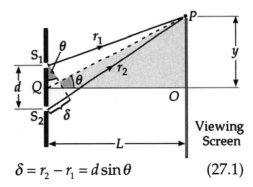

$$\delta = r_2 - r_1 = d \sin \theta \qquad (27.1)$$

$$\tan \theta = \frac{y}{L} \qquad (27.4)$$

Constructive interference (bright fringes) will appear at points on the screen for which the path difference is equal to an integral multiple of the wavelength. The positions of bright fringes can also be located by calculating their distance from the center of the screen (y). In each case, the number m is called the order number of the fringe. *The central bright fringe ($\theta = 0$, $m = 0$) is called the zeroth-order maximum.*

$$\delta = d \sin\theta_{\text{bright}} = m\lambda \qquad (27.2)$$
$$(m = 0, \ \pm 1, \ \pm 2, \ldots)$$

$$y_{\text{bright}} = L \tan\theta_{\text{bright}} \qquad (27.5)$$

$$y_{\text{bright}} = L\left(\frac{m\lambda}{d}\right) \text{ (small angles)} \qquad (27.7)$$

Destructive interference (dark fringes) will appear at points on the screen which correspond to path differences of an odd multiple of half wavelengths. For these points of destructive interference, waves which leave the two slits in phase arrive at the screen 180° out of phase.

$$\delta = d \sin\theta_{\text{dark}} = \left(m + \tfrac{1}{2}\right)\lambda \qquad (27.3)$$
$$(m = 0, \ \pm 1, \ \pm 2, \ldots)$$

$$y_{\text{dark}} = L \tan\theta_{\text{dark}} \qquad (27.6)$$

$$y_{\text{dark}} = L\frac{\left(m + \tfrac{1}{2}\right)\lambda}{d} \text{ (small angles)} \qquad (27.8)$$

The **phase difference** ϕ between the two waves at any point on the screen depends on the path difference at that point.

$$\phi = \frac{2\pi}{\lambda}\delta = \frac{2\pi}{\lambda}d \sin\theta \qquad (27.9)$$

The **average light intensity** (I) at any point P on the screen is proportional to the square of the amplitude of the resultant wave. The average intensity can be written:

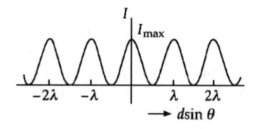

Light intensity from a double slit

• as a **function of phase difference** ϕ;

$$I_{\text{ave}} = I_{\text{max}} \cos^2\left(\frac{\phi}{2}\right)$$

• as a **function of the angle** (θ) subtended by the screen point at the source midpoint; or

$$I = I_{\text{max}} \cos^2\left(\frac{\pi d \sin\theta}{\lambda}\right) \qquad (27.10)$$

- as a **function of the distance** (y) from the center of the screen.

$$I_{ave} \cong I_{max}\cos^2\left(\frac{\pi d}{\lambda L}y\right) \quad \text{(small } \theta)$$

Increasing the number of equally spaced slits in a diffraction grating will increase the number of secondary maxima; the principal maxima will become narrower but remain fixed in position.

The sketches (I vs. sin θ) at right illustrate the effect of the number of slits in a grating on the resulting diffraction pattern. The dashed line shows the decrease in intensity to the left and right of the central maximum.

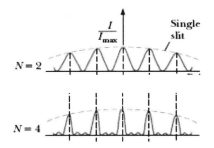

The diffraction pattern on the bottom was obtained using a grating having more slits per unit length than the one on the top.

The wavelength of light in a medium having an index of refraction n is smaller than the wavelength in vacuum. In thin-film interference the wavelength of light within the film is λ_n.

$$\lambda_n = \frac{\lambda}{n} \qquad (27.11)$$

Interference in thin films depends on wavelength, film thickness, and the indices of refraction of the film and surrounding media. *Differences in phase may be due to path difference or phase change upon reflection.* There are two general cases as described below.

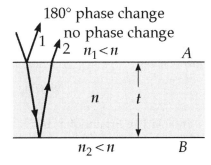

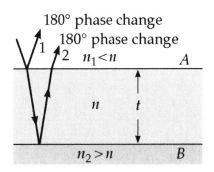

Case (I) Phase change at only one film surface

The indices of refraction of media on both sides of the film are less than that of the film ($n_1 < n$ and $n_2 < n$) as shown in figure above left; or the indices on both sides of the film are greater than that of the film ($n_1 > n$ and $n_2 > n$). This condition is not shown in the figure.

Constructive interference (Case I):

$$2nt = \left(m + \tfrac{1}{2}\right)\lambda \quad m = 0, 1, 2, \ldots$$

(27.13)

Destructive interference (Case I):

$$2nt = m\lambda \quad m = 0, 1, 2, \ldots \quad (24.14)$$

Case (II) Phase changes at either both surfaces or at neither surface

The film is between two media, one of which has an index of refraction greater than that of the film and the other a smaller index ($n_1 < n < n_2$) (as shown in the figure above on right) or (not shown), $n_1 > n > n_2$.

Constructive interference (Case II)
$$2nt = m\lambda \quad m = 0, 1, 2, \ldots$$

Destructive interference (Case II)
$$2nt = \left(m + \tfrac{1}{2}\right)\lambda \quad m = 0, 1, 2, \ldots$$

Note that the roles of the equations for Case (1) and Case (2) are reversed for constructive and destructive interference.

A **single-slit diffraction pattern** consists of a central maximum with a series of alternating bright (maxima) and dark (minima) bands on each side. The intensity at a given point on the screen depends on the angle between the direction to the central maximum and the direction to that point.

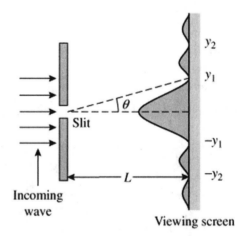

Intensity in Fraunhofer diffraction

The **general condition for destructive interference** in a single-slit diffraction pattern is stated in Equation 27.15.

$$\sin\theta_{\text{dark}} = m\frac{\lambda}{a} \qquad (27.15)$$
$$m = \pm 1, \pm 2, \pm 3, \ldots$$

Rayleigh's criterion states the condition for the resolution of two nearby sources.

For a **slit**, the angular separation between the sources must be greater than the ratio of the wavelength to slit width if the two sources are to be resolved.

$$\theta_{min} = \frac{\lambda}{a} \quad (27.16)$$

The limiting angle of resolution for a slit

For a **circular aperture**, the minimum angular separation depends on D, the diameter of the aperture (or lens).

$$\theta_{min} = 1.22\frac{\lambda}{D} \quad (27.17)$$

A **diffraction grating** (an array of a large number of parallel slits separated by a distance d) will produce an interference pattern in which there is a series of maxima for each wavelength. *Maxima due to wavelengths of different values comprise a spectral order denoted by order number m.* Each spectral order will contain a line characteristic of each wavelength.

$$d\sin\theta_{bright} = m\lambda \quad (27.18)$$
$$m = 0, \pm1, \pm2, \pm3, \ldots$$

Intensity vs. sin θ for a diffraction grating

Bragg's law gives the conditions for **constructive interference** of x-rays reflected from the parallel planes of a crystalline solid separated by a distance d.

$$2d\sin\theta = m\lambda \quad m = 0, 1, 2, 3, \ldots$$
$$(27.19)$$

The angle θ is the angle between the incident beam and the surface.

SUGGESTIONS, SKILLS, AND STRATEGIES

Thin-Film Interference Problems

- Identify the thin film from which interference effects are being observed.

- The type of interference that occurs in a specific problem is determined by the phase relationship between that portion of the wave reflected at the upper surface of the film and that portion reflected at the lower surface of the film.

- Phase differences between the two portions of the wave occur because of differences in the distances traveled by the two portions and by phase changes occurring upon reflection.

Phase Difference due to Path Difference

The wave reflected from the lower surface of the film has to travel a distance equal to twice the thickness of the film before it returns to the upper surface of the film where it interferes with that portion of the wave reflected at the upper surface.

Phase Change due to Reflection

When a wave traveling in a particular medium reflects off a surface having a higher index of refraction than the one it is in, a 180° phase shift occurs. This has the same effect as if the wave had traveled a lesser (or greater) distance of $\frac{1}{2}\lambda$. This effect must be considered in addition to the actual greater path length traveled by one of the waves. The **effective path length** is the combination of the difference distance of travel and any phase changes.

Constructive interference will occur when the *effective path difference* is zero or an integral multiple of λ: $0, \lambda, 2\lambda, 3\lambda, \ldots$

Destructive interference will occur when the *effective path difference* is an odd number of half wavelengths: $\frac{1}{2}\lambda, \frac{3}{2}\lambda, \frac{5}{2}\lambda, \ldots$

REVIEW CHECKLIST

- Describe Young's double-slit experiment to demonstrate the wave nature of light. Account for the phase difference between light waves from the two sources as they arrive at a given point on the screen. State the conditions for constructive and destructive interference in terms of each of the following: path difference, phase difference, distance from the center of the screen, and angle subtended by the observation point at the source mid-point.

- Account for the conditions of constructive and destructive interference in thin films considering both path difference and any expected phase changes due to reflection.

- Determine the positions of the minima in a single-slit diffraction pattern. Determine the positions of the principal maxima in the interference pattern of a diffraction grating.

- Determine whether or not two sources under a given set of conditions are resolvable as defined by Rayleigh's criterion.

ANSWERS TO SELECTED OBJECTIVE QUESTIONS

1. Consider a wave passing through a single slit. What happens to the width of the central maximum of its diffraction pattern as the slit is made half as wide? (a) It becomes one-fourth as wide. (b) It becomes one-half as wide. (c) Its width does not change. (d) It becomes twice as wide. (e) It becomes four times as wide.

Answer The best answer is (d). Equation 27.15 describes the angles at which you get destructive interference; from it, we can obtain the width of the central maximum. For small angles, the equation can be rewritten as

$$\theta_m = \sin^{-1}(m\lambda/a) \approx m\lambda/a$$

Thus, as the width of the slit a is cut in half, the angle of the first destructive interference θ_1 doubles, and the width of the central maximum doubles as well.

□   □   □   □

5. A plane monochromatic light wave is incident on a double slit as illustrated in Active Figure 27.2. **(i)** As the viewing screen is moved away from the double slit, what happens to the separation between the interference fringes on the screen? (a) It increases. (b) It decreases. (c) It remains the same. (d) It may increase or decrease, depending on the wavelength of the light. (e) More information is required. **(ii)** As the slit separation increases, what happens to the separation between the interference fringes on the screen? Select from the same choices.

Answer **(i)** (a) The equations $d \sin\theta_1 = 1\lambda$ and $y_1 = L \tan\theta_1$ describe the location of the first bright fringe to the side of the central bright fringe. As the distance L from the slits to the screen increases, θ_1 is constant and the distance y_1 on the screen gets larger in direct proportion to L. **(ii)** (b) In the same equations think of d getting larger. Then θ_1 gets smaller and all of the fringes get closer together.

□   □   □   □

ANSWERS TO SELECTED CONCEPTUAL QUESTIONS

2. What is the necessary condition on the path length difference between two waves that interfere (a) constructively and (b) destructively?

Answer (a) Two waves interfere constructively if their path difference is either zero or some integral multiple of the wavelength, that is, if the path difference equals $m\lambda$. (b) Two waves interfere destructively if their path difference is an odd multiple of one-half of a wavelength, that is, if the path difference equals

$$\left(m+\frac{1}{2}\right)\lambda.$$

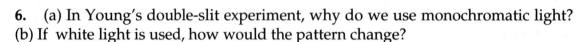

6. (a) In Young's double-slit experiment, why do we use monochromatic light? (b) If white light is used, how would the pattern change?

Answer (a) Suppose we use a mixture of colors. Then each color produces its own pattern, with a spacing between the maxima that is characteristic of the wavelength. With several colors, the patterns are superimposed, and it can become difficult to pick out a single maximum. Using monochromatic light eliminates this problem.

(b) With white light, the central maximum is white. The first side maximum is a full spectrum, with violet on the inside and red on the outside. The second side maximum is a full spectrum also, but red in the second maximum overlaps the violet in the third maximum. At larger angles, the light soon starts mixing to white again, but it may be so faint that you would call it gray.

8. Why can you hear around corners, but not see around corners?

Answer Audible sound has wavelengths on the order of meters or centimeters, while visible light has wavelengths on the order of half a micron. In this world of breadbox-size objects, λ is comparable to the object size a for sound, and sound diffracts around walls and through doorways. But λ/a is much smaller for visible light passing ordinary-size objects or apertures, so light diffracts only through very small angles.

Another way of answering this question would be as follows. We can see by a small angle around a small obstacle or around the edge of a small opening. The side fringes in Figure 27.10 and the Arago spot in the center of Figure 27.11 (in the textbook) show this diffraction. Conversely, we cannot always hear around corners. Out-of-doors, away from reflecting surfaces, have someone a few meters distant face away from you and whisper. The high-frequency, short-wavelength, information-carrying components of the sound do not diffract around his head enough for you to understand his words.

Suppose an opera singer loses the tempo and cannot immediately get it from the orchestra conductor. Then the prompter may make rhythmic kissing noises with her lips and teeth. Try it—you will sound like a birdwatcher trying to lure out a curious bird. This sound is clear on the stage but does not diffract around the prompter's box enough for the audience to notice it.

SOLUTIONS TO SELECTED END-OF-CHAPTER PROBLEMS

1. Young's double-slit experiment is performed with 589-nm light and a distance of 2.00 m between the slits and the screen. The tenth interference minimum is observed 7.26 mm from the central maximum. Determine the spacing of the slits.

Solution

Conceptualize: For the situation described, the observed interference pattern is very narrow (the minima are less than 1 mm apart when the screen is 2 m away). In fact, the minima and maxima are so close together that it would probably be difficult to resolve adjacent maxima, so the pattern might look like a uniform blur to the naked eye. Since the angular spacing of the pattern is inversely proportional to the slit width, we should expect that for this narrow pattern, the space between the slits will be larger than the typical fraction of a millimeter, and certainly much greater than the wavelength of the light ($d \gg \lambda = 589$ nm).

Categorize: Since we are given the location of the tenth minimum for this interference pattern, we should use the equation for **destructive interference** from a double slit. The figure for Problem 13 shows the critical variables for this problem.

Analyze: In the equation $d \sin \theta = \left[m + \dfrac{1}{2} \right] \lambda$, the first minimum is described by $m = 0$ and the tenth by $m = 9$. So $d \sin \theta = \lambda \left[9 + \dfrac{1}{2} \right]$.

Also, $\tan \theta = y/L$, but for small θ, $\sin \theta \approx \tan \theta$. Thus, the distance between the slits is

$$d = \frac{9.5\lambda}{\sin\theta} = \frac{9.5\lambda L}{y} = \frac{9.5(589 \times 10^{-9} \text{ m})(2.00 \text{ m})}{7.26 \times 10^{-3} \text{ m}}$$

$$= 1.54 \times 10^{-3} \text{ m} = 1.54 \text{ mm} \qquad \blacksquare$$

Finalize: The spacing between the slits is relatively large, as we expected (about 3 000 times greater than the wavelength of the light). In order to more clearly distinguish between maxima and minima, the pattern could be expanded by increasing the distance to the screen. However, as L is increased, the overall pattern would be less bright as the light expands over a larger area, so that beyond some distance, the light would be too dim to see.

3. Two radio antennas separated by $d = 300$ m as shown in Figure P27.3 simultaneously broadcast identical signals at the same wavelength. A car travels due north along a straight line at position $x = 1\,000$ m from the center point between the antennas, and its radio receives the signals. (a) If the car is at the position of the second maximum after that at point O when it has traveled a distance $y = 400$ m northward, what is the wavelength of the signals? (b) How much farther must the car travel to encounter the next minimum in reception? *Note:* Do not use the small-angle approximation in this problem.

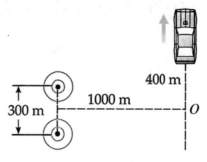

Figure P27.3

Solution

Conceptualize: The car is farther from the south transmitter than from the north transmitter, by two wavelengths. We estimate around 100 m for the wavelength. If the small-angle approximation were good, we would have maxima at 0, at 200 m, at 400 m, and at 600 m along the northward road. The next minimum would be at 500 m, which is 100 m of extra travel. Probably the accurate answer is somewhat more than 100 m.

Categorize: The problem suggests that with the conditions given, the small-angle approximation does not work well. That is, $\sin\theta$, $\tan\theta$, and θ are significantly different. We use the Fraunhofer interference model, treating the waves from the two sources as moving along essentially parallel rays.

Analyze:

(a) At the $m = 2$ maximum, from $\tan\theta = y/L$ we have

$$\theta = \tan^{-1}\frac{y}{L} = \tan^{-1}\frac{400 \text{ m}}{1\,000 \text{ m}} = 21.8°$$

so $\quad d\sin\theta = m\lambda$

gives $\quad \lambda = \dfrac{d\,\sin\theta}{m} = \dfrac{(300 \text{ m})(\sin 21.8°)}{2} = 55.7$ m. ∎

(b) The next minimum encountered is the $m = 2$ minimum, which is the third minimum away from the central maximum.

In that direction $\quad d\sin\theta = \left[m + \tfrac{1}{2}\right]\lambda: \quad d\sin\theta = \dfrac{5}{2}\lambda,$

so $\quad \theta = \sin^{-1}\left(\dfrac{5\lambda}{2d}\right) = \sin^{-1}\left[\dfrac{5(55.7 \text{ m})}{2(300 \text{ m})}\right] = \sin^{-1}(0.464) = 27.7°,$

and $\quad y = L\tan\theta = (1\,000 \text{ m})\tan 27.7° = 524$ m.

Therefore, the car must travel an additional 124 m. ■

Finalize: Our estimates were good in order-of-magnitude terms. If we considered Fresnel interference, we would more precisely find

(a) $\lambda = \dfrac{1}{2}\left(\sqrt{550^2 + 1\,000^2} - \sqrt{250^2 + 1\,000^2}\right) = 55.2$ m

(b) $\Delta y = 123$ m

Note well that interference of waves from side-by-side sources works in the same way for all kinds of waves. Observing how signals from side-by-side sources add together can be thought of as a definitive test for whether the signals are waves or classical particles.

---

7. In Figure P27.7 (not to scale), let $L = 1.20$ m and $d = 0.120$ mm and assume the slit system is illuminated with monochromatic 500-nm light. Calculate the phase difference between the two wave fronts arriving at P when (a) $\theta = 0.500°$ and (b) $y = 5.00$ mm. (c) What is the value of θ for which the phase difference is 0.333 rad? (d) What is the value of θ for which the path difference is $\lambda/4$?

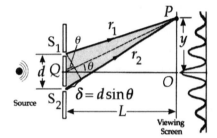

Figure P27.7 (modified)

Solution

Conceptualize: In parts (a) and (b), the phase differences will likely be between 0 and 4π. In parts (c) and (d), the angles will likely be between 0 and 2°. After doing some calculations, we will be able to sort the four cases into order as describing points P from the one closest to O to the farthest.

Categorize: Instead of the usual $d\sin\theta = m\lambda$, we will use more basic equations describing Young's experiment, from which $d\sin\theta = m\lambda$ is derived.

Analyze:

(a) The path difference is

$$\delta = d\sin\theta = (0.120 \times 10^{-3}\ \text{m})(\sin 0.500°) = 1.05 \times 10^{-6}\ \text{m}$$

The phase difference is $\quad \phi = \dfrac{2\pi\delta}{\lambda} = \dfrac{2\pi\left(1.05 \times 10^{-6}\ \text{m}\right)}{500 \times 10^{-9}\ \text{m}} = 13.2$ rad. ■

This produces the same effect as $\phi = 13.159 - 4\pi = 0.593$ rad $= 34.0°$. ■

(b) $\quad \tan\theta = \dfrac{y}{L} = \dfrac{5.00 \times 10^{-3}\ \text{m}}{1.20\ \text{m}} \approx \sin\theta$

$$\phi = \frac{2\pi d \sin\theta}{\lambda} = \frac{2\pi(0.120 \times 10^{-3} \text{ m})(4.17 \times 10^{-3})}{500 \times 10^{-9} \text{ m}} = 2\pi \text{ rad} = 0 \qquad \blacksquare$$

(c) $\phi = \dfrac{2\pi d \sin\theta}{\lambda}$: $\theta = \sin^{-1}\left[\dfrac{\lambda\phi}{2\pi d}\right] = \sin^{-1}\dfrac{(500 \times 10^{-9} \text{ m})(0.333)}{2\pi(1.20 \times 10^{-4} \text{ m})} = 0.012\,7° \qquad \blacksquare$

(d) $\dfrac{\lambda}{4} = d\sin\theta$: $\theta = \sin^{-1}\left[\dfrac{\lambda}{4d}\right] = \sin^{-1}\left[\dfrac{(500 \times 10^{-9} \text{ m})}{4(1.20 \times 10^{-4} \text{ m})}\right] = 0.059\,7° \qquad \blacksquare$

Finalize: In order of distance away from point O, the points considered in this problem are c, d, b, and a. Points c and d are between the zero-order maximum and the first minimum. Point b is at the first side maximum, and point a is a bit beyond the second side maximum.

---

13. A pair of narrow, parallel slits separated by 0.250 mm are illuminated by green light ($\lambda = 546.1$ nm). The interference pattern is observed on a screen 1.20 m away from the plane of the parallel slits. Calculate the distance (a) from the central maximum to the first bright region on either side of the central maximum and (b) between the first and second dark bands in the interference pattern.

Solution

Conceptualize: The spacing between adjacent maxima and minima should be fairly uniform across the pattern as long as the width of the pattern is much less than the distance to the screen (so that the small angle approximation is valid). The separation between fringes should be at least a millimeter if the pattern can be easily observed with a naked eye.

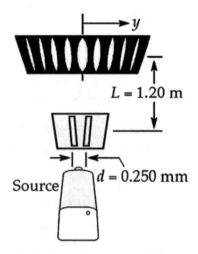

Categorize: The bright regions are areas of constructive interference and the dark bands are destructive interference, so the corresponding double-slit equations will be used to find the y distances.

It can be confusing to keep track of four different symbols for distances. Three are shown in the drawing. Note that:

d is the distance between the adjacent wave sources, here the slits.

L is the distance from the sources to the screen where interference fringes are observed.

y is the unknown distance from the bright central maximum ($m = 0$) to another maximum or minimum on either side of the center of the interference pattern.

λ is the wavelength of the light, determined by the source. The interference maxima are *not* wave crests. The stationary bright fringes are separated by a distance y of a few millimeters while the wave crests are individually unobservable, always moving, and separated by the distance $\lambda = 546.1$ nm.

Analyze:

(a) For any size θ we have $\tan\theta = y/L$.

For **very small** θ we have $\sin\theta \approx \tan\theta$ and the equation for constructive interference $d\sin\theta = m\lambda$ becomes $dy/L \approx m\lambda$ or $y_{\text{bright}} \approx (\lambda L/d)m$.

Substituting values, $y_{\text{bright}} = \dfrac{(546.1\times10^{-9}\text{ m})(1.20\text{ m})}{0.250\times10^{-3}\text{ m}}(1) = 2.62$ mm.   ∎

(b) If you have trouble remembering whether the equation with $m\lambda$ or the equation with $\left(m+\dfrac{1}{2}\right)\lambda$ applies to a particular situation, you can remember that a zero-order bright band is in the center, and dark bands are halfway between bright bands. Thus, the made-up equation $d\sin\theta = (count)\lambda$ describes them all, with $count = 0, 1, 2, \ldots$ for bright bands, and with $count = 0.5, 1.5, 2.5, \ldots$ for dark bands.

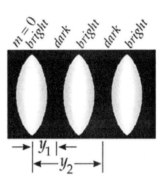

The dark band version of the equation $y_{\text{bright}} \approx (\lambda L/d)m$ is simply

$$y_{\text{dark}} = \frac{\lambda L}{d}\left[m+\frac{1}{2}\right]$$

Then the distance requested in the problem is

$$\Delta y_{\text{dark}} = \frac{\lambda L}{d}\left[1+\frac{1}{2}\right] - \frac{\lambda L}{d}\left[0+\frac{1}{2}\right] = \frac{\lambda L}{d} = 2.62 \text{ mm} \qquad ∎$$

Finalize: This spacing is large enough for easy resolution of adjacent fringes. The distance between minima is the same as the distance between maxima. We expected this equality since the angles are small:

$$\theta = (2.62 \text{ mm})/(1.20 \text{ m}) = 0.002\ 18 \text{ rad} = 0.125°$$

Only when the angular spacing exceeds about $3°$ does $\sin\theta$ differ from $\tan\theta$ when they are written to three significant figures.

22. An oil film (n = 1.45) floating on water is illuminated by white light at normal incidence. The film is 280 nm thick. Find (a) the wavelength and color of the light in the visible spectrum most strongly reflected and (b) the wavelength and color of the light in the spectrum most strongly transmitted. Explain your reasoning.

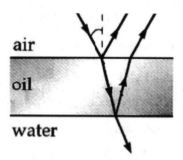

Solution

Conceptualize: We cannot guess the answers. The answers to (a) and (b) need not be colors that an artist would think of as opposites, but their wavelengths should be related by a simple fraction like 5/4. Light of one color must go through a half-integer number of oscillations in the same geometrical space that light of the other color goes through an integer number of oscillations.

Categorize: For thin-film interference, there are too many cases and variations to memorize equations for all of them. We will puzzle it through, thinking of all of the phase shifts of light reflected from the bottom of the film versus light reflected from the top. Table 24.1 in the textbook lists names for colors of various wavelengths.

Analyze: The light reflected from the top of the oil film undergoes phase reversal. Since 1.45 > 1.33, the light reflected from the bottom undergoes no reversal. For constructive interference of reflected light, we then have

$$2nt = \left[m + \frac{1}{2} \right] \lambda \quad \text{or} \quad \lambda_m = \frac{2nt}{m + \dfrac{1}{2}} = \frac{2(1.45)(280 \text{ nm})}{m + \dfrac{1}{2}}$$

(a) Substituting for m, we have

$$m = 0: \lambda_0 = 1\,624 \text{ nm (infrared)}$$

$$m = 1: \lambda_1 = 541 \text{ nm (green)} \qquad \blacksquare$$

$$m = 2: \lambda_2 = 325 \text{ nm (ultraviolet)}$$

Both infrared and ultraviolet light are invisible to the human eye, so the dominant color is green. ∎

(b) Any light that is not reflected is transmitted. To find the color transmitted most strongly, we find the wavelengths reflected least strongly. According to the condition for destructive interference, $2nt = m\lambda$.

Therefore, $\lambda = \dfrac{2nt}{m} = \dfrac{2(1.45)(280 \text{ nm})}{m}$.

For $m = 1$, $\lambda = 812$ nm (near infrared);

$m = 2$, $\lambda = 406$ nm (violet); ∎

$m = 3$, $\lambda = 271$ nm (ultraviolet).

Thus, violet is the only visible color not attenuated by reflection, and the dominant color in the transmitted light. ∎

Finalize: The ratio of the strongly and weakly reflected wavelengths is 541 nm/406 nm = 1.33 = 4/3, a ratio of integers as predicted. In case you did not recognize it, $2nt$ is the *optical path length* down and back through the film. Because we include the index of refraction here, every wavelength we write down is a wavelength in vacuum.

---

25. An air wedge is formed between two glass plates separated at one edge by a very fine wire as shown in Figure P27.25. When the wedge is illuminated from above by 600-nm light and viewed from above, 30 dark fringes are observed. Calculate the diameter d of the wire.

Figure P27.25

Solution

Conceptualize: The diameter of the wire is probably less than 0.1 mm since it is described as a "very fine wire."

Categorize: Light reflecting from the bottom surface of the top plate undergoes no phase shift, while light reflecting from the top surface of the bottom plate is shifted by π, and also has to travel an extra distance $2t$, where t is the local thickness of the air wedge. Then …

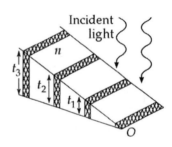

Analyze: … for destructive interference,

$$2t = m\lambda \quad m = 0, 1, 2, 3, \ldots$$

The first dark fringe appears where $m = 0$ at the line of contact between the plates. The 30th dark fringe gives for the diameter of the wire $2d = 29\lambda$, and $d = 14.5\lambda$. The diameter of the wire is then

$$d = 14.5\lambda = 14.5\left(600 \times 10^{-9}\text{m}\right) = 8.70 \ \mu\text{m} \qquad ∎$$

Finalize: This wire is not only less than 0.1 mm thick; it is much thinner than a typical human hair (~50 μm).

---

29. A screen is placed 50.0 cm from a single slit, which is illuminated with light of wavelength 690 nm. If the distance between the first and third minima in the diffraction pattern is 3.00 mm, what is the width of the slit?

Solution

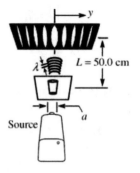

Conceptualize: We estimate between 0.1 mm and 1 mm. The width of the slit will be much greater than the wavelength of the light, but small enough to reveal that light is not a stream of classical particles.

Categorize: The textbook's analysis of single-slit diffraction will give us the answer. We will have to apply it both to the first and to the third minimum.

Analyze: In the equation for single-slit diffraction minima at small angles,

$$\frac{y}{L} \approx \sin\theta_{\text{dark}} = \frac{m\lambda}{a}$$

we take differences between the first and third dark fringes, to see that

$$\frac{\Delta y}{L} = \frac{\Delta m\lambda}{a} \quad \text{with } \Delta y = 3.00 \times 10^{-3} \text{ m} \quad \text{and} \quad \Delta m = 3 - 1 = 2.$$

The width of the slit is then $\quad a = \dfrac{\lambda L \Delta m}{\Delta y} = \dfrac{\left(690 \times 10^{-9} \text{ m}\right)(0.500 \text{ m})(2)}{3.00 \times 10^{-3} \text{ m}}.$

$$a = 2.30 \times 10^{-4} \text{ m} \qquad\qquad\blacksquare$$

Finalize: It is 0.230 mm; our estimate was good. It is the light that goes through the opening unimpeded that diffracts. We are not studying reflection from the edges of the slit. In sharp contrast to Newton's first law, the light does not travel just in a straight line. Diffraction can be thought of as an aspect of the basic dynamic of wave motion.

====

35. A helium–neon laser emits light that has a wavelength of 632.8 nm. The circular aperture through which the beam emerges has a diameter of 0.500 cm. Estimate the diameter of the beam 10.0 km from the laser.

Solution

Conceptualize: A typical laser pointer makes a spot about 5 cm in diameter at 100 m, so the spot size at 10 km would be about 100 times bigger, or about 5 m across. Assuming that this HeNe laser is similar, we could expect a comparable beam diameter.

Categorize: We assume that the light is parallel and not diverging before it passes through and fills the circular aperture. However, after the light passes through the circular aperture, it will spread from diffraction according to Equation 27.17.

Analyze: The beam spreads into a cone of half-angle

$$\theta_{min} = 1.22\frac{\lambda}{D} = 1.22\left(\frac{632.8\times10^{-9}\text{ m}}{0.005\,00\text{ m}}\right) = 1.54\times10^{-4}\text{ rad}$$

The radius of the beam ten kilometers away is, from the definition of radian measure,

$$r_{beam} = \theta_{min}\left(1.00\times10^{4}\text{ m}\right) = 1.54\text{ m}$$

and its diameter is $d_{beam} = 2r_{beam} = 3.09$ m. ∎

Finalize: The beam is several meters across as expected, and is about 600 times larger than the laser aperture. Since most HeNe lasers are low power units in the mW range, the beam at this range would be so spread out that it would be too dim to see on a screen.

---

40. White light is spread out into its spectral components by a diffraction grating. If the grating has 2 000 grooves per centimeter, at what angle does red light of wavelength 640 nm appear in first order?

Solution

Conceptualize: We expect an angle on the order of ten degrees, easily measurable on a tabletop and very precisely measurable with a spectrometer.

Categorize: We use the grating equation. Note that it is identical in form to the equation describing constructive interference with two side-by-side sources.

Analyze: The ruling engine that cut the diffraction grating (or the aluminum plate from which the gelatin or plastic was cast) sliced each centimeter into two thousand divisions. So the grating spacing is

$$d = \frac{1.00\times10^{-2}\text{ m}}{2\,000} = 5.00\times10^{-6}\text{ m}$$

The light is deflected according to $d\sin\theta = m\lambda$:

$$\theta = \sin^{-1}\frac{m\lambda}{d} = \sin^{-1}\left[\frac{1(640\times10^{-9}\text{m})}{5.00\times10^{-6}\text{m}}\right] = \sin^{-1}0.128 = 7.35°$$ ∎

Finalize: As shown in Example 27.6, the grating spacing is the reciprocal of the number of grooves per unit width. Just think about a pie being cut into eighths for an eight-person family. A spotlight or the beam from a slide projector shining through a diffraction grating, onto a classroom projection screen, makes a beautiful display. An individual student can see the beauty of the visible spectrum by using a spectrometer with an ordinary light bulb.

---

42. The hydrogen spectrum includes a red line at 656 nm and a blue-violet line at 434 nm. What are the angular separations between these two spectral lines for all visible orders obtained with a diffraction grating that has 4 500 grooves/cm?

Solution

Conceptualize: Many diffraction gratings yield several spectral orders within the viewing range of 90° to either side. So the angle between red and blue lines is probably 10° to 30°. (We call the color of the 434-nm line blue, but note that it is not the blue-green line of hydrogen at 486.1 nm.)

Categorize: The angular separation is the difference between the angles corresponding to the red and blue wavelengths for each visible spectral order according to the diffraction grating equation, $d \sin\theta = m\lambda$.

Analyze: The grating spacing is

$$d = 1.00 \times 10^{-2} \text{ m/4 500 grooves} = 2.22 \times 10^{-6} \text{ m}$$

In the first-order spectrum ($m = 1$), the angles of diffraction are given by $\sin\theta = \lambda/d$. We have

$$\theta_{1r} = \sin^{-1}\left(\frac{\lambda_r}{d}\right) = \sin^{-1}\left(\frac{656 \times 10^{-9} \text{ m}}{2.22 \times 10^{-6} \text{ m}}\right) = \sin^{-1}(0.295) = 17.17°$$

and for the light in the blue spectral line

$$\theta_{1b} = \sin^{-1}\left(\frac{\lambda_b}{d}\right) = \sin^{-1}\left(\frac{434 \times 10^{-9} \text{ m}}{2.22 \times 10^{-6} \text{ m}}\right) = \sin^{-1}(0.195) = 11.26°$$

The angular separation is $\Delta\theta = \theta_{1r} - \theta_{1b} = 17.17° - 11.26° = 5.91°$.   ■

In the second order of interference, with $m = 2$, similarly,

$$\Delta\theta_2 = \sin^{-1}\left(\frac{2\lambda_r}{d}\right) - \sin^{-1}\left(\frac{2\lambda_b}{d}\right) = 13.2°$$ ■

In the third order ($m = 3$), $\Delta\theta_3 = \sin^{-1}\left(\frac{3\lambda_r}{d}\right) - \sin^{-1}\left(\frac{3\lambda_b}{d}\right) = 26.5°$.   ■

Examining the fourth order, we find the red line is not visible:

$$\theta_{4r} = \sin^{-1}\left(\frac{4\lambda_r}{d}\right) = \sin^{-1}(1.18)$$

does not exist, so the answer is complete with the angular separations in the first three orders.

Finalize: The full spectrum is visible in the first 3 orders with this diffraction grating, and the fourth is partially visible. We can also see that the pattern is dispersed more for higher spectral orders so that the angular separation between the red and blue lines increases as m increases. It is also worth noting that the spectra of different orders can overlap, as the graphical display indicates. For example, the red line in second order is at a larger angle than the blue line in

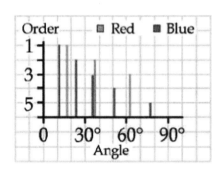

third order. This effect can make the pattern look confusing if you do not know what you are looking for

---

44. A grating with 250 grooves/mm is used with an incandescent light source. Assume the visible spectrum to range in wavelength from 400 nm to 700 nm. In how many orders can one see (a) the entire visible spectrum and (b) the short-wavelength region of the visible spectrum?

Solution

Conceptualize: Compare this problem with Problem 42. With 250 grooves per millimeter instead of 450 grooves per millimeter, the grating is coarser and the angles for constructive interference will be smaller. The red will be visible (if it is bright enough) in more than three orders and the blue in more than five orders.

Categorize: We use the grating equation, this time thinking with some care about the order of interference m as the unknown.

Analyze: The grating spacing is $d = \dfrac{1.00 \text{ mm}}{250} = 4.00 \times 10^{-6}$ m.

(a) In each order of interference m, red light diffracts at a larger angle than the other colors with shorter wavelengths. We find the largest integer m satisfying

$$d \sin\theta = m\lambda \quad \text{with} \quad \lambda = 700 \text{ nm}.$$

With $\sin\theta$ having its largest possible value, we have

$$m = \frac{d \sin\theta}{\lambda} = \frac{(4.00 \times 10^{-6}\text{m}) \sin 90°}{700 \times 10^{-9}\text{m}} = 5.71$$

Thus, the red light cannot be seen in the sixth order, and the full visible spectrum appears in only five orders.   ∎

(b) Now consider light at the boundary between violet and ultraviolet.

$$m = \frac{d \sin\theta}{\lambda} = \frac{(4.00 \times 10^{-6} \text{ m}) \sin 90°}{400 \times 10^{-9} \text{ m}} = 10.0$$

and $\qquad$ $m = 10$. $\qquad$ ∎

Finalize: Seeing more orders is not an advantage. It is natural to want a grating with closely spaced grooves so that the angles to be measured are nice and large for good precision. But note that the grating only works if the distance between grooves is larger than the wavelength.

"Order of interference" has a physical meaning: When you are looking at a particular color in third order, you are just three wavelengths farther from one grating groove than from the adjacent groove.

---

48. If the spacing between planes of atoms in an NaCl crystal is 0.281 nm, what is the predicted angle at which 0.140-nm x-rays are diffracted in a first-order maximum?

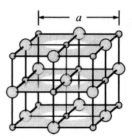

Solution

Conceptualize: X-rays have very short wavelength compared to visible light. It is convenient that a crystal forms a natural diffraction grating to prove that an x-ray moves as a wave, and to make its wavelength measurable. We expect an angle on the order of ten degrees, easily measurable on a divided circle.

Figure 27.25

Categorize: The atomic planes in this crystal are shown in Figure 27.25 of the text. The diffraction they produce is described by Bragg's law, …

Analyze: … which is $\qquad$ $2d \sin\theta = m\lambda$.

Solving for the angle gives

$$\theta = \sin^{-1} \frac{m\lambda}{2d} = \sin^{-1}\left[\frac{1(0.140 \times 10^{-9} \text{ m})}{2(0.281 \times 10^{-9} \text{ m})}\right] = \sin^{-1}(0.249) = 14.4° \qquad ∎$$

Finalize: Note that this is a grazing angle, an angle between the incident beam and the surface. The angle of incidence is 90°–14.4°.

In the same way that regularly stacked atoms in a crystal diffract x-rays according to Bragg's law, soldiers standing in ranks and files on a parade ground can diffract sound. On a bed of nails, regularly spaced nails can diffract microwaves. Think up some more examples yourself.

---

58. The condition for constructive interference by reflection from a thin film in air as developed in Section 27.5 assumes nearly normal incidence. **What If?** Suppose the light is incident on the film at a nonzero angle θ_1 (relative to the normal). The index of refraction of the film is n and the film is surrounded by vacuum. Find the condition for constructive interference that relates the thickness t of the film, the index of refraction n of the film, the wavelength λ of the light, and the angle of incidence θ_1.

Solution

Conceptualize: We need to analyze the textbook's Figure 27.8 without the simplifying assumption of normal incidence.

Categorize: We draw the diagram as shown. The ray reflected from the top of the film undergoes phase reversal. The ray reflected from the bottom surface has no phase change on reflection. We must figure out geometrically the extra distance it travels relative to the upper ray.

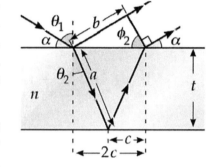

Analyze: Because refraction out of the film is the opposite of refraction into the film, the ray reflected from the bottom surface leaves the film parallel to the ray from the top surface. Then, since the angles marked α are equal, $\phi_2 + \alpha + 90° = 180°$.

From the angle of incidence, $\phi_2 + \alpha = 90°$,

so $\phi_2 = \theta_1$.

Before they head off together along parallel paths, one beam travels a distance b and undergoes phase reversal on reflection; the other beam travels a distance $2a$ inside the film, where its wavelength is λ/n. The number of cycles it completes along this path is

$$\frac{2a}{\lambda/n} = \frac{2na}{\lambda}$$

The optical path length in the film is $2na$. Then the shift between the two outgoing rays is

$$\delta = 2na - b - \frac{\lambda}{2}$$

where a and b are as shown in the ray diagram, n is the index of refraction, and the term $\lambda/2$ is due to phase reversal at the top surface. For constructive interference, $\delta = m\lambda$, where m has integer values.

This condition becomes $2na - b = \left[m + \dfrac{1}{2}\right]\lambda.$ **[1]**

From the figure's geometry, $a = \dfrac{t}{\cos\theta_2}$ and $c = a\sin\theta_2 = \dfrac{t\sin\theta_2}{\cos\theta_2}.$

Further, $b = 2c\sin\theta_1 = \dfrac{2t\sin\theta_2}{\cos\theta_2}\left(\sin\theta_1\right).$

Also, from Snell's law, $\sin\theta_1 = n\sin\theta_2.$

so $b = \dfrac{2nt\sin^2\theta_2}{\cos\theta_2}.$

With these results, the condition for constructive interference given in equation [1] becomes

$$2n\left(\dfrac{t}{\cos\theta_2}\right) - \dfrac{2nt\sin^2\theta_2}{\cos\theta_2} = \left(\dfrac{2nt}{\cos\theta_2}\right)\left(1 - \sin^2\theta_2\right) = \left[m + \dfrac{1}{2}\right]\lambda$$

or $\qquad 2nt\cos\theta_2 = \left[m + \dfrac{1}{2}\right]\lambda.$

Finally, we apply the law of refraction at entry, $\sin\theta_1 = n\sin\theta_2,$ as

$$2nt\cos\theta_2 = 2nt\sqrt{1 - \sin^2\theta_2} = 2nt\sqrt{1 - \left(\sin^2\theta_1\right)/n^2}$$

so the reflected light will show constructive interference provided that

$$2nt\sqrt{1 - \left(\sin^2\theta_1\right)/n^2} = \left(m + \dfrac{1}{2}\right)\lambda, \qquad \text{where } m = 0,\ 1,\ 2,\ldots \qquad \blacksquare$$

Finalize: If the light comes in perpendicular to the film, $\theta_1 = 0$ and the equation we proved reduces to $2nt\left(m + \dfrac{1}{2}\right)\lambda,$ in agreement with Equation 27.13. As you turn the film, the color you see at one spot changes. The equation describes this as λ for constructive interference changing when θ_1 changes. This iridescence effect, characteristic of interference colors, is very different from the way pigment colors behave. Abalone shells show beautiful iridescence of thin sheets.

---

65. Light of wavelength 500 nm is incident normally on a diffraction grating. If the third-order maximum of the diffraction pattern is observed at 32.0°, (a) what is the number of rulings per centimeter for the grating? (b) Determine the total number of primary maxima that can be observed in this situation.

Solution

Conceptualize: The diffraction pattern described in this problem seems to be similar to previous problems about diffraction gratings with 2 000 to 5 000 grooves/cm. With the third-order maximum at 32°, there are probably 5 or 6 maxima on each side of the central bright fringe, for a total of 11 or 13 primary maxima.

Categorize: The diffraction grating equation can be used to find the grating spacing and the angles of the other maxima that should be visible within the viewing range of 90° on both sides of the central maximum.

Analyze:

(a) We use the grating equation $d \tan\theta = m\lambda$:

$$d = \frac{m\lambda}{\sin\theta} = \frac{3\left(5.00\times10^{-7}\ \text{m}\right)}{\sin 32.0°} = 2.83\times10^{-6}\ \text{m}$$

Thus the grating gauge is $\dfrac{1}{d} = 3.53\times10^{5}$ grooves/m = 3 530 grooves/cm. ■

(b) For any interference maximum for this light going through this grating,

$$\sin\theta = m\left(\frac{\lambda}{d}\right) = \frac{m\left(5.00\times10^{-7}\ \text{m}\right)}{2.83\times10^{-6}\ \text{m}} = m(0.177)$$

For $\sin\theta \le 1$, we require that $m(0.177) \le 1$ or $m \le 5.65$. Because m must be an integer, its maximum value is really 5. Therefore, the total number of maxima is $2m + 1 = 11$. ■

Finalize: The results agree with our predictions, and there are 5 maxima on either side of the central maximum. If more maxima were desired, a coarser grating with **fewer** grooves/cm would be required; however, this could impede our ability to resolve the angles between exiting light beams that appear close together.

Chapter 28
Quantum Physics

Section 28.1 Blackbody Radiation and Planck's Theory

A **black body** is an ideal body that absorbs all radiation incident on it. Any body at some temperature T emits thermal radiation, which is characterized by the properties of the body and its temperature. The spectral distribution of blackbody radiation at various temperatures is shown in the figure. As the temperature increases, the total power of the emitted radiation (area under the curve) increases, while the peak of the distribution shifts to shorter wavelengths.

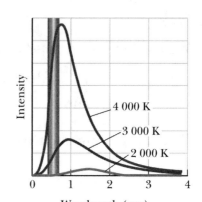

Intensity of blackbody radiation versus wavelength at three temperatures

Classical theories failed to explain blackbody radiation. A new theory, proposed by Max Planck, is consistent with this distribution at all wavelengths. Planck made two basic assumptions in the development of this result:

- The oscillators emitting the radiation can only have **discrete energies** given by $E_n = nhf$, where n is a quantum number ($n = 1, 2, 3, ...$), f is the oscillator frequency, and h is Planck's constant.

- These oscillators can emit or absorb energy in discrete units called **quanta** (or photons), where the energy of a light quantum obeys the relation $E = hf$.

Subsequent developments showed that the quantum concept was necessary in order to explain several phenomena at the atomic level, including the photoelectric effect, the Compton effect, and atomic spectra.

Section 28.2   The Photoelectric Effect

When light is incident on certain metallic surfaces, electrons can be emitted from the surfaces. This is called the photoelectric effect, discovered by Hertz. Several features of the photoelectric effect can not be explained with classical physics or with the wave theory of light. In 1905, Einstein provided a successful explanation of the photoelectric effect by extending Planck's quantum concept to include electromagnetic fields.

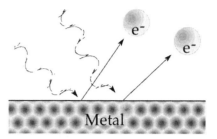

When photons with a frequency beyond the cutoff frequency are incident on certain surfaces photoelectrons are emitted.

Observed features of the photoelectric effect can be explained and understood on the basis of the photon theory of light. These features include:

- **Photoelectrons have a maximum kinetic energy that is independent of light intensity** (number of photons incident per unit area of the photo surface per second). This is confirmed experimentally by showing that the photocurrent becomes zero at the same stopping potential for different light intensities.

- **Photoelectrons are emitted almost instantaneously** by photons with a frequency above the cutoff frequency (a threshold or minimum value characteristic of a particular material). This instantaneous emission occurs even at very low light intensity.

- **No photoelectrons are emitted when the frequency of incident light is below the cutoff frequency** characteristic of the material being illuminated. This is true regardless of the degree of light intensity.

- **The maximum kinetic energy of photoelectrons increases with increasing light frequency.** This is seen by examining the photoelectric equation (Equation 28.11) which states that $K_{max} = hf - \phi$. The work function (ϕ) represents the minimum energy with which an electron is bound in the surface of a metal. The cutoff frequency is $f_c = \phi/h$.

Section 28.3   The Compton Effect

The Compton effect involves the scattering of x-rays by electrons. In the quantum model illustrated in the figure an x-ray photon collides with an electron and is scattered at an angle θ relative to the incident direction. The scattered photon experiences an increase in wavelength ($\Delta\lambda$) called the Compton shift.

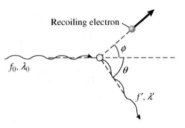

The quantum model for x-ray scattering from an electron

Section 28.4   Photons and Electromagnetic Waves

The results of some experiments are better described based on the particle nature of light; other experimental outcomes are better described in terms of the wave properties of light. The photoelectric effect and the Compton effect show light as being composed of photons with "particle-like" energy and momentum; interference and diffraction are explained on the basis of light acting as a wave The experimental evidence regarding the nature of light cannot be described by a single model; the particle model and the wave model of light are complementary.

Section 28.5   The Wave Properties of Particles

De Broglie postulated that a particle in motion has wave properties and a corresponding wavelength inversely proportional to the particle's momentum. The wave nature of particles was confirmed in an experiment in which the wavelength of electrons was measured after scattering from a crystal.

Section 28.8   The Uncertainty Principle

It is physically impossible to simultaneously measure the exact position and exact momentum of a particle. The uncertainties in position and momentum are due to the quantum structure of matter. Another form of the uncertainty principle places a similar limitation on the simultaneous measurement of energy and time.

Section 28.9   An Interpretation of Quantum Mechanics

The wave function is a complex-valued quantity. The square of the absolute value of the wave function gives the probability per volume of finding a particle at a given point in the volume. The wave function contains all the information that can be known about the particle. The **Schrödinger equation** describes the manner in which matter waves change in time and space. The average experimental value of a quantity such as position or energy is called the **expectation value** of the quantity.

Section 28.10   A Particle in a Box
Section 28.12   The Schrödinger Equation

The time independent Schrödinger equation can be applied to the case of a particle of constant energy E confined to a one-dimensional region of space (a "particle in a box") where the potential U is zero inside the box and infinite outside. The equation can be solved to find the wave functions and the corresponding probability densities as shown in the figure below.

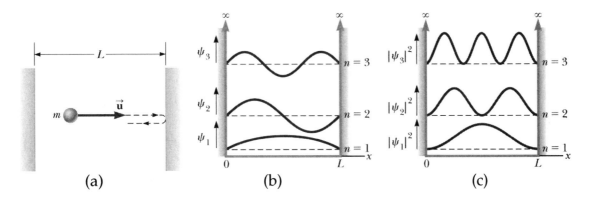

(a) Infinite square well (b) Wave functions (c) Probability densities for a particle in a box.

Section 28.13   Tunneling Through a Potential Energy Barrier

When a particle is incident onto a potential energy barrier, the height of which is greater than the energy of the particle, there is a finite probability that the particle will penetrate the barrier. In this process, called **tunneling**, part of the incident wave is transmitted and part is reflected.

EQUATIONS AND CONCEPTS

The **distribution of black body radiation** varies with temperature and wavelength.

$$P = \sigma A e T^4 \qquad (28.1)$$

$$\lambda_{max} T = 2.898 \times 10^{-3} \text{ m} \cdot \text{K} \qquad (28.2)$$

σ = the Stefan-Boltzan constant

e = emissivity of the surface

Stefan's law: the total power of emitted radiation increases with temperature.

Wien displacement law: the peak wavelength of the distribution shifts to shorter wavelengths as the temperature increases. *The total radiation emitted is the area under the temperature vs. wavelength curve.*

Discrete energy values of an atomic oscillator are determined by a quantum number, n. Each discrete energy value corresponds to a quantum state.

$$E_n = nhf \qquad n = 1, 2, 3, \ldots \quad (28.4)$$

$$h = 6.626 \times 10^{-34} \text{ J} \cdot \text{s} \qquad (28.7)$$

The **energy of a quantum** or photon (a discrete unit of energy) corresponds to the energy difference between initial and final quantum states. *An oscillator emits or absorbs energy only when there is a transition between quantum states.*

$$E = hf \qquad (28.5)$$

The **maximum kinetic energy of a photoelectron** depends on the frequency of the incident light and the work function of the metal, ϕ, which is typically a few eV. This model is in excellent agreement with experimental results.

$$K_{max} = hf - \phi \qquad (28.11)$$

The **cutoff wavelength** (and corresponding frequency) depends on the value of the work function of a specific surface; for wavelengths greater than λ_c no photoelectric effect will be observed. *The work function represents the minimum energy with which an electron is bound in a metal.*

$$\lambda_c = \frac{hc}{\phi} \qquad (28.12)$$

$$hc = 1\,240 \text{ eV} \cdot \text{nm}$$

The **Compton shift** is the change in wavelength of an x-ray when scattered from an electron. *The scattered x-ray makes an angle θ with the direction of the incident x-ray.* The figure at right represents the data for Compton scattering of x-rays from graphite at $\theta = 90.0°$.

$\lambda_C = h / m_e c = 0.002\ 43$ nm is called the Compton wavelength.

$$\lambda' - \lambda_0 = \frac{h}{m_e c}(1 - \cos\theta) \qquad (28.13)$$

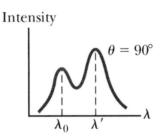

Scattered x-ray intensity versus wavelength. The wavelength of the incident beam is λ_0 and the scattered beam has wavelength λ'.

The **de Broglie wavelength** of a particle is inversely proportional to the momentum of the particle; and a particle has a **frequency** proportional to its energy. These statements reflect the dual nature of matter.

$$\lambda = \frac{h}{p} = \frac{h}{mu} \qquad (28.15)$$

$$f = \frac{E}{h} \qquad (28.16)$$

The **wave function of a free particle** moving along the *x*-axis is a sinusoidal wave. The wave representing the particle has a constant amplitude, A.

$$\psi(x) = Ae^{ikx} \qquad (28.26)$$

$k = (2\pi / \lambda)$ is the angular wave number
A = constant amplitude

The **probability**, P_{ab}, **of finding a particle within an arbitrary interval** $a \le x \le b$ equals the area under the curve of $|\psi|^2$ vs. *x* between the end points of the interval.

$$P_{ab} = \int_a^b |\psi|^2 dx \qquad (28.29)$$

$|\psi|^2$ is the relative probability of finding the particle at a given point along the interval.

The **normalization condition** is a statement of the requirement that the particle exists at some point (along the *x* axis in the one-dimensional case) at all times.

$$\int_{-\infty}^{\infty} |\psi|^2 dx = 1 \qquad (28.28)$$

The **expectation value for the position** is the expected average value of the coordinate. *This is the average of the values of the position if calculated from the known wave function. To find the expectation value of any function f(x), replace x in Equation 28.30 with f(x).*

$$\langle x \rangle \equiv \int_{-\infty}^{\infty} \psi^* x \psi \, dx \qquad (28.30)$$

$\psi^* = $ complex conjugate of ψ

The **Heisenberg uncertainty principle** can be stated in two forms involving uncertainties in the simultaneous measurement of:
(1) position and momentum and (2) energy and time.

$$\Delta x \Delta p_x \geq \frac{\hbar}{2} \qquad (28.22)$$

$$\Delta E \Delta t \geq \frac{\hbar}{2} \qquad (28.23)$$

For a **particle-in-a-box of width L** in one dimensional motion:
- The **wave function** can be represented as a sinusoidal function.

$$\psi(x) = A \sin\left(\frac{n\pi x}{L}\right) \quad n = 1, 2, 3, \ldots$$
$$(28.34)$$

- The **energy of the particle** is quantized. The ground state energy corresponds to $n = 1$ and is the lowest energy state.

$$E_n = \left(\frac{h^2}{8mL^2}\right) n^2 \quad n = 1, 2, 3, \ldots$$
$$(28.35)$$

The **time-independent Schrödinger equation** for a particle confined to moving along the x axis (with constant total energy E) allows in principle the determination of the wave functions and energies of the allowed states if the potential energy function, U, is known.

$$\frac{-\hbar^2}{2m}\frac{d^2\psi}{dx^2} + U\psi = E\psi \qquad (28.36)$$

The **normalized wave function** for a particle confined to a one-dimensional box of width L is an example of applying the Schrödinger equation to a specific problem.

$$\psi_n(x) = \sqrt{\frac{2}{L}} \sin\left(\frac{n\pi x}{L}\right) \qquad (28.41)$$

REVIEW CHECKLIST

- Describe the formula for blackbody radiation proposed by Planck, the assumption made in formulating this formula, and the related experimental results.

- Describe the Einstein model for the photoelectric effect and the important experimental results. Make calculations using the photoelectric equation.

- Describe the Compton effect (the scattering of x-rays by electrons) and be able to use the formula for the Compton shift. Recognize that the Compton effect can only be explained using the photon concept.

- Discuss the wave properties of particles, the de Broglie wavelength concept, and the dual nature of both matter and light. Make calculations using the de Broglie wavelength equation.

- Describe the concept of wave function for the representation of matter waves and state in equation form the normalization condition and expectation value of the coordinate.

- Make calculations using both forms of the Heisenberg uncertainty principle.

ANSWERS TO SELECTED OBJECTIVE QUESTIONS

4. The probability of finding a certain quantum particle in the section of the x axis between $x = 4$ nm and $x = 7$ nm is 48%. The particle's wave function $\psi(x)$ is constant over this range. What numerical value can be attributed to $\psi(x)$, in units of $\text{nm}^{-1/2}$? (a) 0.48 (b) 0.16 (c) 0.12 (d) 0.69 (e) 0.40

Answer The square of the wave function at point x gives the probability per unit length of finding the particle at point x. The average value of $|\psi|^2$ multiplied by the width between two points gives the probability of finding the particle in that range. For ψ with a constant value, we have $\psi^2 (3 \text{ nm}) = 0.48$ for the probability of finding the particle in the range stated. Then solving for ψ gives $\psi = \sqrt{0.16 / \text{nm}} = 0.4 / \sqrt{\text{nm}}$. This is answer (e).

□   □   □   □

7. In a certain experiment, a filament in an evacuated lightbulb carries a current I_1 and you measure the spectrum of light emitted by the filament, which behaves as a black body at temperature T_1. The wavelength emitted with highest intensity (symbolized by λ_{max}) has the value λ_1. You then increase the potential difference across the filament by a factor of 8, and the current increases by a factor of 2. **(i)** After this change, what is the new value of the temperature of the filament? (a) $16T_1$ (b) $8T_1$ (c) $4T_1$ (d) $2T_1$ (e) still T_1 **(ii)** What is the new value of the wavelength emitted with highest intensity? (a) $4\lambda_1$ (b) $2\lambda_1$ (c) λ_1 (d) $\frac{1}{2}\lambda_1$ (e) $\frac{1}{4}\lambda_1$

Answer The temperature T_1 is no doubt well above room temperature, and the temperature T_2 with higher voltage will be higher still. So we might expect the resistance of the filament to increase, and that is why the current may increase only by a factor of 2 as the potential difference becomes 8 times larger. Taking the product, we find the electric power $P = I\Delta V$ delivered to the filament is 16 times higher at the final setting. With a steady-state temperature in a vacuum, the filament must radiate all of this power as electromagnetic waves, according to $P = 1A\sigma T^4$. Thermal expansion in surface area has a negligible effect, so the final temperature is then given by $P_2/P_1 = (T_2/T_1)^4$, so $T_2 = T_1(P_2/P_1)^{1/4} = T_1(16)^{1/4} = 2T_1$. For question **(i)** this is answer (d). Now Wien's displacement law, that the wavelength of strongest emission is inversely proportional to the absolute temperature, says that **(ii)** the new wavelength where the spectrum intensity peaks is $\lambda_1/2$, answer (d).

□ □ □ □

12. An x-ray photon is scattered by an originally stationary electron. Relative to the frequency of the incident photon, is the frequency of the scattered photon (a) lower, (b) higher, or (c) unchanged?

Answer (a) The x-ray photon transfers some of its energy to the electron. Thus, its energy, and therefore its frequency, must be decreased.

□ □ □ □

ANSWERS TO SELECTED CONCEPTUAL QUESTIONS

5. Discuss the relationship between ground-state energy and the uncertainty principle.

Answer Consider a particle bound to a restricted region of space. If its minimum energy were zero, then the particle could have zero momentum and zero uncertainty in its momentum. At the same time, the uncertainty in its position would not be infinite, but equal to the width of the region. In such a case, the uncertainty product $\Delta x \Delta p_x$ would be zero, violating the uncertainty principle. This contradiction proves that the minimum energy of the particle is not zero.

□   □   □   □

7. If the photoelectric effect is observed for one metal, can you conclude that the effect will also be observed for another metal under the same conditions? Explain.

Answer No. Suppose that the incident light frequency at which you first observed the photoelectric effect is above the cutoff frequency of the first metal, but less than the cutoff frequency of the second metal. In that case, the photoelectric effect would not be observed at all in the second metal.

□   □   □   □

9. Why does the existence of a cutoff frequency in the photoelectric effect favor a particle theory for light over a wave theory?

Answer A theory modeling light as a classical wave would predict that the photoelectric effect should occur at any frequency, provided that the light intensity is high enough. As is implied by the question, this is in contradiction to experimental results. Light with a frequency below the cutoff cannot knock electrons out of a particular target at all.

□   □   □   □

19. If matter has a wave nature, why is this wave-like characteristic not observable in our daily experiences?

Answer For any object that we can perceive directly, the de Broglie wavelength $\lambda = h/mu$ is too small to be measured by any means; therefore, no wavelike characteristics can be observed. The object will not diffract noticeably when it goes through an aperture. It will not show resolvable interference maxima and minima when it goes through two openings. It will not show resolvable nodes and antinodes if it is in resonance.

□   □   □   □

SOLUTIONS TO SELECTED END-OF-CHAPTER PROBLEMS

4. **(i)** Calculate the energy, in electron volts, of a photon whose frequency is (a) 620 THz, (b) 3.10 GHz, and (c) 46.0 MHz. **(ii)** Determine the corresponding wavelengths for the photons listed in part (i) and **(iii)** state the classification of each on the electromagnetic spectrum.

Solution

Conceptualize: From part (a) through (b) to (c) the frequency of the electromagnetic wave under discussion goes down by on the order of ten million times. So the photon energy, starting from a few electronvolts, will go down by ~10^7 times to a fraction of a microelectronvolt. The wavelength will go up by this factor, from something small compared to your height to something large.

Categorize: We use just Planck's equation for the energy of a photon and the relationship between wavelength and frequency for any continuous wave. 1 eV = 1.60×10^{-19} J is a unit of energy we can use for anything if we want to.

Analyze: **(i)** Planck's equation is $E = hf$. The photon energies are

(a) $E = (6.63 \times 10^{-34} \text{ J} \cdot \text{s})(6.20 \times 10^{14} \text{ Hz}) = 4.11 \times 10^{-19} \text{ J} = 2.57 \text{ eV}$ ∎

(b) $E = (6.63 \times 10^{-34} \text{ J} \cdot \text{s})(3.10 \times 10^{9} \text{ Hz}) = 2.06 \times 10^{-24} \text{ J} = 12.8 \text{ } \mu\text{eV}$ ∎

(c) $E = (6.63 \times 10^{-34} \text{ J} \cdot \text{s})(46.0 \times 10^{6} \text{ Hz}) = 3.05 \times 10^{-26} \text{ J} = 1.91 \times 10^{-7} \text{ eV}$ ∎

(ii)   The wavelengths are

(a) $\lambda = \dfrac{c}{f} = \dfrac{3.00 \times 10^{8} \text{ m/s}}{6.20 \times 10^{14} \text{ s}^{-1}} = 4.84 \times 10^{-7} \text{ m} = 484 \text{ nm}$ ∎

(b) $\lambda = \dfrac{c}{f} = \dfrac{3.00 \times 10^{8} \text{ m/s}}{3.10 \times 10^{9} \text{ s}^{-1}} = 0.096 \text{ 8 m} = 9.68 \text{ cm}$ ∎

(c) $\lambda = \dfrac{c}{f} = \dfrac{3.00 \times 10^{8} \text{ m/s}}{46.0 \times 10^{6} \text{ s}^{-1}} = 6.52 \text{ m}$ ∎

(iii)   These wavelengths correspond to (a) blue light, (b) microwave radiation, and (c) radio waves in the public LO band.   ∎

Finalize: Make sure you never confuse a photon with a proton. A photon is a quantum particle with zero mass, a parcel of electromagnetic radiation that cannot be subdivided in absorption or emission. It carries energy like a classical particle, with the amount given by $E = hf$; and it moves as a continuous wave, described by $\lambda = v/f$, just as classical waves do. It is neither a classical particle nor a classical wave, but a quantum particle.

---

15. Two light sources are used in a photoelectric experiment to determine the work function for a particular metal surface. When green light from a mercury lamp ($\lambda = 546.1$ nm) is used, a stopping potential of 0.376 V reduces the photocurrent to zero. (a) Based on this measurement, what is the work function for this metal? (b) What stopping potential would be observed when using the yellow light from a helium discharge tube ($\lambda = 587.5$ nm)?

Solution

Conceptualize: According to Table 28.1, the work function for most metals is on the order of a few eV, so this metal is probably similar. We can expect the stopping potential for the yellow light to be slightly lower than 0.376 V since the yellow light has a longer wavelength (lower frequency) and therefore less energy per photon than the green light.

Categorize: In this photoelectric experiment, the green light has sufficient energy hf to overcome the work function of the metal ϕ so that the ejected electrons have a maximum kinetic energy of 0.376 eV. With this information, we can use the photoelectric effect equation to find the work function, which can then be used to find the stopping potential for the less energetic yellow light.

Analyze:

(a) Einstein's photoelectric effect equation is $K_{max} = hf - \phi$ and the energy required to raise an electron through a 1-V potential is 1 eV, so that

$$K_{max} = e\Delta V_s = 0.376 \text{ eV}$$

The energy of a photon from the mercury lamp is:

$$hf = \frac{hc}{\lambda} = \frac{(6.626 \times 10^{-34} \text{ J} \cdot \text{s})(2.998 \times 10^8 \text{ m/s})}{546.1 \times 10^{-9} \text{ m}} \left(\frac{1 \text{ eV}}{1.602 \times 10^{-19} \text{ J}} \right)$$

$$= \frac{1\ 240 \text{ eV} \cdot \text{nm}}{546.1 \text{ nm}} = 2.27 \text{ eV}$$

Therefore, the work function for this metal is:

$$\phi = hf - K_{max} = 2.27 \text{ eV} - 0.376 \text{ eV} = 1.89 \text{ eV} \qquad \blacksquare$$

(b) For the yellow light, λ = 587.5 nm and the photon energy is

$$hf = \frac{hc}{\lambda} = \frac{1\ 240 \text{ eV} \cdot \text{nm}}{587.5 \times 10^{-9} \text{ m}} = 2.11 \text{ eV}$$

Therefore the maximum energy that can be given to an ejected electron is

$$K_{max} = hf - \phi = 2.11 \text{ eV} - 1.89 \text{ eV} = 0.216 \text{ eV}$$

so the stopping voltage is $\quad \Delta V_s = 0.216 \text{ V}.$ $\qquad \blacksquare$

Finalize: The work function for this metal is lower than we expected and does not correspond with any of the values in Table 28.1. Further examination in the *CRC Handbook of Chemistry and Physics* (online at http://www.hbcpnetbase.com/) reveals that all of the metal elements have work functions between 2 and 6 eV. However, a single metal's work function may vary by about 1 eV depending on impurities in the metal, so it is just barely possible that a metal might have a work function of 1.89 eV.

The stopping potential for the yellow light is indeed lower than for the green light as we expected. An interesting calculation is to find the wavelength for the lowest energy light that will eject electrons from this metal. That threshold wavelength for $K_{max} = 0$ is 654 nm, which is red light in the visible portion of the electromagnetic spectrum.

==========

21. A 0.001 60-nm photon scatters from a free electron. For what (photon) scattering angle does the recoiling electron have kinetic energy equal to the energy of the scattered photon?

Solution

Conceptualize: Scattering angles can lie between 0° and 180°. If photon and electron end up with equal energies in this Compton scattering process, the photon must give just half its original energy to the originally stationary electron, and the photon's wavelength must double. It is not clear that this is allowed ...

Categorize: ... by Compton's equation but, if it is, that equation will relate the change in wavelength to the scattering angle of the x-ray photon.

Analyze: The energy of the incoming photon is

$$E_0 = \frac{hc}{\lambda_0} = \frac{\left(6.626 \times 10^{-34} \text{ J} \cdot \text{s}\right)\left(3.00 \times 10^8 \text{ m/s}\right)}{0.001\ 60 \times 10^{-9} \text{ m}} = 1.24 \times 10^{-13} \text{ J}$$

The outgoing photon and the electron share equally in this energy. The kinetic energy of the electron and the energy of the scattered photon are each one-half of E_0.

$$E' = 6.22 \times 10^{-14} \text{ J} \quad \text{and} \quad \lambda' = \frac{hc}{E'} = 3.20 \times 10^{-12} \text{ m}$$

The shift in wavelength is $\Delta\lambda = \lambda' - \lambda_0 = 1.60 \times 10^{-12}$ m.

But by Equation 28.13, $\Delta\lambda = \lambda_C(1 - \cos\theta)$, where λ_C is the Compton wavelength.

Then, $\cos\theta = 1 - \dfrac{\Delta\lambda}{\lambda_C}$ implies

$$\theta = \cos^{-1}\left(1 - \frac{\Delta\lambda}{\lambda_C}\right) = \cos^{-1}\left(1 - \frac{1.60 \times 10^{-12} \text{ m}}{2.43 \times 10^{-12} \text{ m}}\right) = \cos^{-1}(0.342) = 70.0° \quad \blacksquare$$

Finalize: Compton's equation says that the smallest possible change in photon wavelength is zero, occurring if the photon barely grazes the electron and its scattering angle is zero. The largest possible change in wavelength is $2\lambda_C = 4.86$ pm, if the photon recoils straight back from a head-on collision. So the 1.60-pm wavelength change in this problem is allowed.

==========

23. Review. A helium–neon laser produces a beam of diameter 1.75 mm, delivering 2.00×10^{18} photons/s. Each photon has a wavelength of 633 nm. Calculate the amplitudes of (a) the electric field and (b) the magnetic field inside the beam. (c) If the beam shines perpendicularly onto a perfectly reflecting surface, what force does it exert on the surface? (d) If the beam is absorbed by a block of ice at 0°C for 1.50 h, what mass of ice is melted?

Solution

Conceptualize: We are reviewing several properties of light. It is a wave of electric field and magnetic field; we expect the amplitudes for this bright light to be several newtons per coulomb and a small fraction of a tesla. It carries momentum, and will exert a tiny force on the mirror. It carries energy and will melt a bit of ice, perhaps a few grams.

Categorize: The photon stream is the light beam. We will find its intensity. We have studied how the intensity is related to both of the field amplitudes, to the radiation pressure, and to the energy transported in a certain time interval.

Analyze: The energy of one photon is

$$E = \frac{hc}{\lambda} = \frac{\left(6.63 \times 10^{-34}~\text{J}\cdot\text{s}\right)\left(3.00 \times 10^{8}~\text{m/s}\right)}{633 \times 10^{-9}~\text{m}} = 3.14 \times 10^{-19}~\text{J}$$

The power carried by the beam is

$$P = \left(2 \times 10^{18}~\text{photons/s}\right)\left(3.14 \times 10^{-19}~\text{J/photon}\right) = 0.628~\text{W}$$

Its intensity is the average Poynting vector

$$I = S_{\text{avg}} = \frac{P}{A} = \frac{P}{\pi r^2} = \frac{0.628~\text{W}\,(4)}{\pi\left(1.75 \times 10^{-3}\text{m}\right)^2} = 2.61 \times 10^{5}~\text{W/m}^2$$

From Chapter 24, the intensity is related to the field amplitudes by

$$S_{\text{avg}} = \left|\frac{1}{\mu_0}\vec{\mathbf{E}} \times \vec{\mathbf{B}}\right|_{\text{av}} = \frac{1}{\mu_0}E_{\text{rms}}B_{\text{rms}}\sin 90° = \frac{1}{\mu_0}\frac{E_{\text{max}}}{\sqrt{2}}\frac{B_{\text{max}}}{\sqrt{2}}$$

We also have $E_{\text{max}} = B_{\text{max}}c.$

(a) So $S_{\text{avg}} = \dfrac{E_{\text{max}}^2}{2\mu_0 c}$ and the electric field amplitude is

$$E_{\text{max}} = \left(2\mu_0 c S_{\text{av}}\right)^{1/2}$$

$$= \left[2\left(4\pi \times 10^{-7}~\text{T}\cdot\text{m/A}\right)\left(3.00 \times 10^{8}~\text{m/s}\right)\left(2.61 \times 10^{5}~\text{W/m}^2\right)\right]^{1/2}$$

$$= 1.40 \times 10^{4}~\text{N/C} \qquad\blacksquare$$

(b) The magnetic field amplitude is $B_{max} = \dfrac{E_{max}}{c} = \dfrac{1.40 \times 10^4 \text{ N/C}}{3 \times 10^8 \text{ m/s}} = 4.68 \times 10^{-5}$ T. ∎

(c) Each photon carries momentum $\dfrac{E}{c}$. The beam transports momentum at the rate $\dfrac{P}{c}$.

It would impart momentum to an absorbing surface at this rate, and it bounces back to give momentum to a perfectly reflecting surface at the rate

$$\dfrac{2P}{c} = \text{force} = \dfrac{2(0.628 \text{ W})}{3 \times 10^8 \text{ m/s}} = 4.19 \times 10^{-9} \text{ N}$$ ∎

(d) The block of ice absorbs energy $L\Delta m = P\Delta t$, melting the mass

$$\Delta m = \dfrac{P\Delta t}{L} = \dfrac{(0.628 \text{ W})(1.5 \times 3\,600 \text{ s})}{3.33 \times 10^5 \text{ J/kg}} = 1.02 \times 10^{-2} \text{ kg}$$ ∎

Finalize: This problem involves lots of review. You might think of it as a fine point, but it is useful to know that the rms value is $1/\sqrt{2}$ times the maximum value for a sinusoidally varying signal, and it is the rms value that tells most directly about energy transport. It is a very fundamental idea that force is the time rate of momentum change. The laser in this problem is more powerful than a typical student-laboratory laser, at 628 milliwatts instead of on the order of one milliwatt. But it would be challenging to make direct measurements of the field amplitudes and the force on the mirror. It would be challenging to keep the ice block insulated well enough to see clearly that the light melts ten grams in ninety minutes. And it would be especially challenging to demonstrate the photon-granularity of the energy transport.

---

25. The resolving power of a microscope depends on the wavelength used. If you wanted to "see" an atom, a wavelength of approximately 1.00×10^{-11} m would be required. (a) If electrons are used (in an electron microscope), what minimum kinetic energy is required for the electrons? (b) **What If?** If photons are used, what minimum photon energy is needed to obtain the required resolution?

Solution

Conceptualize: Higher energy goes with shorter wavelength, both for photons and for electrons, but with quite different patterns. An electron with a few thousand eV's of kinetic energy may have wavelength 10^{-11} m. It will take a hard x-ray with much higher energy to get an electromagnetic wave with the same wavelength.

Categorize: Think of this as two separate problems. Quite different equations describe the wavelength of a quantum particle with mass (de Broglie's equation) and the wavelength of a photon.

Analyze:

(a) Since the de Broglie wavelength is $\lambda = h/p$, the electron momentum is

$$p_e = \frac{h}{\lambda} = \frac{6.63 \times 10^{-34} \text{ J} \cdot \text{s}}{1.00 \times 10^{-11} \text{ m}} = 6.63 \times 10^{-23} \text{ kg} \cdot \text{m/s}$$

and its kinetic energy is

$$K = \frac{p_e^2}{2m_e} = \frac{\left(6.63 \times 10^{-23} \text{ kg} \cdot \text{m/s}\right)^2}{2\left(9.11 \times 10^{-31} \text{ kg}\right)} = 2.41 \times 10^{-15} \text{ J} = 15.1 \text{ keV} \quad \blacksquare$$

For better accuracy, you can use the relativistic equation

$$\left(m_e c^2 + K\right)^2 = p_e^2 c^2 + m_e^2 c^4 \quad \text{to find} \quad K = 14.9 \text{ keV.} \quad \blacksquare$$

(b) For photons the energy is

$$E = hf = \frac{hc}{\lambda} = \frac{\left(6.63 \times 10^{-34} \text{ J} \cdot \text{s}\right)\left(3.00 \times 10^8 \text{ m/s}\right)}{1.00 \times 10^{-11} \text{ m}}$$

$$= 1.99 \times 10^{-14} \text{ J} = 124 \text{ keV} \quad \blacksquare$$

For the photon, this wavelength $\lambda = 10$ pm is in the x-ray range of the electromagnetic spectrum.

Finalize: The photon has on the order of ten times more energy than an electron with the same wavelength.

Never try to use $\lambda = h/mu$ for a photon. A photon has no mass. (But $\lambda = h/p$ does apply to photons, where $p = E/c$ is the momentum a photon transports, as evidenced by its radiation pressure.) Good advice is not to use $v = f\lambda$ for electrons or other quantum particles with mass. The v in this equation could not be interpreted as the speed of the particle. On the other hand, 1 eV is a convenient unit for energies of photons, protons, and other quantum particles and atomic systems, not just for electrons.

---

28. The nucleus of an atom is on the order of 10^{-14} m in diameter. For an electron to be confined to a nucleus, its de Broglie wavelength would have to be on this order of magnitude or smaller. (a) What would be the kinetic energy of an electron confined to this region? (b) Make an order-of-magnitude estimate of the electric potential energy of a system of an electron inside an atomic nucleus. (c) Would you expect to find an electron in a nucleus? Explain.

Solution

Conceptualize: The de Broglie wavelength of a normal ground-state orbiting electron is on the order of 10^{-10} m (the diameter of a hydrogen atom) so with a shorter wavelength, the electron would have more kinetic energy if confined inside the nucleus. If the kinetic energy is much greater than the potential energy characterizing its attraction with the positive nucleus, then the electron will escape from its electrostatic potential well.

Categorize: If we try to calculate the velocity of the electron from the de Broglie wavelength as

$$u = \frac{h}{m_e \lambda} = \frac{6.626 \times 10^{-34} \text{ J} \cdot \text{s}}{(9.11 \times 10^{-31} \text{ kg})(10^{-14} \text{ m})} = 7.27 \times 10^{10} \text{ m/s}$$

we find a value which is not possible since it exceeds the speed of light. Therefore, we must use the relativistic energy expression to find the kinetic energy of this fast-moving electron.

Analyze:

(a) We find the momentum of the particle:

$$p = \frac{h}{\lambda} = \frac{6.626 \times 10^{-34} \text{ J} \cdot \text{s}}{10^{-14} \text{ m}} = 6.63 \times 10^{-20} \text{ N} \cdot \text{s}$$

We find the particle's relativistic total energy from $E^2 = (pc)^2 + (mc^2)^2$ as

$$E = \sqrt{(1.99 \times 10^{-11} \text{ J})^2 + (8.19 \times 10^{-14} \text{ J})^2} = 1.99 \times 10^{-11} \text{ J}$$

Its relativistic kinetic energy is $K = E - mc^2$.

$$K = \frac{1.99 \times 10^{-11} \text{ J} - 8.19 \times 10^{-14} \text{ J}}{1.60 \times 10^{-19} \text{ J/eV}} = 124 \text{ MeV} \sim 100 \text{ MeV} \qquad \blacksquare$$

(b) The electrostatic potential energy of an electron-proton system with a separation of 10^{-14} m is

$$U = -\frac{k_e e^2}{r} = -\frac{(8.99 \times 10^9 \text{ N} \cdot \text{m}^2/\text{C}^2)(1.60 \times 10^{-19} \text{ C})^2}{10^{-14} \text{ m}}$$
$$= -2.30 \times 10^{-14} \text{ J} \sim -0.1 \text{ MeV} \qquad \blacksquare$$

(c) Since the kinetic energy is nearly 1000 times greater than the magnitude of the potential energy, the electron would immediately escape the proton's attraction and would not be confined to the nucleus. $\qquad \blacksquare$

Finalize: It is also interesting to notice in the above calculations that the rest energy of the electron is negligible compared to the momentum contribution to the total energy.

31. Consider a freely moving quantum particle with mass m and speed u. Its energy is $E = K = \frac{1}{2}mu^2$. (a) Determine the phase speed of the quantum wave representing the particle and (b) show that it is different from the speed at which the particle transports mass and energy.

Solution

Conceptualize: Section 28.6 in the textbook demonstrates how the group speed of a wave packet is the speed u of the quantum particle with mass that the wave packet represents. A point of constant phase, such as a crest of the wave within the packet, can move at a different speed. Its speed, called the phase speed v_{phase}, can in principle be larger or smaller than the group speed; this problem identifies the phase speed.

Categorize: Note that Chapters 9 and 28 use u to represent the speed of a particle with mass, so that v is available for the speeds of waves and reference frames. We are to determine the phase speed $v_{\text{phase}} = \omega/k$ of the wave representing the particle. We can use the Planck and de Broglie equations $E = hf$ and $\lambda = h/p$ that relate the energy and momentum of the quantum particle to the wavelength and frequency of the wave that represents it.

Analyze: The particle is freely moving, so we attribute no potential energy to it. Its energy is $E = K = \frac{1}{2}mu^2 = hf = (h/2\pi)(2\pi f) = \hbar\omega$.

For its momentum we have $p = mu = h/\lambda = (h/2\pi)(2\pi/\lambda) = \hbar k$.

Thus $\omega = K/\hbar$ and $k = p/\hbar$. Then (a) the phase speed is

$$v_{\text{phase}} = f\lambda = (2\pi f)\left(\frac{\lambda}{2\pi}\right) = \frac{\omega}{k} = \left(\frac{K}{\hbar}\right)\left(\frac{\hbar}{p}\right) = \frac{K}{p} = \frac{\frac{1}{2}mu^2}{mu} = \frac{u}{2} \qquad \blacksquare$$

(b) We see that the phase speed is only one-half of the experimentally measurable speed u at which the quantum particle transports mass, energy, and momentum. In the textbook's Active Figure 28.17, individual wave crests would move forward more slowly than their envelope moves forward, so individual crests would appear to move backward relative to the packet containing them. $\qquad \blacksquare$

Finalize: By contrast, a photon is a quantum particle with no mass. It transports energy and momentum, related by $E = pc$. Its energy is not kinetic energy, but energy in electric and magnetic fields. For a photon in vacuum the same procedure identifies the phase speed as

$$v_{\text{phase}} = f\lambda = (2\pi f)(\lambda/2\pi) = \omega/k = (E/\hbar)(\hbar/p) = E/p = pc/p = c$$

Its phase speed is the same as its group speed, the speed of light.

33. Neutrons traveling at 0.400 m/s are directed through a pair of slits separated by 1.00 mm. An array of detectors is placed 10.0 m from the slits. (a) What is the de Broglie wavelength of the neutrons? (b) How far off axis is the first zero-intensity point on the detector array? (c) When a neutron reaches a detector, can we say which slit the neutron passed through? Explain.

Solution

Conceptualize: The momentum of each neutron is small, but Planck's constant is so very small that the de Broglie wavelength of the neutrons will be small compared to a millimeter. Still the two-slit diffraction pattern may be observable, with the answer to (b) being on the order of a millimeter.

Categorize: We use de Broglie's equation and the waves in interference model.

Analyze:

(a) The wavelength of the neutrons is

$$\lambda = \frac{h}{mu} = \frac{6.626 \times 10^{-34} \text{ J} \cdot \text{s}}{(1.67 \times 10^{-27} \text{ kg})(0.400 \text{ m/s})} = 9.93 \times 10^{-7} \text{ m} \qquad \blacksquare$$

(b) The condition for destructive interference in a multiple-slit experiment is

$$d \sin \theta = \left(m + \frac{1}{2} \right) \lambda, \text{ with } m = 0 \text{ for the first minimum.}$$

$$\text{Then} \quad \theta = \sin^{-1}\left(\frac{\lambda}{2d}\right) = \sin^{-1}\left(\frac{9.93 \times 10^{-7} \text{ m}}{2 \times 10^{-3} \text{ m}}\right) = \sin^{-1}\left(4.96 \times 10^{-4}\right) = 0.028 \ 4°.$$

And the geometry of a double-slit experiment implies that the distance to the first minimum on the screen is given by

$$\frac{y}{L} = \tan \theta: \quad y = L \tan \theta = (10.0 \text{ m}) \tan(0.028 \ 4°) = 4.96 \text{ mm} \qquad \blacksquare$$

(c) We cannot say the neutron passed through one slit. We can only say it passed through the pair of slits, as a water wave does to produce an interference pattern. $\qquad \blacksquare$

Finalize: Experimentally it would be hard to set up a monoenergetic beam of neutrons moving as slowly as 40 cm/s, but essentially this experiment has been done at higher energies with finer-grained diffracting targets, and the results agree with the theory of waves in interference. If we set up any kind of detector to see which slit a particular neutron passed through, the interference pattern would disappear.

37. An electron and a 0.020 0 kg bullet each have a velocity of magnitude 500 m/s, accurate to within 0.010 0%. Within what lower limit could we determine the position of each object along the direction of the velocity?

Solution

Conceptualize: It seems reasonable that a tiny particle like an electron could be located within a narrower region than a big object like a bullet, but we may find that the realm of the very small does not obey common sense.

Categorize: Heisenberg's uncertainty principle can be used to find the uncertainty in position from the uncertainty in the momentum.

Analyze: The uncertainty principle states $\Delta x \Delta p_x \geq \dfrac{\hbar}{2}$,

where $\Delta p_x = m \Delta u$ and $\hbar \equiv h/2\pi$.

Both the electron and bullet have a velocity uncertainty

$$\Delta u = (0.000\ 100)(500\ \text{m/s}) = 0.050\ 0\ \text{m/s}$$

For the electron, the minimum uncertainty in position is

$$\Delta x = \frac{h}{4\pi m \Delta u} = \frac{6.626 \times 10^{-34}\ \text{J} \cdot \text{s}}{4\pi \left(9.11 \times 10^{-31}\ \text{kg}\right)(0.050\ 0\ \text{m/s})} = 1.16\ \text{mm} \qquad \blacksquare$$

For the bullet,

$$\Delta x = \frac{h}{4\pi m \Delta u} = \frac{6.626 \times 10^{-34}\ \text{J} \cdot \text{s}}{4\pi (0.020\ 0\ \text{kg})(0.050\ 0\ \text{m/s})} = 5.28 \times 10^{-32}\ \text{m} \qquad \blacksquare$$

Finalize: Our intuition did not serve us well here, since the position of the center of the larger bullet can be determined much more precisely than the electron. Quantum mechanics describes all objects, but the quantum fuzziness in position is too small to observe for the bullet. It is large for the small-mass electron.

---

41. A free electron has a wave function

$$\psi(x) = Ae^{i\left(5.00 \times 10^{10}\, x\right)}$$

where x is in meters. Find its (a) de Broglie wavelength, (b) momentum, and (c) kinetic energy in electron volts.

Solution

Conceptualize: We cannot draw a simple graph of Ae^{ikx}, because it has both real and imaginary parts. The statement of the problem does not include the time variation of the wave function, but it would be as in

$$\psi(x,t) = Ae^{ikx-iwt} = A\cos(kx - \omega t) + Ai\sin(kx - \omega t)$$

In a word, this wave function is for the wave motion of a quantum particle moving toward the right with a particular wave number.

Categorize: The wave function represents the state of the electron, so it contains information about its wavelength, momentum, and energy that we can read out.

Analyze:

(a) The wave function, $\psi(x) = Ae^{i(5\times 10^{10}x)} = A\cos(5\times 10^{10}x) + iA\sin(5\times 10^{10}x)$,

will go through one full cycle between $x_1 = 0$ and $(5.00 \times 10^{10})x_2 = 2\pi$. The wavelength is then

$$\lambda = x_2 - x_1 = \frac{2\pi}{5.00\times 10^{10}\ \text{m}^{-1}} = 1.26\times 10^{-10}\ \text{m} \qquad \blacksquare$$

To say the same thing, we can inspect $Ae^{i(5\times 10^{10}x)}$ to see that the wave number is $k = 5.00 \times 10^{10}\ \text{m}^{-1} = 2\pi/\lambda$.

(b) Since $\lambda = h/p$, the momentum is

$$p = \frac{h}{\lambda} = \frac{6.626\times 10^{-34}\ \text{J}\cdot\text{s}}{1.26\times 10^{-10}\ \text{m}} = 5.27\times 10^{-24}\ \text{kg}\cdot\text{m/s} \qquad \blacksquare$$

(c) The electron's kinetic energy is

$$K = \frac{1}{2}mu^2 = \frac{p^2}{2m}:$$

$$K = \frac{\left(5.27\times 10^{-24}\ \text{kg}\cdot\text{m/s}\right)^2}{2\left(9.11\times 10^{-31}\ \text{kg}\right)}\left(\frac{1\ \text{eV}}{1.60\times 10^{-19}\ \text{J}}\right) = 95.3\ \text{eV} \qquad \blacksquare$$

[We use u to represent the speed of a particle with mass in chapters 9 and 28.]

Finalize: Its relativistic total energy is 511 keV + 95.3 eV. This electron could be coasting along in a field-free vacuum tube after being fired from a 95.3-volt electron gun. Its wavelength is suitable for being diffracted by a crystal in a Davisson-Germer experiment.

---

43. An electron is contained in a one-dimensional box of length 0.100 nm. (a) Draw an energy-level diagram for the electron for levels up to $n = 4$. (b) Photons are emitted by the electron making downward transitions that could eventually carry it from the $n = 4$ state to the $n = 1$ state. Find the wavelengths of all such photons.

Solution

Conceptualize: This problem is about a one-dimensional model for an atom or for a quantum dot. We will see whether the photon wavelengths are around the visible part of the electromagnetic spectrum.

Categorize: In part (a), we can draw a diagram that parallels our treatment of mechanical waves under boundary conditions. In each standing-wave state, we measure the distance d from one node to the next (N to N), and base our solution upon that. In part (b) we identify the loss of energy by the electron with the Planck energy of an emitted photon.

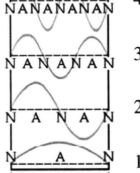

Analyze:

(a) Since $d_{\text{N to N}} = d = \dfrac{\lambda}{2}$ and $\lambda = \dfrac{h}{p}$, the electron's

momentum in each state is $p = \dfrac{h}{\lambda} = \dfrac{h}{2d}$.

The electron's energy is $K = \dfrac{p^2}{2m_e} = \dfrac{h^2}{8m_e d^2} = \left(\dfrac{1}{d^2}\right)\left[\dfrac{\left(6.63 \times 10^{-34} \text{ J} \cdot \text{s}\right)^2}{8\left(9.11 \times 10^{-31} \text{ kg}\right)}\right].$

Evaluating, $K = \dfrac{6.03 \times 10^{-38} \text{ J} \cdot \text{m}^2}{d^2} = \dfrac{3.77 \times 10^{-19} \text{ eV} \cdot \text{m}^2}{d^2}.$

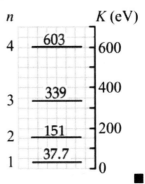

In state 1, $d = 1.00 \times 10^{-10}$ m $\qquad K_1 = 37.7$ eV

In state 2, $d = 5.00 \times 10^{-11}$ m $\qquad K_2 = 151$ eV

In state 3, $d = 3.33 \times 10^{-11}$ m $\qquad K_3 = 339$ eV

In state 4, $d = 2.50 \times 10^{-11}$ m $\qquad K_4 = 603$ eV

These energy levels are shown in the diagram. ∎

(b) When the charged, massive electron inside the box makes a downward transition from one energy level to another, a chargeless, massless photon comes out of the box, carrying the difference in energy ΔE. Its wavelength is

$$\lambda = \dfrac{c}{f} = \dfrac{hc}{\Delta E} = \dfrac{\left(6.626 \times 10^{-34} \text{ J} \cdot \text{s}\right)\left(3.00 \times 10^8 \text{ m/s}\right)}{\Delta E\left(1.602 \times 10^{-19} \text{ J/eV}\right)} = \dfrac{1.24 \times 10^{-6} \text{ eV} \cdot \text{m}}{\Delta E}$$

For the downward tumble from state 4 to state 2, for example, the photon energy is $(603 - 151)$ eV $= 452$ eV and the photon wavelength is

$$\lambda = \frac{hc}{\Delta E} = \frac{1.24 \times 10^{-6} \text{ eV} \cdot \text{m}}{452 \text{ eV}} = 2.74 \text{ nm}$$

The wavelengths of light emitted in each transition are given in the table.   ■

Transition	$4 \rightarrow 3$	$4 \rightarrow 2$	$4 \rightarrow 1$	$3 \rightarrow 2$	$3 \rightarrow 1$	$2 \rightarrow 1$
ΔE (eV)	264	452	565	188	302	113
Wavelength (nm)	4.71	2.74	2.20	6.59	4.12	11.0

Finalize: This is a good problem to understand thoroughly. Note which equations for wavelength and energy apply to the electron and which equations for wavelength and energy apply to the photon. The size of the box is too small to model the space available to an outer electron in a real atom, so the photon wavelengths are in the ultraviolet or x-ray part of the electromagnetic spectrum. We could say the box models the space available to an inner electron in a multi-electron atom.

---

47. A quantum particle in an infinitely deep square well has a wave function

$$\psi_n(x) = \sqrt{\frac{2}{L}} \sin\left(\frac{\pi x}{L}\right)$$

for $0 \le x \le L$ and is zero otherwise. (a) Determine the probability of finding the particle between $x = 0$ and $x = L/3$. (b) Use the result of this calculation and a symmetry argument to find the probability of finding the particle between $x = L/3$ and $x = 2L/3$. Do not re-evaluate the integral.

Solution

Conceptualize: The textbook's Active Figure 28.22, at the bottom of column (b), shows the probability density $|\psi_1|^2$ for finding the particle at different points along the length of the well. We reproduce this graph here. Consider the area under the curve between $x = 0$ and $x = L/3$. The size of this area represents the probability of finding the particle in this section of its range. Because the curve is low near the boundary and high at the center, the probability is much less than one-third; we might estimate about 0.2.

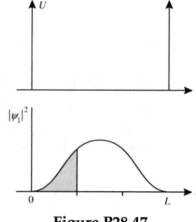

Figure P28.47

Categorize: We must square the given wave function and integrate it over x between the limits 0 and $L/3$ to find the probability. That is all we have to do, because the coefficient $\sqrt{2/L}$ has been chosen "for normalization" to make the whole area under the graph line equal to 1, representing the 100% probability of finding the particle anywhere.

Analyze:

(a) The probability is

$$P = \int_0^{L/3} |\psi|^2 \, dx = \int_0^{L/3} \frac{2}{L} \sin^2\left(\frac{\pi x}{L}\right) dx = \frac{2}{L} \int_0^{L/3} \left[\frac{1}{2} - \frac{1}{2}\cos\left(\frac{2\pi x}{L}\right)\right] dx$$

$$= \frac{1}{L} \int_0^{L/3} dx - \frac{1}{L} \int_0^{L/3} \frac{L}{2\pi}\left[\cos\left(\frac{2\pi x}{L}\right)\right] \frac{2\pi dx}{L}$$

$$P = \frac{x}{L} - \frac{1}{2\pi}\sin\left(\frac{2\pi x}{L}\right)\Big|_0^{L/3} = \left[\frac{1}{3} - \frac{1}{2\pi}\sin\left(\frac{2\pi}{3}\right)\right] = \frac{1}{3} - \frac{\sqrt{3}}{4\pi} = 0.196 \qquad \blacksquare$$

(b) The probability density is symmetric about $x = \dfrac{L}{2}$. Thus, the probability of finding the particle between $x = \dfrac{2L}{3}$ and $x = L$ is the same 0.196. Therefore, the probability of finding it in the range $\dfrac{L}{3} \le x \le \dfrac{2L}{3}$ is

$$P = 1.00 - 2(0.196) = 0.609 \qquad \blacksquare$$

Finalize: Our 20% estimate for part (a) was right on. Classically, the particle would move back and forth with constant speed between the walls, and the probability of finding the particle is the same for all points between the walls. Thus, the classical probability of finding the particle in any range equal to one-third of the available space is $P_{\text{classical}} = \dfrac{1}{3}$. The result of part (a) is significantly smaller and the result of (b) is much larger, because of the curvature of the graph of the probability density. We could say, because the quantum particle moves as a wave.

***

58. An electron with kinetic energy $E = 5.00$ eV is incident on a barrier of width $L = 0.200$ nm and height $U = 10.0$ eV (Fig. P28.58). What is the probability that the electron (a) tunnels through the barrier? (b) Is reflected?

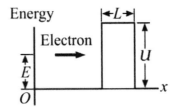

Figure P28.58

Solution

Conceptualize: Since the barrier energy is higher than the kinetic energy of the electron, transmission is not likely, but should be possible since the barrier is not infinitely high or wide.

Categorize: The probability of transmission is found from Equations 28.42 and 28.43.

Analyze: The decay constant for the wave function inside the barrier is:

$$C = \frac{\sqrt{2m(U-E)}}{\hbar}$$

$$= \frac{\sqrt{2(9.11 \times 10^{-31} \text{ kg})(10.0 \text{ eV} - 5.00 \text{ eV})(1.60 \times 10^{-19} \text{ J/eV})}}{6.626 \times 10^{-34} \text{ J} \cdot \text{s}/2\pi}$$

$$= 1.14 \times 10^{10} \text{ m}^{-1}$$

(a) The approximate probability of transmission is

$$T \approx e^{-2CL} = e^{-2(1.14 \times 10^{10} \text{ m}^{-1})(2.00 \times 10^{-10} \text{ m})} = 0.010\ 3 \qquad \blacksquare$$

(b) If the electron does not tunnel, it is reflected, with probability

$$1 - 0.010\ 3 = 0.990 \qquad \blacksquare$$

Finalize: Our expectation was correct; there is only a 1% chance that the electron will penetrate the barrier. This tunneling probability would be greater if the barrier were narrower, shorter, or if the kinetic energy of the electron were greater.

Related Comment: A typical scanning tunneling electron microscope (STM) can be built for less than $5 000, and can be tied directly to a home computer. The STM essentially consists of a needle (1) that is mounted on a few piezoelectric crystals (2). When a voltage is applied across the piezoelectric crystals, the crystals change their shape, and the needle moves.

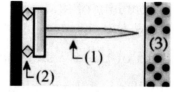

The STM charges the needle, and moves the tip of the needle to within a few tenths of a nanometer of the test sample (3); the remaining gap provides the energy barrier that is required for tunneling. When the electron cloud of one of the needle's atoms is close to the electron cloud of one of the sample's atoms, a relatively large number of electrons tunnel from one to the other, constituting a current. The current is measured by a computer. As the needle scans across the surface of the sample, an image of the atoms is generated.

72. For a quantum particle described by a wave function $\psi(x)$, the expectation value of a physical quantity $f(x)$ associated with the particle is defined by

$$\langle f(x) \rangle \equiv \int_{-\infty}^{\infty} \psi^* f(x)\psi \, dx$$

For a particle in an infinitely deep one-dimensional box extending from $x = 0$ to $x = L$, show that

$$\langle x^2 \rangle = \frac{L^2}{3} - \frac{L^2}{2n^2\pi^2}$$

Solution

Conceptualize: An example in the chapter text proved that the average experimental value of position x for a particle in this state is $L/2$. The average value of x^2 over many experimental trials will not be just $L^2/4$, but something larger than that, because of the uncertainty spread in the particle's position.

Categorize: We note that the wave function for a quantum particle in a box is

$$\psi_n(x) = A\sin\left(\frac{n\pi x}{L}\right), \quad \text{where} \quad A = \sqrt{2/L}.$$

Substituting x^2 for $f(x)$ and noting that $\psi^*\psi = |\psi|^2$, we are to evaluate

$$\langle x^2 \rangle = \int_{-\infty}^{\infty} x^2 |\psi|^2 \, dx.$$

Analyze: Performing the integral,

$$\langle x^2 \rangle = \left(\frac{2}{L}\right)\int_0^L x^2 \sin^2\left(\frac{n\pi x}{L}\right) dx$$

$$= \left(\frac{1}{L}\right)\int_0^L x^2\left[1 - \cos\left(\frac{2n\pi x}{L}\right)\right] dx$$

$$= \left(\frac{1}{L}\right)\int_0^L x^2 \, dx - \left(\frac{1}{L}\right)\int_0^L x^2 \cos\left(\frac{2n\pi x}{L}\right) dx$$

$$= \left(\frac{1}{L}\right)\frac{x^3}{3}\bigg|_0^L - \left(\frac{1}{L}\right)\left[x^2 \frac{L}{2n\pi}\sin\left(\frac{2n\pi x}{L}\right)\right]_0^L$$

$$+ \left(\frac{1}{L}\right)\int_0^L 2x\frac{L}{2n\pi}\sin\left(\frac{2n\pi x}{L}\right) dx$$

$$= \frac{L^2}{3} - 0 + \frac{1}{n\pi}\left[-x\frac{L}{2n\pi}\cos\left(\frac{2n\pi x}{L}\right)\right]_0^L + \frac{1}{n\pi}\int_0^L \frac{L}{2n\pi}\cos\left(\frac{2n\pi x}{L}\right) dx$$

$$= \frac{L^2}{3} + \frac{1}{n\pi}\left[-L\frac{L}{2n\pi}1\right] + 0 = \frac{L^2}{3} - \frac{L^2}{2n^2\pi^2} \quad \blacksquare$$

Finalize: In the ground state $n = 1$ we have

$$\langle x^2 \rangle = \frac{L^2}{3} - \frac{L^2}{2\pi^2} = 0.283L^2$$

In higher and higher states as $n \to \infty$ the expectation value approaches $0.333\ L^2$. An elementary theorem in statistics says that the root-mean-square uncertainty in x is given by $\Delta x = \sqrt{\langle x^2 \rangle - \langle x \rangle^2}$. This quantity is smallest in the ground state, where it is $0.181\ L$.

---

74. An electron is represented by the time-independent wave function

$$\psi(x) = \begin{cases} Ae^{-\alpha x} & \text{for } x > 0 \\ Ae^{+\alpha x} & \text{for } x < 0 \end{cases}$$

(a) Sketch the wave function as a function of x. (b) Sketch the probability density representing the likelihood that the electron is found between x and $x + dx$. (c) Only an infinite value of potential energy could produce the discontinuity in the derivative of the wave function at $x = 0$. Aside from this feature, argue that $\psi(x)$ can be a physically reasonable wave function. (d) Normalize the wave function. (e) Determine the probability of finding the electron somewhere in the range

$$-\frac{1}{2\alpha} \leq x \leq \frac{1}{2\alpha}$$

Solution

Conceptualize: This could be a model for an electron tied to the center of a very tiny attractor, forming a one-dimensional quantum dot.

Categorize: Think of α as some known number such as $5/\text{nm}$. The sketches emphasize that the electron state is spread out in space. The probability in part (e) will be notably less than 1. The probability in part (d) will be 1 because we think of the probability of finding the electron anywhere at all, from $-\infty$ to $+\infty$ on the x axis.

Analyze:

(a) The diagram below shows the wave function.

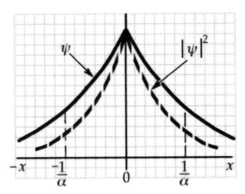

(b) For $x > 0$, the probability of finding the particle between coordinate x and coordinate $x + dx$ is

$$P_{x \text{ to } x+dx} = P(x)dx = |\psi|^2 \, dx = A^2 e^{-2\alpha x} dx \qquad \blacksquare$$

For $x < 0$ this probability is

$$P_{x \text{ to } x+dx} = P(x)dx = |\psi|^2 \, dx = A^2 e^{+2\alpha x} dx \qquad \blacksquare$$

The same diagram shows the probability distribution function as $|\psi|^2$. It shows exponential decays steeper than those of ψ, because it has 2α instead of α in the exponent.

(c) This can be reasonable since (1) ψ is continuous, (2) $\psi \to 0$ as $x \to \pm\infty$, and (3) the waveform can represent an electron bound by an infinitely deep, infinitely narrow potential well at $x = 0$. The wave function's derivative has no single value at the origin, but the wave function need not be differentiable at a point where the potential is infinite. (4) The wave function is integrable and can be normalized, as we show in part (d). $\qquad \blacksquare$

(d) Because ψ is symmetric, $\displaystyle\int_{-\infty}^{\infty} |\psi|^2 \, dx = 2\int_{0}^{\infty} |\psi|^2 \, dx = 1,$

or the normalization requirement is $2A^2 \displaystyle\int_{0}^{\infty} e^{-2\alpha x} \, dx = 1.$

Integrating,

$$\frac{2A^2}{-2\alpha} \int_{0}^{\infty} e^{-2\alpha x}(-2\alpha \, dx) = \frac{2A^2}{-2\alpha} e^{-2\alpha x} \Big|_{0}^{\infty} = \left[\frac{2A^2}{-2\alpha}\right]\left[e^{-\infty} - e^0\right] = A^2/\alpha = 1$$

which gives $A = \sqrt{\alpha}.$ $\qquad \blacksquare$

(e) The probability of finding the particle between $-1/2\alpha$ and $+1/2\alpha$ is

$$P_{-1/2\alpha \text{ to } 1/2\alpha} = 2P_{0 \text{ to } 1/2\alpha} = 2\int_{0}^{1/2\alpha} |\psi|^2 \, dx = 2\int_{0}^{1/2\alpha} \left(\sqrt{\alpha}\right)^2 e^{-2\alpha x} \, dx$$

$$= \left[\frac{2\alpha}{-2\alpha}\right]\left[e^{-2\alpha/2\alpha} - 1\right] = \left[1 - e^{-1}\right] = 0.632 \qquad \blacksquare$$

Finalize: This can be another good problem to understand thoroughly. The wave function contains all the information about the particle's state, so it is possible to find lots of things from the wave function. The problem may be more tangible to you if you assume $\alpha = 5/\text{nm}$ from the start so that definite answers show up as definite numbers. The graph of ψ would then start from $2.24/\sqrt{\text{nm}}$ at $x = 0$ and the graph of $|\psi|^2$ would start from $5/\text{nm}$. Extra practice: Taking $\alpha = 5/\text{nm}$, substitute the wave function into the Schrödinger equation, with also the assumption that $U = 4.00$ eV for $x > 0$ and for $x < 0$. Prove that it satisfies the Schrödinger equation at all points other than $x = 0$, and evaluate the particle's energy. The answer is $E = 3.05$ eV.

Chapter 29
Atomic Physics

NOTES FROM SELECTED CHAPTER SECTIONS

Section 29.1   Early Structural Models of the Atom

A need for modification of the Bohr theory became apparent when improved spectroscopic techniques were used to examine the spectral lines of hydrogen. It was found that many of the lines in the Balmer and other series were not single lines at all. Instead, each was a group of lines spaced very close together. An additional difficulty arose when it was observed that, in some situations, certain single spectral lines were split into three closely spaced lines when the atoms were placed in a strong magnetic field.

Atomic spectra and the Bohr theory of the hydrogen atom are described in Chapter 11, Section 5. It would be useful to review this discussion of spectral lines, Bohr's basic assumptions, and energy level diagrams.

Section 29.4   Physical Interpretation of the Quantum Numbers

The possible stationary energy states of an electron in an atom are determined by the values of four quantum numbers.

- **Principle quantum** number (n): integer values from 1 to ∞

   The principal quantum number follows from the concept of quantization of angular momentum and specifies the radii of the non-radiating orbits. Electron energy states with the same principal quantum number form a shell identified by the letters K, L, M . . . corresponding to $n = 1, 2, 3 \ldots$

- **Orbital quantum number** (ℓ): integer values from 0 to $(n-1)$

  An electron in a given allowed energy state may exist in different elliptical orbits determined by the value of ℓ. For each value of n there are n possible orbits corresponding to different values of ℓ. Energy states with given values of n and ℓ form a subshell identified by the letters s, p, d, f, . . . corresponding to $\ell = 0, 1, 2, 3, \ldots$. The maximum number of electrons allowed in any subshell is $2(2\ell + 1)$.

- **Orbital magnetic quantum number** (m_ℓ): integer values from -1 to $+1$

  The orbital magnetic quantum number accounts for the observed Zeeman effect; when a gas is placed in an external magnetic field single spectral lines are split into several lines.

- **Spin magnetic quantum number** (m_s): values of $-1/2$ and $+1/2$

  The spin magnetic quantum number m_s accounts for the two closely spaced energy states in spectral lines ("doublets") corresponding to the two possible orientations of electron spin ($+1/2$ is "up" spin and $-1/2$ is "down" spin).

Section 29.5   The Exclusion Principle and the Periodic Table

The **Pauli exclusion principle** states that no two electrons can exist in identical quantum states. This means that no two electrons in a given atom can be characterized by the same set of quantum numbers at the same time.

Hund's rule states that when an atom has orbitals of equal energy, the order in which they are filled by electrons is such that a maximum number of electrons will have unpaired spins.

Section 29.6   More on Atomic Spectra: Visible and X-Ray

An atom will emit electromagnetic radiation if the atom in an excited state makes a transition to a lower energy state. The set of wavelengths observed for a species by such processes is called an **emission spectrum**.

An atom can also absorb electromagnetic radiation at specific wavelengths resulting in a transition from a lower energy state to a higher energy state. These transitions produce **absorption spectra;** these spectra can be used to identify the elements in gases.

Since the orbital angular momentum of an atom changes when a photon is emitted or absorbed (as a result of a transition); and since angular momentum must be conserved, the photon involved in the process must carry angular momentum.

The x-ray spectrum of a metal target consists of a broad continuous spectrum (**bremsstrahlung**) of wavelengths from some minimum value to infinity produced by slowing down of electrons as they strike the target. Superimposed on this continuum is a series of sharp lines (**characteristic x-rays**) which are emitted by atoms when an electron undergoes a transition from an outer shell into an electron vacancy in one of the inner shells; these x-rays have wavelengths characteristic of the target. Transitions into a vacant state in the K shell give rise to the K series of spectral lines, transitions into a vacant state in the L shell create the L series of lines, and so on. Lines within a series are given a notation to designate the shell from which the transition originates. Examples from the K and L series (in order of increasing energy and decreasing intensity) are:

- L_α line: an electron drops from the L shell to the K shell

- L_β line: an electron drops from the M shell to the K shell

- K_α line: an electron drops from the M shell to the L shell

- K_β line: an electron drops from the N shell to the L shell

EQUATIONS AND CONCEPTS

The **potential energy** of the hydrogen atom (electron-proton system) is negative and depends only on the distance of the electron from the nucleus. The potential energy is zero when r is at infinity and negative otherwise.

$$U(r) = -k_e \frac{e^2}{r} \qquad (29.1)$$

Quantized energy level values can be expressed in units of electron volts (eV). The lowest allowed energy state or ground state corresponds to the principal quantum number $n = 1$. *The absolute value of the ground state energy is equal to the ionization energy of the atom.* The energy level approaches $E = 0$ as the electron radius, r, approaches infinity.

$$E_n = -\left(\frac{k_e e^2}{2a_0}\right)\frac{1}{n^2} = -\frac{13.606 \text{ eV}}{n^2}$$
$$n = 1, 2, 3, \dots \qquad (29.2)$$

The **ground state wave function for hydrogen** (1s state) depends only on the radial distance r. The parameter a_0 is the Bohr radius. *All s states have spherical symmetry.*

$$\psi_{1s}(r) = \frac{1}{\sqrt{\pi a_0^3}} e^{-r/a_0} \qquad (29.3)$$

The **probability density** is the square of the absolute value of the wave function. The probability density of the 1s state is given by Equation 29.4 when ψ is normalized.

$$|\psi_{1s}|^2 = \left(\frac{1}{\pi a_0^3}\right) e^{-\frac{2r}{a_0}} \qquad (29.4)$$

The **radial probability density** for the 1s state of hydrogen is defined as the probability per unit radial distance of finding the electron in a spherical shell of radius r and thickness dr.
As shown in the figure, the value of $P_{1s}(r)$ is maximum when $r = a_0$.

$$P_{1s}(r) = \left(\frac{4r^2}{a_0^3}\right) e^{-2r/a_0} \qquad (29.7)$$

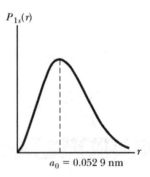

$P_{1s}(r)$

$a_0 = 0.052\,9 \text{ nm}$

The **normalized wave function for the 2s state** in hydrogen ($n = 2$, $\ell = 0$) corresponds to the energy level representing the first excited state.
ψ_{2s} is spherically symmetric.

$$\psi_{2s}(r) = \frac{1}{4\sqrt{2\pi}}\left(\frac{1}{a_0}\right)^{3/2}\left[2 - \frac{r}{a_0}\right]e^{-r/2a_0}$$
$$(29.8)$$

Quantum numbers determine the allowed wave functions and energy levels for the hydrogen atom.

The **orbital quantum number** (ℓ) determines discrete values of the magnitude of the angular momentum (L). *There are n different possible values of ℓ. All states having the same values of n and ℓ form a sub-shell.*

$$L = \sqrt{\ell(\ell+1)}\,\hbar \qquad (29.9)$$
$$\ell = 0, 1, 2, 3, \ldots, n-1$$

The **orbital magnetic quantum number** (m_ℓ) specifies the allowed values of the z component of the orbital momentum vector ($\bar{\mathbf{L}}$). *When an atom is placed in an external magnetic field, the vector $\bar{\mathbf{L}}$ lies on the surface of a cone and precesses about the z axis.* This effect is referred to as **space quantization**.

$$L_z = m_\ell\,\hbar \qquad (29.10)$$
$$m_\ell = -\ell, -\ell+1, -\ell+2, \ldots, \ell$$

The **Zeeman effect** ("splitting" of spectral lines) is due to the quantization of the angle θ that $\bar{\mathbf{L}}$ makes with the direction of a weak external magnetic field. $\bar{\mathbf{L}}$ can never be parallel or antiparallel to the z axis (θ cannot be zero).

$$\cos\theta = \frac{L_z}{L} = \frac{m_\ell}{\sqrt{\ell(\ell+1)}} \qquad (29.11)$$

The **spin quantum number** (s) is related to the intrinsic angular momentum or spin angular momentum (S) of the electron spinning on its axis.

$$S = \sqrt{s(s+1)}\,\hbar = \frac{\sqrt{3}}{2}\hbar \qquad (29.12)$$
$$s = \tfrac{1}{2}$$

The **spin magnetic quantum number** (m_s) specifies the space orientation, relative to a z axis, of the spin angular momentum vector.

$$S_z = m_s\hbar = \pm\frac{1}{2}\hbar \qquad (29.13)$$
$$m_s = \pm\tfrac{1}{2}$$

The **spin magnetic moment** ($\vec{\mu}_{\text{spin}}$) is related to the spin angular momentum.

$$\vec{\mu}_{\text{spin}} = -\frac{e}{m_e}\vec{\mathbf{S}} \qquad (29.14)$$

The **z-component of the spin magnetic moment** is quantized; there are two allowed values.

$$\vec{\mu}_{\text{spin, }z} = \pm\frac{e\hbar}{2m_e} \qquad (29.15)$$

The quantity $\mu_B = \dfrac{e\hbar}{2m_e}$ is the **Bohr magneton.**

$$\mu_B = 9.274 \times 10^{-24} \text{ J/T}$$

The **exclusion principle** dictates that no two electrons in the same atom can have the exact same set of quantum numbers.

Selection rules govern the allowed changes in the orbital quantum number and the orbital magnetic quantum number when an electron undergoes a transition between stationary states.

$$\Delta\ell = \pm 1 \text{ and } \Delta m_\ell = 0 \text{ or } \pm 1 \quad (29.16)$$

Energy levels of multielectron atoms are calculated by taking into account the shielding effect of the nuclear charge by inner-core electrons. *The atomic number is replaced by an effective atomic number Z_{eff}, which depends on the value of n and ℓ.*

$$E_n \approx -\frac{(13.6 \text{ eV})Z_{\text{eff}}^2}{n^2} \quad (29.18)$$

SUGGESTIONS, SKILLS, AND STRATEGIES

Before reading this chapter, it may be helpful to review Section 5 of Chapter 11. This section deals with the Bohr model of the atom, and makes it easier to understand the modern concept of the atomic structure.

After reading the chapter in your text, review the significance of each of the quantum numbers that is used to describe the various electronic states of electrons in an atom. Next, review the set of allowed values for each of the quantum numbers.

In addition to the principal quantum number n (which can range from 1 to ∞), other quantum numbers are necessary to specify completely the possible energy levels in the hydrogen atom and also in more complex atoms.

All energy states with the same principal quantum number, n, form a shell. These shells are identified by the spectroscopic notation K, L, M, . . . corresponding to $n = 1, 2, 3, \ldots$.

The orbital quantum number ℓ [which can range from 0 to $(n - 1)$], determines the allowed value of orbital angular momentum. All energy states having the same values of n and ℓ form a subshell. The letter designations s, p, d, f, ... correspond to values of $\ell = 0, 1, 2, 3, \ldots$

The magnetic orbital quantum number m_ℓ (which can range from $-\ell$ to ℓ) determines the possible orientations of the electron's orbital angular momentum vector in the presence of an external magnetic field.

The spin magnetic quantum number m_s can have only two values, $m_s = -\dfrac{1}{2}$ and $m_s = \dfrac{1}{2}$, which in turn correspond to the two possible directions of the electron's intrinsic spin.

REVIEW CHECKLIST

- Understand the significance of the wave function and the associated radial probability density for the ground state of hydrogen.

- For each of the quantum numbers, n (the principal quantum number), ℓ (the orbital quantum number), m_ℓ (the orbital magnetic quantum number), and m_s (the spin magnetic quantum number):

  o qualitatively describe what each implies concerning atomic structure;

  o state the allowed values which may be assigned to each, and the number of allowed states that may exist in a particular atom corresponding to each quantum number.

- Associate the customary shell and subshell spectroscopic notations with allowed combinations of quantum numbers n and ℓ. Calculate the possible values of the orbital angular momentum, L, corresponding to a given value of the principal quantum number.

- Describe how allowed values of the magnetic orbital quantum number, m_ℓ, may lead to a restriction on the orientation of the orbital angular momentum vector in an external magnetic field. Find the allowed values for L_z (the component of the angular momentum along the direction of an external magnetic field) for a given value of L.

- State the Pauli exclusion principle and describe its relevance to the periodic table of the elements. Show how the exclusion principle leads to the known electronic ground state configuration of the light elements.

ANSWER TO AN OBJECTIVE QUESTION

7. Which of the following electronic configurations are *not* allowed for an atom? Choose all correct answers. (a) $2s^2 2p^6$ (b) $3s^2 3p^7$ (c) $3d^7 4s^2$ (d) $3d^{10} 4s^2 4p^6$ (e) $1s^2 2s^2 2d^1$

Answer In (b) it is impossible to have seven electrons in a p subshell. With $\ell = 1$, the possibilities for m_ℓ are –1, 0, and +1, and the possibilities for m_s are +1/2 and –1/2. So only six electrons can fit into a p subshell. In (e) it is impossible for a $2d$ state to exist. With $n = 2$, the possible values for l are 0 and 1. The shell with $n = 2$ has just a $2s$ and a $2p$ subshell.

ANSWERS TO SELECTED CONCEPTUAL QUESTIONS

5. Could the Stern–Gerlach experiment be performed with ions rather than neutral atoms? Explain.

Answer Practically speaking, no. Since ions have a net charge, the magnetic force $q\vec{v} \times \vec{B}$ would deflect the beam, making it very difficult to separate ions with different magnetic-moment orientations.

☐ ☐ ☐ ☐

7. Why do lithium, potassium, and sodium exhibit similar chemical properties?

Answer The three elements have similar electronic configurations, with filled inner shells plus a single outer electron in an s orbital. Since atoms typically interact through their unfilled outer shells, and the outer shell of each of these atoms is similar, the chemical interactions of the three atoms are also similar.

☐ ☐ ☐ ☐

10. Discuss some consequences of the exclusion principle.

Answer If the Pauli exclusion principle were not valid, the elements and their chemical behavior would be grossly different because every electron would end up in the lowest energy level of the atom. All substances would be nearly alike in their chemistry. Atoms would be only weakly bound in molecules, and complicated molecules might not exist. Most materials would have a much higher density, and the spectra of atoms would be very simple. There would be much less color in the world.

☐ ☐ ☐ ☐

SOLUTIONS TO SELECTED END-OF-CHAPTER PROBLEMS

2. According to classical physics, a charge e moving with an acceleration a radiates energy at a rate

$$\frac{dE}{dt} = -\frac{1}{6\pi \,\epsilon_0} \frac{e^2 a^2}{c^3}$$

(a) Show that an electron in a classical hydrogen atom (see Fig. 29.3) spirals into the nucleus at a rate

$$\frac{dr}{dt} = -\frac{e^4}{12\pi^2 \,\epsilon_0^2 \, m_e^2 c^3} \frac{1}{r^2}$$

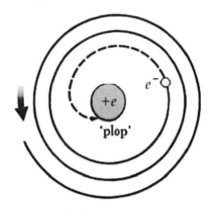

Figure 29.3

(b) Find the time interval over which the electron reaches $r = 0$, starting from $r_0 = 2.00 \times 10^{-10}$ m.

Solution

Conceptualize: This problem emphasizes that a theory has a limited range of applicability. Classical mechanics does not describe an electron.

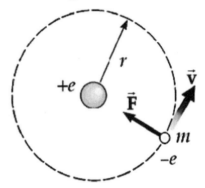

Categorize: We imagine that Newton's second law describes a single electron moving around a proton. As its energy decreases according to the given radiation equation, its radius must decrease.

Analyze: According to a classical model, the electron moving as a particle in uniform circular motion about the proton in the hydrogen atom experiences a force $k_e e^2 / r^2$; and from Newton's second law, $F = ma$, its acceleration is $k_e e^2 / m_e r^2$.

(a) Using the fact that the Coulomb constant is $k_e = \dfrac{1}{4\pi \,\epsilon_0}$,

we have the electron's acceleration $a = \dfrac{v^2}{r} = \dfrac{k_e e^2}{m_e r^2} = \dfrac{e^2}{4\pi \,\epsilon_0 m_e r^2}$. [1]

From the Bohr structural model of the atom, we can write the total energy of the atom as

$$E = -\frac{k_e e^2}{2r} = -\frac{e^2}{8\pi \,\epsilon_0 r} \quad \text{so} \quad \frac{dE}{dt} = \frac{e^2}{8\pi \,\epsilon_0 r^2} \frac{dr}{dt} = -\frac{1}{6\pi \,\epsilon_0} \frac{e^2 a^2}{c^3}. \quad [2]$$

Substituting [1] into [2] for a, solving for dr/dt, and simplifying gives

$$\frac{dr}{dt} = -\frac{4r^2}{3c^3}\left(\frac{e^2}{4\pi\,\epsilon_0\,m_e r^2}\right)^2 = -\frac{e^4}{12\pi^2\,\epsilon_0^2\,r^2 m_e^2 c^3}$$ ∎

(b) We can evaluate the constant to express dr/dt in a simpler form:

$$\frac{dr}{dt} = -\frac{A}{r^2} = -\frac{3.15 \times 10^{-21}\ \text{m}^3}{r^2}\frac{}{\text{s}}$$

Thus the time required to spiral in to the proton is described by

$$-\int_{2.00 \times 10^{-10}\ \text{m}}^{0} r^2 dr = 3.15 \times 10^{-21}\int_{0}^{T} dt$$

Integration gives

$$T = \left(3.17 \times 10^{20}\ \frac{\text{s}}{\text{m}^3}\right)\frac{r^3}{3}\Big|_{0}^{2.00 \times 10^{-10}\ \text{m}} = 8.46 \times 10^{-10}\ \text{s} = 0.846\ \text{ns}$$ ∎

Finalize: We know that atoms last much longer than eight tenths of a nanosecond. Thus, classical physics does not hold (fortunately) for atomic systems.

—————————

9. A general expression for the energy levels of one-electron atoms and ions is

$$E_n = -\frac{\mu k_e^2 q_1^2 q_2^2}{2\hbar^2 n^2}$$

Here μ is the reduced mass of the atom, given by $\mu = m_1 m_2/(m_1 + m_2)$, where m_1 is the mass of the electron and m_2 is the mass of the nucleus; k_e is the Coulomb constant; and q_1 and q_2 are the charges of the electron and the nucleus, respectively. The wavelength for the $n = 3$ to $n = 2$ transition of the hydrogen atom is 656.3 nm (visible red light). What are the wavelengths for this same transition in (a) positronium, which consists of an electron and a positron, and (b) singly ionized helium? *Note:* A positron is a positively charged electron.

Solution

Conceptualize: The reduced mass of positronium is **less** than hydrogen, so the photon energy will be **less** for positronium than for hydrogen. This means that the wavelength of the emitted photon will be **longer** than 656.3 nm. On the other hand, helium has about the same reduced mass but more charge than hydrogen, so its transition energy will be **larger**, corresponding to a wavelength **shorter** than 656.3 nm.

Categorize: All the factors in the above equation are constant for this problem except for the reduced mass and the nuclear charge. Therefore, the wavelength corresponding ·to the energy difference for the transition can be found simply from the ratio of mass and charge variables.

Analyze: For hydrogen, $\mu = \dfrac{m_p m_e}{m_p + m_e} \approx m_e$.

The photon energy is $\Delta E = E_3 - E_2$.

Its wavelength is $\lambda = 656.3$ nm, where $\lambda = \dfrac{c}{f} = \dfrac{hc}{\Delta E}$.

(a) For positronium, $\mu = \dfrac{m_e m_e}{m_e + m_e} = \dfrac{m_e}{2}$,

so the energy of each level is one half as large as in hydrogen, for which we could invent the name "protonium." The photon energy is inversely proportional to its wavelength, so for positronium,

$$\lambda_{32} = 2(656.3 \text{ nm}) = 1.31 \ \mu\text{m (in the infrared region)}. \qquad \blacksquare$$

(b) For He$^+$, $\mu \approx m_e$, $q_1 = e$, and $q_2 = 2e$, so the transition energy is $2^2 = 4$ times as large as in hydrogen. Then,

$$\lambda_{32} = \left(\frac{656}{4}\right) \text{nm} = 164 \text{ nm (in the ultraviolet region)} \qquad \blacksquare$$

Finalize: As expected, the wavelengths for positronium and helium are respectively larger and smaller than for hydrogen. Other energy transitions should have wavelength shifts consistent with this pattern. It is important to remember that the reduced mass is not the total mass, but is generally close in magnitude to the smaller mass of the system (hence the name **reduced** mass).

10. An electron of momentum p is at a distance r from a stationary proton. The electron has kinetic energy $K = p^2/2m_e$. The atom has potential energy $U = -k_e e^2/r$ and total energy $E = K + U$. If the electron is bound to the proton to form a hydrogen atom, its average position is at the proton but the uncertainty in its position is approximately equal to the radius r of its orbit. The electron's average vector momentum is zero, but its average squared momentum is approximately equal to the squared uncertainty in its momentum as given by the uncertainty principle. Treating the atom as a one-dimensional system, (a) estimate the uncertainty in the electron's momentum in terms of r. Estimate the electron's (b) kinetic energy and (c) total energy in terms of r. The actual value of r is the one that *minimizes the total energy*, resulting in a stable atom. Find (d) that value of r and (e) the resulting total energy. (f) State how your answers compare with the predictions of the Bohr theory.

Solution

Conceptualize: Bohr's model makes an arbitrary and incorrect assumption that the angular momentum of a hydrogen atom in its ground state is $1\hbar$ instead of zero. This problem shows how important the Heisenberg uncertainty principle is, by using it instead of Bohr's quantization of angular momentum to determine the size of a hydrogen atom in its ground state, and the energy of the atom.

Categorize: We follow the step-by-step directions in the problem.

Analyze:

(a) The uncertainty principle is represented by $\Delta x \Delta p \geq \dfrac{\hbar}{2}$.

Thus, if $\Delta x = r$, $\Delta p \geq \dfrac{\hbar}{2r}$. ∎

(b) The minimum uncertainty would be attained only if the wave function had a particular (gaussian) waveform. We assume that the momentum uncertainty is just twice as large as its minimum possible value: $\Delta p = \dfrac{\hbar}{r}$. Then the kinetic

energy is $K = \dfrac{(\Delta p)^2}{2m_e} = \dfrac{\hbar^2}{2m_e r^2}$. ∎

(c) The electric potential energy is $U = -\dfrac{k_e e^2}{r}$ so the total energy is

$$E = \frac{\hbar^2}{2m_e r^2} - \frac{k_e e^2}{r}$$ ∎

(d) To minimize E as a function of r, we require

$$\frac{dE}{dr} = -\frac{\hbar^2}{m_e r^3} + \frac{k_e e^2}{r^2} = 0$$

Solving for r gives $r = \dfrac{\hbar^2}{m_e k_e e^2}$, which is just the Bohr radius. ∎

(e) Then the energy is

$$E = \frac{\hbar^2}{2m_e}\left(\frac{m_e k_e e^2}{\hbar^2}\right)^2 - k_e e^2 \left(\frac{m_e k_e e^2}{\hbar^2}\right) = -\frac{m_e k_e^2 e^4}{2\hbar^2} = -13.6 \text{ eV}$$ ∎

(f) The radius and this energy have the same values as those predicted by the Bohr theory for the ground state of hydrogen. ∎

Finalize: In the Schrödinger theory the angular momentum of a hydrogen atom in its ground state is in fact zero, and not $1\hbar$ as in the Bohr theory; and the zero value is confirmed by experiment. The argument given in this problem comes out so neatly partly because we assumed $\Delta p = \dfrac{\hbar}{r}$ rather than $\Delta p = \dfrac{\hbar}{2r}$ or some other multiple of the minimum value required by the uncertainty principle. The argument given here cannot be generalized to excited states of the atom.

═══════════════

16. For a spherically symmetric state of a hydrogen atom, the Schrödinger equation in spherical coordinates is

$$-\frac{\hbar^2}{2m_e}\left(\frac{d^2\psi}{dr^2} + \frac{2}{r}\frac{d\psi}{dr}\right) - \frac{k_e e^2}{r}\psi = E\psi$$

(a) Show that the 1s wave function for an electron in hydrogen,

$$\psi_{1s}(r) = \frac{1}{\sqrt{\pi a_0^{\,3}}}e^{-r/a_0}$$

satisfies the Schrödinger equation. (b) What is the energy of the atom for this state?

Solution

Conceptualize: The potential energy function $k_e e^2/r$ makes this Schrödinger equation describe an electron in hydrogen. With this particular functional form of the potential energy term, the wave function must have another particular functional form, which we confirm.

Categorize: We use the quantum particle under boundary conditions model. We substitute the wave function and its derivatives into the Schrödinger equation, and then simplify the resulting equation. If we find that the resulting equation is true, then we know that the Schrödinger equation is satisfied.

Analyze: We evaluate the derivatives of the proposed wave function solution:

$$\frac{d\psi}{dr} = \frac{1}{\sqrt{\pi a_0^{\,3}}}\frac{d}{dr}\left(e^{-r/a_0}\right) = -\frac{1}{\sqrt{\pi a_0^{\,3}}}\left(\frac{1}{a_0}\right)e^{-r/a_0} = -\frac{\psi}{a_0} \qquad [1]$$

Differentiating again, $\dfrac{d^2\psi}{dr^2} = -\dfrac{1}{\sqrt{\pi a_0^5}}\dfrac{d}{dr}\left(e^{-r/a_0}\right) = \dfrac{1}{\sqrt{\pi a_0^7}}e^{-r/a_0} = \dfrac{1}{a_0^2}\psi.$ $\qquad [2]$

Substituting [1] and [2] into the Schrödinger equation, and noting that m is the electron's mass m_e, we have

$$-\frac{\hbar^2}{2m_e}\left(\frac{1}{a_0^{\,2}} - \frac{2}{a_0 r}\right)\psi - \frac{k_e e^2}{r}\psi = E\psi \qquad [3]$$

Substituting $\hbar^2 = m_e k_e e^2 a_0$ and canceling m_e and ψ, we have

$$-\frac{k_e e^2 a_0}{2}\left(\frac{1}{a_0^2} - \frac{2}{a_0 r}\right) - \frac{k_e e^2}{r} = E$$

We find that the second and third terms add to zero, leaving the equation $E = -\dfrac{k_e e^2}{2a_0}$. This is a true statement of the ground state energy of hydrogen, so (a) the Schrödinger equation is satisfied, with (b) the value of energy just stated. ∎

Finalize: The equation we were testing had functions of r on both sides. An important step in the proof was showing that they amounted to the same function of r, when we divided out the ψ.

---

25. How many sets of quantum numbers are possible for a hydrogen atom for which (a) $n = 1$, (b) $n = 2$, (c) $n = 3$, (d) $n = 4$, and (e) $n = 5$?

Solution

Conceptualize: Several state functions, with different angular momentum values, different z components of angular momentum, or different orientations of electron spin, have the same energy.

Categorize: We follow the rules about possible values for ℓ, for m_ℓ, and for m_s to tabulate the states for each value of the principal quantum number n. We will check our results to show that they agree with the general rule that the number of sets of quantum numbers for a shell is equal to $2n^2$.

Analyze:

(a) For $n = 1$, we have $\ell = 0$, $m_\ell = 0$, $m_s = \pm\frac{1}{2}$.

n	ℓ	m_ℓ	m_s
1	0	0	−1/2
1	0	0	+1/2

This yields $2n^2 = 2(1)^2 = 2$ sets. ∎

(b) For $n = 2$, we have

n	ℓ	m_ℓ	m_s
2	0	0	$\pm 1/2$
2	1	-1	$\pm 1/2$
2	1	0	$\pm 1/2$
2	1	$+1$	$\pm 1/2$

This yields $2n^2 = 2(2)^2 = 8$ sets. ∎

Note that the number is twice the number of m_ℓ values. Also, for each ℓ there are $(2\ell + 1)$ different m_ℓ values. Finally, l can take on values ranging from 0 to $n - 1$.

So the general expression is $\text{number} = \sum_{\ell=0}^{n-1} 2(2\ell + 1)$.

The series is an arithmetic progression like $2 + 6 + 10 + 14$.

The sum is $\sum_{0}^{n-1} 4\ell + \sum_{0}^{n-1} 2 = 4\left[\dfrac{n^2 - n}{2}\right] + 2n = 2n^2$.

(c) $n = 3$, $2(1) + 2(3) + 2(5) = 2 + 6 + 10 = 18$: $2n^2 = 2(3)^2 = 18$ ∎

(d) $n = 4$, $2(1) + 2(3) + 2(5) + 2(7) = 32$: $2n^2 = 2(4)^2 = 32$ ∎

(e) $n = 5$, $32 + 2(9) = 32 + 18 = 50$: $2n^2 = 2(5)^2 = 50$ ∎

Finalize: We have computed the number of states in two different ways for each of the assigned n values, and have shown that the results agree.

⸻

32. Scanning through Figure 29.12 in order of increasing atomic number, notice that the electrons usually fill the subshells in such a way that those subshells with the lowest values of $n + \ell$ are filled first. If two subshells have the same value of $n + \ell$, the one with the lower value of n is generally filled first. Using these two rules, write the order in which the subshells are filled through $n + \ell = 7$.

Solution

Conceptualize: There is a pattern to the ground-state electron configurations of the elements. Writing down the whole configuration for the heaviest element, number 110, and then crossing out the appropriate number of electrons at the end, will give a good estimate of the configuration of any other element.

Categorize: We follow the directions given in the problem, remembering that the possible values for ℓ range from zero to $n - 1$.

Analyze:

$n + \ell$	1	2	3	4	5	6	7
subshell	$1s$	$2s$	$2p, 3s$	$3p, 4s$	$3d, 4p, 5s$	$4d, 5p, 6s$	$4f, 5d, 6p, 7s$

■

Finalize: Checking against the periodic table in Figure 29.12 shows that the prediction made from the two rules given in the problem works well for the ground-state electron configurations of atoms of most elements. But for one exception, consider silver (Ag). For it the rules suggest a configuration ending in $\ldots 5s^2 4d^9$. But this atom happens to have lower energy in the configuration $\ldots$ $5s^1 4d^{10}$. The energy advantage for having the full d subshell must be greater than for having a full s subshell.

═══════════════

35. Use the method illustrated in Example 29.5 to calculate the wavelength of the x-ray emitted from a molybdenum target ($Z = 42$) when an electron moves from the L shell ($n = 2$) to the K shell ($n = 1$).

Solution

Conceptualize: Most electrons in a multielectron atom are feeling forces due to one another as well as due to the nucleus, and their wave functions are quite complex. "Shells" interpenetrate in space. But an inner electron can be modeled as moving in the field of just the nucleus and few or no electrons below it, and in fact the Bohr theory gives a fair estimate for its energy.

Categorize: Following Example 29.5, we suppose the electron is originally in the L shell with just one other electron in the K shell between it and the nucleus…

Analyze: … so it moves in a field of effective charge $(42 - 1)e$. Its energy is then $E_L = -(42 - 1)^2 (13.6 \text{ eV}/4) = -5.72$ keV.

In the electron's final state we choose to estimate the charge holding it in orbit as $42e$, so its energy is $E_K = -(42)^2 (13.6 \text{ eV}) = -24.0$ keV.

The photon energy emitted is the absolute value of the difference,

$$E_\gamma = 24.0 \text{ keV} - 5.72 \text{ keV} = 18.3 \text{ keV} = 2.92 \times 10^{-15} \text{ J}$$

Then $f = \dfrac{E}{h} = 4.41 \times 10^{18} \text{ Hz}$ and $\lambda = \dfrac{c}{f} = 68 \text{ pm}$. ∎

Finalize: In effect, we visualize the shells as like hollow nesting dolls. We imagine this particular electron as making a transition from the inner surface of the larger shell where it starts, to the inner surface of the smaller shell where it ends up. It is a little remarkable that this method works as well as it does, giving an estimate good to within a few percent.

---

43. The first quasar to be identified and the brightest found to date, 3C 273 in the constellation Virgo, was observed to be moving away from the Earth at such high speed that the observed blue 434-nm H_γ line of hydrogen is Doppler-shifted to 510 nm, in the green portion of the spectrum (Fig. P29.43). (a) How fast is the quasar receding? (b) Edwin Hubble discovered that all objects outside the local group of galaxies are moving away from us, with speeds v proportional to their distances R. Hubble's law is expressed as $v = HR$, where Hubble's constant has the approximate value $H \approx 22 \times 10^{-3} \text{ m/s} \cdot \text{ly}$. Determine the distance from the Earth to this quasar.

Solution

Conceptualize: The problem states that the quasar is moving very fast, and since there is a significant red shift of the light, the quasar must be moving away from Earth at a relativistic speed ($v > 0.1c$). Quasars are very distant astronomical objects, and since our universe is estimated to be about 14 billion years old, we should expect this quasar to be some billions of light-years away.

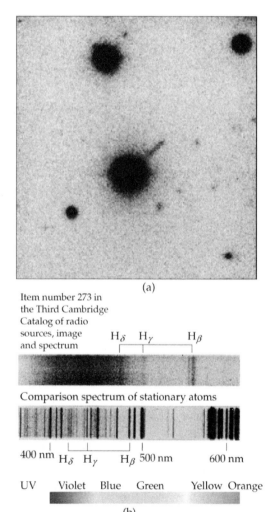

(a)

Item number 273 in the Third Cambridge Catalog of radio sources, image and spectrum

H_δ H_γ H_β

Comparison spectrum of stationary atoms

400 nm H_δ H_γ H_β 500 nm   600 nm

UV   Violet   Blue   Green   Yellow Orange

(b)

Maarten Schmidt/Palomar Observatory/California Institute of Technology

Figure P29.43

Categorize: We use the treatment of the relativistic Doppler effect in Chapter 24 to find the speed of the quasar from the Doppler red shift, and this speed can then be used to find the distance using Hubble's law.

Analyze:

(a) We use primed symbols to represent observed Doppler-shifted values and unprimed symbols to represent values as they would be measured by an observer stationary relative to the source. Doppler-shift equations from Chapter 13 do not apply to electromagnetic waves, because the speed of source or observer relative to some medium cannot be defined for these waves. Instead, we use Equation 24.23, expressing it as

$$f' = \frac{c}{\lambda'} = \sqrt{\frac{1+v/c}{1-v/c}}\, f = \sqrt{\frac{1+v/c}{1-v/c}}\left(\frac{c}{\lambda}\right)$$

where v is the velocity of mutual approach. Then we have

$$\frac{\lambda'}{\lambda} = \sqrt{\frac{1-v/c}{1+v/c}}$$

Squaring both sides, and solving,

$$\left(\frac{\lambda'}{\lambda}\right)^2 = \frac{1-v/c}{1+v/c}$$

$$\left(\frac{\lambda'}{\lambda}\right)^2 + \left(\frac{\lambda'}{\lambda}\right)^2 \frac{v}{c} = 1 - \frac{v}{c}$$

$$\left(\frac{\lambda'}{\lambda}\right)^2 - 1 = -\frac{v}{c}\left[\left(\frac{\lambda'}{\lambda}\right)^2 + 1\right]$$

$$\frac{v}{c} = -\frac{(\lambda'/\lambda)^2 - 1}{(\lambda'/\lambda)^2 + 1} = -\frac{(510\text{ nm}/434\text{ nm})^2 - 1}{(510\text{ nm}/434\text{ nm})^2 + 1} = \frac{(1.18)^2 - 1}{(1.18)^2 + 1}$$

$$= -\frac{1.381 - 1}{1.381 + 1} = -0.160$$

The negative sign indicates that the quasar is moving away from us, or us from it. The speed of recession that the problem asks for is then

$$v = 0.160c \quad \text{(or 16.0\% of the speed of light).} \qquad \blacksquare$$

(b) Hubble's law asserts that the universe is expanding at a constant rate so that the speeds of galaxies are proportional to their distance R from Earth, as described by $v = HR$.

So, $$R = \frac{v}{H} = \frac{0.160(3.00 \times 10^8\text{ m/s})}{2.2 \times 10^{-2}\text{ m/s} \cdot \text{ly}} = 2.18 \times 10^9\text{ ly.} \qquad \blacksquare$$

Finalize: The speed and distance of this quasar are consistent with our predictions. It appears that this quasar is quite far from Earth, but the universe was comparable to its present age when the quasar was radiating the light we now receive.

---

50. (a) Use Bohr's model of the hydrogen atom to show that when the electron moves from the state n to the state $n - 1$, the frequency of the emitted light is

$$f = \left(\frac{2\pi^2 m_e k_e^2 e^4}{h^3} \right) \frac{2n - 1}{n^2(n - 1)^2}$$

(b) Bohr's correspondence principle claims that quantum results should reduce to classical results in the limit of large quantum numbers. Show that as $n \to \infty$, this expression varies as $1/n^3$ and reduces to the classical frequency one expects the atom to emit. *Suggestion:* To calculate the classical frequency, note that the frequency of revolution is $v / 2\pi r$, where v is the speed of the electron and r is given by Equation 11.22.

Solution

Conceptualize: "Correspondence" refers to a prediction of quantum theory agreeing with an expectation from classical physics, as we can expect if the quantum number is very large.

Categorize: We follow the directions given in the problem.

Analyze:

(a) Combining Equations 11.23 and 29.2, the energy of a state of the hydrogen atom is

$$E_n = -\frac{m_e k_e^2 e^4}{2n^2 \hbar^2}$$

We use $\hbar = h/2\pi$ and identify the photon energy emitted in a transition between adjacent states as

$$hf = \Delta E = \frac{4\pi^2 m_e k_e^2 e^4}{2h^2} \left(\frac{1}{(n - 1)^2} - \frac{1}{n^2} \right)$$

which reduces to

$$f = \frac{2\pi^2 m_e k_e^2 e^4}{h^3} \left(\frac{2n - 1}{(n - 1)^2 n^2} \right) \qquad \blacksquare$$

(b) As $n \to \infty$, the "–1" terms lose importance and drop out, leaving the right-hand factor equal to

$$\lim_{n \to \infty} \left(\frac{2n - 1}{(n-1)^2 n^2} \right) = \frac{2n}{n^4} = \frac{2}{n^3}$$

Thus the emission frequency becomes for n large $f = \dfrac{4\pi^2 m_e k_e^2 e^4}{h^3} \left(\dfrac{1}{n^3} \right)$.   ∎

But in the Bohr model, $v^2 = \dfrac{k_e e^2}{m_e r}$ and $r = \dfrac{n^2 h^2}{4\pi^2 m_e k_e e^2}$,

so the classical frequency is $f = \dfrac{v}{2\pi r} = \dfrac{1}{2\pi r^{3/2}} \sqrt{\dfrac{k_e e^2}{m_e}}$,

or $f = \dfrac{1}{2\pi} \left(\sqrt{\dfrac{k_e e^2}{m_e}} \right) \left(\dfrac{4\pi^2 m_e k_e e^2}{n^2 h^2} \right)^{3/2}$.

And this is identical to $f = \dfrac{4\pi^2 m_e k_e^2 e^4}{h^3} \left(\dfrac{1}{n^3} \right)$.   ∎

Finalize: Experiments on macroscopic objects showed that an oscillating charge radiates electromagnetic waves at its own frequency. We have shown in the case of an electron orbiting a proton that this idea agrees with the quantum description of large orbits.

––––––––––

51. Suppose a hydrogen atom is in the 2s state, with its wave function given by Equation 29.8. Taking $r = a_0$, calculate values for (a) $\psi_{2s}(a_0)$, (b) $|\psi_{2s}(a_0)|^2$, and (c) $P_{2s}(a_0)$.

Solution

Conceptualize: Schrödinger's theory is a quantitative theory. The wave function tells probabilities of finding the particle in particular places.

Categorize: We evaluate the wave function as stated in the problem.

Analyze: The wave function for the 2s state is given by

$$\psi_{2s}(r) = \frac{1}{4\sqrt{2\pi}} \left(\frac{1}{a_0} \right)^{3/2} \left(2 - \frac{r}{a_0} \right) e^{-r/2a_0}$$

(a) Substituting $r = a_0 = 0.529 \times 10^{-10}$ m, we find

$$\psi_{2s}(a_0) = \frac{1}{4\sqrt{2\pi}} \left(\frac{1}{0.529 \times 10^{-10} \text{ m}} \right)^{3/2} (2 - 1) e^{-1/2} = 1.57 \times 10^{14} \text{ m}^{-3/2}$$ ∎

(b) $\left|\psi_{2s}\left(a_0\right)\right|^2 = \left(1.57 \times 10^{14} \text{ m}^{-3/2}\right)^2 = 2.47 \times 10^{28} \text{ m}^{-3}$ ∎

(c) Using Equation 29.6 and the results to (b) gives

$$P_{2s}\left(a_0\right) = 4\pi a_0^2 \left|\psi_{2s}\left(a_0\right)\right|^2 = 8.69 \times 10^8 \text{ m}^{-1}$$ ∎

Finalize: The numbers are large because we are describing the probability *per volume* of finding the electron, and the electron really is in a small parcel of volume around the atomic nucleus. Scattering experiments can show the "shape" of an atomic electron cloud, and give results agreeing with the theory.

---

56. We wish to show that the most probable radial position for an electron in the 2*s* state of hydrogen is $r = 5.236a_0$. (a) Use Equations 29.6 and 29.8 to find the radial probability density for the 2*s* state of hydrogen. (b) Calculate the derivative of the radial probability density with respect to *r*. (c) Set the derivative in (b) equal to zero and identify three values of *r* that represent minima in the function. (d) Find two values of *r* that represent maxima in the function. (e) Identify which of the values in part (d) represents the highest probability.

Solution

Conceptualize: An example in the chapter found the most probable radial position for an electron in the ground state of hydrogen. Here we expect a larger modal position for an excited state. The textbook Active Figure 29.6 shows the radial probability distribution.

Categorize: We are using the quantum particle model. We differentiate to find the most probable radius.

Analyze: We use Equation 29.8:

$$\psi_{2s}(r) = \frac{1}{4\sqrt{2\pi}}\left(\frac{1}{a_0}\right)^{3/2}\left[2 - \frac{r}{a_0}\right]e^{-r/2a_0}$$

(a) By Equation 29.6, the radial probability distribution function is

$$P(r) = 4\pi r^2 \psi^2 = \frac{1}{8}\left(\frac{r^2}{a_0^3}\right)\left(2 - \frac{r}{a_0}\right)^2 e^{-r/a_0}$$ ∎

(b) Its extremes are given by

$$\frac{dP(r)}{dr} = \frac{1}{8}\left[\frac{2r}{a_0^3}\left(2 - \frac{r}{a_0}\right)^2 - \frac{2r^2}{a_0^3}\left(\frac{1}{a_0}\right)\left(2 - \frac{r}{a_0}\right) - \frac{r^2}{a_0^3}\left(2 - \frac{r}{a_0}\right)^2\left(\frac{1}{a_0}\right)\right]e^{-r/a_0}$$

$$= 0$$

We factor in the following manner:

$$\frac{1}{8}\left(\frac{r}{a_0^3}\right)\left(2-\frac{r}{a_0}\right)\left[2\left(2-\frac{r}{a_0}\right)-\frac{2r}{a_0}-\frac{r}{a_0}\left(2-\frac{r}{a_0}\right)\right]e^{-r/a_0}=0$$

The left-hand side of this equation is the derivative requested.   ∎

(c) The roots of $dP/dr = 0$ at $r = 0$, $r = 2a_0$, and $r = \infty$ are minima, because $P(r) = 0$, as shown in Active Figure 29.6.   ∎

Therefore, we focus instead on the roots given by

$$2\left(2-\frac{r}{a_0}\right)-2\frac{r}{a_0}-\frac{r}{a_0}\left(2-\frac{r}{a_0}\right)=4-\frac{6r}{a_0}+\left(\frac{r}{a_0}\right)^2=0$$

(d) This equation has solutions $r = \left(3\pm\sqrt{5}\right)a_0$.   ∎

We substitute two roots into $P(r)$:

When $r = \left(3-\sqrt{5}\right)a_0 = 0.764a_0$, then $P(r) = \dfrac{0.051\,9}{a_0}$. This is the smaller peak in the textbook graph.

(e) When $r = \left(3+\sqrt{5}\right)a_0 = 5.24a_0$, then $P(r) = \dfrac{0.191}{a_0}$. This is the higher peak in the textbook graph. The most probable value of r is $\left(3+\sqrt{5}\right)a_0 = 5.24a_0$.   ∎

Finalize: Finding the most probable radial position told us the least probable radii as well. In particular, and as Active Figure 29.6 shows, in this state the electron is exactly never at $r = 2a_0$. So then, how can it be sometimes in closer to the nucleus and most of the time farther away? The answer: The electron moves as a wave. It is in a standing-wave state with a nodal surface at this radius. Its motion is not that of a classical particle, moving through a succession of points one after another. In particular, the electron does not speed up to infinite speed to jump across the forbidden radius.

---

57. For hydrogen in the 1s state, what is the probability of finding the electron farther than $2.50a_0$ from the nucleus?

Solution

Conceptualize: From the red graph shown in Active Figure 29.6, it appears that the probability of finding the electron beyond $2.5a_0$ may be about 20%.

Categorize: The precise probability can be found by integrating the 1s radial probability distribution function from $r = 2.50a_0$ to ∞.

Analyze: The general radial probability distribution function is

$$P(r) = 4\pi r^2 |\psi|^2$$

With $\psi_{1s} = (\pi a_0^3)^{-1/2} e^{-r/a_0}$, it is $P(r) = 4r^2 a_0^{-3} e^{-2r/a_0}$.

The required probability is then $P = \int_{2.50a_0}^{\infty} P(r)\,dr = \int_{2.50a_0}^{\infty} \frac{4r^2}{a_0^3} e^{-2r/a_0}\,dr$.

Let $z = 2r/a_0$ and $dz = 2dr/a_0$. Then we want $P = \frac{1}{2}\int_{5.00}^{\infty} z^2 e^{-z}\,dz$.

Performing this integration by parts,

$$P = -\frac{1}{2}\left(z^2 + 2z + 2\right)e^{-z}\Big|_{5.00}^{\infty}$$

$$P = -\frac{1}{2}(0) + \frac{1}{2}(25.0 + 10.0 + 2.00)e^{-5.00} = \left(\frac{37}{2}\right)(0.006\ 74) = 0.125 \qquad \blacksquare$$

Finalize: The probability of 12.5% is less than the 20% we estimated, but close enough to be a reasonable result. As Active Figure 29.6 suggests, the 1s probability density function peaks more sharply and closer to the origin than other states.

---

61. In the technique known as electron spin resonance (ESR), a sample containing unpaired electrons is placed in a magnetic field. Consider a situation in which a single electron (*not* contained in an atom) is immersed in a magnetic field. In this simple situation only two energy states are possible, corresponding to $m_s = \pm\frac{1}{2}$. In ESR, the absorption of a photon causes the electron's spin magnetic moment to flip from the lower energy state to the higher energy state. According to the result of Problem 24 in Chapter 22, the change in energy is $2\mu_B B$. (The lower energy state corresponds to the case in which the z component of the magnetic moment $\vec{\mu}_{\text{spin}}$ is aligned with the magnetic field, and the higher energy state corresponds to the case in which the z component of $\vec{\mu}_{\text{spin}}$ is aligned opposite to the field.) What is the photon frequency required to excite an ESR transition in a 0.350-T magnetic field?

Solution

Conceptualize: Application of the magnetic field splits an atomic energy level into two. The separation between the levels can be adjusted by changing the field. This action will change the frequency of the photons that are absorbed as they boost the atom from the lower to the higher state, and the photons that are emitted as the atom tumbles back down again. In this problem we model the experimental situation by ignoring the rest of the atom.

Categorize: The magnetic moment feels torque $\vec{\tau} = \vec{\mu} \times \vec{B}$ in an external field. The electron spin vector cannot be geometrically parallel or antiparallel to the field. But its z component must be either parallel or antiparallel to any applied field.

Analyze: To evaluate the energy difference, we imagine the z component of the electron's magnetic moment as continuously variable. In turning it from alignment with the field to the opposite direction, the field does work according to Equation 10.31,

$$W = \int dW = \int_0^{180°} \tau \, d\theta = \int_0^{\pi} \mu B \sin\theta \, d\theta = -\mu B \cos\theta \big|_0^{\pi} = 2\mu B$$

To make the electron flip, the photon must carry energy $\Delta E = 2\mu_B B = hf$.

Therefore,

$$f = \frac{2\mu_B B}{h} = \frac{2(9.27 \times 10^{-24} \text{ J/T})(0.350 \text{ T})}{6.626 \times 10^{-34} \text{ J} \cdot \text{s}} = 9.79 \times 10^9 \text{ Hz} \qquad \blacksquare$$

Finalize: This is a microwave photon. Experimentally we identify a resonance phenomenon, because the sample will absorb at just this frequency, and will not absorb microwaves of slightly lower or higher frequency.

---

65. An electron in chromium moves from the $n = 2$ state to the $n = 1$ state without emitting a photon. Instead, the excess energy is transferred to an outer electron (one in the $n = 4$ state), which is then ejected by the atom. In this Auger (pronounced "ohjay") process, the ejected electron is referred to as an Auger electron. Use the Bohr theory to find the kinetic energy of the Auger electron.

Solution

Conceptualize: The chromium atom with nuclear charge $Z = 24$ starts with one vacancy in the $n = 1$ shell, perhaps produced by the absorption of an x-ray, which ionized the atom. An electron from the $n = 2$ shell tumbles down to fill the vacancy.

Categorize: We use the same ideas as for estimating a wavelength in the x-ray spectrum.

Analyze: We suppose that the electron that makes the transition is shielded from the electric field of the full nuclear charge by the one K-shell electron originally below it. With $Z = 24$, its original energy is

$$E = -(Z - 1)^2 (13.6 \text{ eV}) \left(\frac{1}{2^2} \right) = -1.80 \text{ keV}$$

Its final energy is $E = -Z^2(13.6 \text{ eV})\left(\dfrac{1}{1^2}\right) = -7.83 \text{ keV}$.

The magnitude of the electron's energy loss is 7.83 keV – 1.80 keV = 6.04 keV

Then, instead of coming out as an x-ray photon, this +6.04 keV can be transferred to the single 4s electron. Suppose that it is shielded by the 22 electrons in the K, L, and M shells. To break the outermost electron out of the atom, producing a Cr^{2+} ion, requires an energy investment of

$$E_{\text{ionize}} = \frac{(Z - 22)^2(13.6 \text{ eV})}{4^2} = \frac{2^2(13.6 \text{ eV})}{16} = 3.40 \text{ eV}$$

Then the remaining energy that can appear as kinetic energy is

$$K = |\Delta E| - E_{\text{ionize}} = 6\,035 \text{ eV} - 3.4 \text{ eV} = 6.03 \text{ keV} \qquad \blacksquare$$

Because of conservation of momentum for the ion-electron system and the tiny mass of the electron compared to that of the Cr^{2+} ion, almost all of this kinetic energy will belong to the electron.

Finalize: As evidence that the relatively tiny 3.40 eV can be the right order of magnitude for the energy to remove a second electron from ionized chromium, note that the (first) ionization energy for neutral chromium is tabulated as 6.76 eV.

Chapter 30
Nuclear Physics

Section 30.1   Some Properties of Nuclei

Important quantities in the description of nuclear properties are:

- the **atomic number**, Z, which equals the number of protons in the nucleus;

- the **neutron number**, N, which equals the number of neutrons in the nucleus;

- the **mass number**, A, which equals the number of nucleons (neutrons plus protons) in the nucleus.

Nuclei can be represented symbolically as $^A_Z X$, where X designates the chemical symbol for a specific element. The nuclei of all atoms of a particular element contain the same number of protons but may contain different numbers of neutrons. Nuclei that are related in this way are called **isotopes.** *The isotopes of an element have the same Z value but different N and A values.* The **unified mass unit, u,** is defined such that the mass of the isotope ^{12}C is exactly 12 u ($1\,u = 1.669\,559 \times 10^{-27}$ kg).

Based on nuclear scattering and other experiments, it is possible to conclude the following:

- Most nuclei are approximately spherical and all have nearly the same density.

- Nucleons combine to form a nucleus as though they were tightly packed spheres.

- The stability of nuclei is due to a short-range attractive nuclear force.

- The nuclear force is approximately the same for an interaction between any pair of nucleons (p-p, n-n, or p-n).

Nuclei have **intrinsic angular momentum** which is quantized by the **nuclear spin quantum number** (*I*) and may be an integer or a half integer.

The **magnetic moment of the nucleus** is measured in terms of the **nuclear magneton.** When placed in an external magnetic field, nuclear magnetic moments precess with a frequency called the **Larmor precessional frequency.**

Section 30.2   Nuclear Binding Energy

The total mass of a nucleus is always less than the sum of the masses of its individual nucleons. This difference in mass is the origin of the **nuclear binding energy** and represents the energy that would be required to separate the nucleus into neutrons and protons

Section 30.3   Radioactivity

The decay of radioactive substances can be accompanied by the emission of three forms of radiation that vary in ability to penetrate shielding materials:

- **alpha** (α) **particles** — helium nuclei (^4_2He) which can barely penetrate a sheet of paper;

- **beta** (β) **particles** — either electrons (e⁻) or positrons (e⁺) able to penetrate a few millimeters of aluminum;

- **gamma** (γ) **rays** — high-energy photons (γ) capable of penetrating several centimeters of lead.

A positron is the antiparticle of the electron; it has a mass equal to m_e and a charge of $+q_e$.

The decay rate or activity (*R*) of a sample of a radioactive substance is defined as the number of decays occurring per second. The half-life ($T_{1/2}$) of a radioactive substance is the time during which half of an initial number of radioactive nuclei in a sample will decay. A common unit of radioactivity is the curie (Ci) and the SI unit of activity is the **Becquerel** (Bq); 1 Bq = 1 decay/s and $1\,Ci = 3.7 \times 10^{10}\,Bq$.

Section 30.4   The Radioactive Decay Processes

The overall decay process can be represented in equation form as:

$$\underset{\substack{\text{Parent}\\\text{nucleus}}}{X} \rightarrow \underset{\substack{\text{Daughter}\\\text{nucleus}}}{Y} + \underset{\substack{\text{radiation}}}{\text{emitted}}$$

In an equation representing a specific radioactive decay process, the sum of the mass numbers on each side of the equation must be equal and the sum of the atomic numbers on each side of the equation must be equal.

Alpha decay can occur because some nuclei have energy barriers that can be penetrated by the alpha particles (the tunneling process). Alpha decay is energetically more favorable for proton-rich nuclei.

Beta decay is accompanied by the emission of either an electron (e^-) and an antineutrino ($\bar{\nu}$), or by emission of a positron (e^+) and a neutrino (ν). The neutrino and antineutrino shown in the e^- and e^+ decay processes in the table on the following page are required for conservation of energy and momentum.

The neutrino has the following properties:

- zero electric charge

- rest mass much smaller than that of the electron

- spin of $\frac{1}{2}$ (satisfying the law of conservation of angular momentum)

- very weak interaction with matter (therefore very difficult to detect)

In the **electron-capture process** the nucleus of an atom absorbs one of its own electrons (usually from the K shell) and emits a neutrino.

In **gamma decay** a nucleus in an excited state decays to its ground state and emits a gamma ray. The disintegration energy is the energy released as a result of the decay process. The general characteristics of alpha, beta, and gamma decay are summarized below.

Decay	Can be written as	Process
Alpha decay	$^A_Z X \rightarrow ^{A-4}_{Z-2} Y + ^4_2 He$	The parent nucleus emits an alpha particle and loses two protons and two neutrons.
Beta decay	$^A_Z X \rightarrow ^A_{Z+1} Y + e^- + \bar{\nu}$ (electron emission) $^A_Z X \rightarrow ^A_{Z-1} Y + e^+ + \nu$ (positron emission)	A nucleus can undergo beta decay in two ways. The parent nucleus can emit an electron (e^-) and an antineutrino ($\bar{\nu}$) or emit a positron (e^+) and a neutrino (ν).
Gamma decay	$^A_Z X^* \rightarrow ^A_Z X + \gamma$ * parent nucleus in an excited state	A nucleus in an excited state decays to a lower energy state (often the ground state) and emits a gamma ray.

Section 30.5   Nuclear Reactions

Nuclear reactions are events in which collisions between nuclei change the structure or properties of nuclei. The **Q value** is the energy required to balance the overall reaction equation

Exothermic reactions release energy and have positive Q values.
The Q values of endothermic reactions are negative and the minimum energy of the incoming particle for which an endothermic reaction will occur is called the **threshold energy.**

EQUATIONS AND CONCEPTS

The **nuclear radius** is proportional to the cube root of the mass number (the total number of nucleons). *Most nuclei are nearly spherical in shape and all nuclei have nearly the same density.*

$$r = aA^{1/3} \qquad (30.1)$$

$$a = 1.2 \times 10^{-15} \text{ m} = 1.2 \text{ fm}$$

The **femtometer** (fm), sometimes called the **Fermi**, is a convenient unit in which to express nuclear dimensions.

$$1 \text{ fm} = 10^{-15} \text{ m}$$

The **nuclear magneton** (μ_n) is the unit of magnetic moment used to express a property associated with the angular momentum of the nucleus.

$$\mu_n \equiv \frac{e\hbar}{2m_p} = 5.05 \times 10^{-27} \text{ J/T} \quad (30.3)$$

The **binding energy** of any nucleus can be calculated in terms of the mass of a neutral hydrogen atom, $M(\text{H})$, the mass of a neutron, m_n, and the atomic mass of an atom of the isotope associated with the compound nucleus $M\left({}^A_Z X\right)$.

$$E_b(\text{MeV}) = \left[ZM(\text{H}) + Nm_n \right.$$
$$\left. -M\left({}^A_Z X\right) \right]$$
$$\times \left(931.494 \; \frac{\text{MeV}}{\text{u}} \right)$$
$$(30.4)$$

Masses are in atomic mass units (u).

The **decay rate or activity** of a radioactive sample is defined as the number of decays per second. The decay rate is proportional to the number of nuclei present and depends on the **decay constant** which is characteristic of a particular radioactive species.

$$\frac{dN}{dt} = -\lambda N \quad (30.5)$$

λ = decay constant

$$R = \left| \frac{dN}{dt} \right| = N_0 \lambda e^{-\lambda t} = R_0 e^{-\lambda t} \quad (30.7)$$

R_0 is the decay rate at $t = 0$.

The **exponential decay law** is illustrated in the graph of N versus t as illustrated at right. The decay constant λ is characteristic of a specific isotope and is the probability of decay per nucleus per second. N_0 is the number of radioactive nuclei present at time $t = 0$. Time is shown in units of half-life ($T_{1/2}$). See Equation 30.8 below.

$$N = N_0 e^{-\lambda t} \quad (30.6)$$

Plot of the exponential decay law for radioactive nuclei

Units of activity are the becquerel (Bq) and the curie (Ci). *The becquerel is the SI unit and the curie is approximately the activity of 1 gram of radium.*

$$1 \text{ Ci} \equiv 3.7 \times 10^{10} \text{ decays/s}$$

$$1 \text{ Bq} \equiv 1 \text{ decay/s}$$

The **half-life** of a radioactive substance ($T_{1/2}$) is the time required for half of a given number of radioactive nuclei in a sample of the substance to decay. *After an elapsed time of n half-lives, the number of radioactive nuclei remaining will be $N_0/2^n$.*

$$T_{1/2} = \frac{\ln 2}{\lambda} = \frac{0.693}{\lambda} \qquad (30.8)$$

Spontaneous decay of radioactive nuclei proceeds by one of the following processes: alpha decay, beta decay, electron capture, or gamma decay. The overall decay process can be represented as a decay equation. *The sum of the mass numbers on each side of the equation must be equal and the sum of the atomic numbers on each side of the equation must be equal.*

$$\underset{\substack{\text{Parent} \\ \text{nucleus}}}{X} \rightarrow \underset{\substack{\text{Daughter} \\ \text{nucleus}}}{Y} + \underset{\substack{\text{Emitted} \\ \text{radiation}}}{\;}$$

Alpha Decay
When a nucleus decays by alpha emission, the parent nucleus loses two neutrons and two protons. For alpha emission to occur, the mass of the parent nucleus ($^A_Z X$) must be greater than the combined masses of the daughter nuclei ($^{A-4}_{Z-2} Y$) and the emitted alpha particle.

$$^A_Z X \rightarrow \,^{A-4}_{Z-2} Y + \,^4_2 He \qquad (30.9)$$

$$^{238}_{92} U \rightarrow \,^{234}_{90} Th + \,^4_2 He \qquad (30.10)$$

$$^{226}_{88} Ra \rightarrow \,^{222}_{86} Rn + \,^4_2 He \qquad (30.11)$$

The **disintegration energy** Q of the system (parent nucleus, daughter nucleus, and alpha particle) is required in order to conserve energy during spontaneous decay of the isolated parent nucleus. Q appears in the form of kinetic energy of the daughter nucleus and the alpha particle.

$$Q = \left(M_X - M_Y - M_\alpha \right)$$
$$\times \, 931.494 \text{ Mev/u} \quad (30.13)$$

Masses are in atomic mass units (u).

Beta decay

When a radioactive nucleus undergoes beta decay the process is accompanied by a third particle required to conserve energy and momentum: either an antineutrino ($\bar{v}$) with e^- emission or a neutrino (v) with e^+ emission. The equations at right show: (i) The equation representing the process; (ii) an example of the decay mode; and (iii) an equation representing the origin of the electron or positron.

Electron Emission

(i) $_Z^A X \rightarrow\, _{Z+1}^{A} Y + e^- + \bar{v}$ (30.16)

(ii) $_6^{14} C \rightarrow\, _7^{14} N + e^- + \bar{v}$ (30.18)

(iii) $n \rightarrow p + e^- + \bar{v}$

Positron Emission

(i) $_Z^A X \rightarrow\, _{Z-1}^{A} Y + e^+ + v$ (30.17)

(ii) $_7^{12} N \rightarrow\, _6^{12} C + e^+ + v$ (30.19)

(iii) $p \rightarrow n + e^+ + v$

Electron Capture

This decay process competes with positron beta decay and occurs when a parent nucleus captures an orbital electron and emits a neutrino. *It is typically the K-shell electron that is captured by the parent nucleus.*

$$_Z^A X + e^- \rightarrow\, _{Z-1}^{A} Y + v \qquad (30.20)$$

Gamma Decay

Nuclei which undergo alpha or beta decay are often left in an excited energy state (indicated by the * symbol). The nucleus returns to the ground state by emission of one or more photons. The excited state of $^{12}C^*$ is an example of decay by gamma emission. *Gamma decay results in no change in mass number or atomic number.*

$$_Z^A X^* \rightarrow\, _Z^A X + \gamma \qquad (30.21)$$

$$_6^{12} C^* \rightarrow\, _6^{12} C + \gamma \qquad (30.23)$$

Nuclear reactions can occur when target nuclei (X) are bombarded with energetic particles (a) resulting in a daughter or product nucleus (Y) and an outgoing particle (b). The reaction can be represented in a compact form as shown. *In these reactions the structure, identity, or properties of the target nuclei are changed.*

$$a + X \rightarrow Y + b \qquad (30.24)$$

$$X(a, b)Y$$

The **reaction energy** associated with a nuclear reaction is the total change in mass-energy resulting from the reaction.

$$Q = \left(M_a + M_X - M_Y - M_b\right)c^2$$

(30.25)

The **threshold energy** is the minimum energy required for an endothermic reaction to occur.

$Q > 0$, exothermic reaction
$Q < 0$, endothermic reaction

SUGGESTIONS, SKILLS, AND STRATEGIES

The rest energy of a particle is given by $E = mc^2$. It is therefore often convenient to express the unified mass unit in terms of its equivalent energy, $1\,u = 1.660\,559 \times 10^{-27}$ kg or $1\,u = 931.494$ MeV/c^2. When masses are expressed in units of u, energy values are then $E = m(931.494 \text{ MeV/u})$.

Equation 30.6 can be solved for the particular time t after which the number of remaining nuclei will be some specified fraction of the original number N_0. This can be done by taking the natural log of each side of Equation 30.6 to find

$$t = \frac{1}{\lambda}\ln\left(\frac{N_0}{N}\right)$$

REVIEW CHECKLIST

- Use the appropriate nomenclature in describing the static properties of nuclei.

- Discuss nuclear stability in terms of the strong nuclear force and a plot of N vs. Z.

- Account for nuclear binding energy in terms of the Einstein mass-energy relationship. Describe the basis for energy released by fission and fusion in terms of the shape of the curve of binding energy per nucleon vs. mass number.

- Identify each of the components of radiation that are emitted by the nucleus through natural radioactive decay and describe the basic properties of each. Write out generic equations [e.g., Equations 30.9 ($_Z^A X \rightarrow _{Z-2}^{A-4} Y + _2^4 He$), 30.16 ($_Z^A X \rightarrow _{Z+1}^A Y + e^- + \bar{\nu}$), 30.20 ($_Z^A X + e^- \rightarrow _{Z-1}^A Y + \nu$), and 30.19 ($_7^{12}N \rightarrow _6^{12}C + e^+ + \nu$)] that illustrate the conservation of nucleon number and charge in the processes of radioactive decay by alpha, beta and gamma emission and by electron capture. Explain why the neutrino must be considered in the analysis of beta decay.

- State, and apply to the solution of related problems, the formula which expresses decay rate as a function of the decay constant and the number of radioactive nuclei. Describe the process of carbon dating as a means of determining the age of ancient objects.

- Calculate the Q value of given nuclear reactions and determine the threshold energy of endothermic reactions.

ANSWER TO AN OBJECTIVE QUESTION

2. Two samples of the same radioactive nuclide are prepared. Sample G has twice the initial activity of sample H. **(i)** How does the half-life of G compare with the half-life of H? (a) It is two times larger. (b) It is the same. (c) It is half as large. **(ii)** After each has passed through five half-lives, how do their activities compare? (a) G has more than twice the activity of H. (b) G has twice the activity of H. (c) G and H have the same activity. (d) G has lower activity than H.

Answer **(i)** (b) Because they are the same nuclide (samples of the same isotope), they have the same half-life. Nothing in the physical or chemical environment of the sample can alter a nucleus's probability of decay, and certainly the sample size cannot alter it.

(ii) (b) The activities of G and H decrease in time together, but G will always have twice as many remaining parent nuclei, and will always have twice the activity of H. After five half-lives, each has 1/32 as many parent nuclei as it started with, and the ratio of their activities is $(2/32)/(1/32) = 2/1$, as at the start. We ignore statistical fluctuations that would only show up over short periods of observation, and the possibility that the daughter nucleus is also radioactive.

ANSWERS TO SELECTED CONCEPTUAL QUESTIONS

4. Explain why nuclei that are well off the line of stability in Figure 30.4 tend to be unstable.

Answer The nuclear force favors the formation of neutron-proton pairs, so a stable nucleus cannot be too far away from having equal numbers of protons and neutrons. This effect sets the upper boundary of the zone of stability on the neutron-proton diagram. All of the protons repel one another electrically, so a stable nucleus cannot have too many protons. This effect sets the lower boundary of the zone of stability, and also the upper end.

□   □   □   □

6.   Why do nearly all the naturally occurring isotopes lie above the $N = Z$ line in Figure 30.4?

Answer   As Z increases, extra neutrons are required to overcome the increasing electrostatic repulsion of the protons. The extra neutrons attract their neighbors, including protons, with the strong nuclear force.

14. Suppose it could be shown that the cosmic-ray intensity at the Earth's surface was much greater 10 000 years ago. How would this difference affect what we accept as valid carbon-dated values of the age of ancient samples of once-living matter? Explain your answer.

Answer   If the cosmic ray intensity at the Earth's surface was much greater 10 000 years ago, the fraction of the Earth's carbon dioxide with the radioactive nuclide ^{14}C would also be greater at that time than now. Thus, there would initially be a greater fraction of ^{14}C in the organic artifacts, and we would believe that the artifact was more recent than it actually is.

For example, suppose that that actual ratio of atmospheric ^{14}C to ^{12}C, two half-lives (11 460 years) ago, was 2.6×10^{-12}. The current ratio of isotopes in a sample from that time would be $\left(\frac{1}{2}\right)\left(\frac{1}{2}\right)\left(2.6 \times 10^{-12}\right) = 0.65 \times 10^{-12}$. We would measure this ratio today; but, believing the initial ratio to be equal to the present value of 1.3×10^{-12}, we would think that the creature had died only one half-life (5 730 years) ago. We would calculate $\left(\frac{1}{2}\right)\left(1.3 \times 10^{-12}\right) = 0.65 \times 10^{-12}$.

17. If a nucleus such as ^{226}Ra initially at rest undergoes alpha decay, which has more kinetic energy after the decay, the alpha particle or the daughter nucleus? Explain your answer.

Answer   The alpha particle and the daughter nucleus carry momenta of equal magnitudes in opposite directions. Since kinetic energy can be written as $p^2/2m$, the less massive alpha particle has much more of the decay energy than the recoiling nucleus.

SOLUTIONS TO SELECTED END-OF-CHAPTER PROBLEMS

1. (a) Use energy methods to calculate the distance of closest approach for a head-on collision between an alpha particle having an initial energy of 0.500 MeV and a gold nucleus (^{197}Au) at rest. Assume the gold nucleus remains at rest during the collision. (b) What minimum initial speed must the alpha particle have to approach as close as 300 fm to the gold nucleus?

Solution

Conceptualize: The positively charged alpha particle ($q = +2e$) will be repelled by the positive gold nucleus ($Q = +79e$), so that the particles probably will not touch each other in the electrostatic "collision" of part (a). The closest the alpha particle can get to the gold nucleus would be if the two nuclei did touch, in which case the distance between their centers would be about 9 fm (using $r = aA^{1/3}$ for the radius of each nucleus). To get this close or even within 300 fm, the alpha particle must be traveling very fast, probably close to the speed of light (but of course v must be less than c).

Categorize: The initial kinetic energy is equal to the electric potential energy of the alpha-particle-gold-nucleus system at the distance of closest approach, r_{min}.

Analyze:

(a) $K_\alpha + 0 = 0 + U = k_e \dfrac{qQ}{r_{min}}$

$$r_{min} = k_e \frac{qQ}{K_\alpha} = \frac{(8.99 \times 10^9 \text{ N} \cdot \text{m}^2 / \text{C}^2)(2)(79)(1.60 \times 10^{-19} \text{ C})^2}{(0.500 \text{ MeV})(1.60 \times 10^{-13} \text{ J} / \text{MeV})} = 455 \text{ fm}$$ ∎

(b) Since $K_\alpha = \dfrac{1}{2}mv^2 = k_e \dfrac{qQ}{r_{min}}$, we find that

$$v = \sqrt{\frac{2k_e qQ}{mr_{min}}}$$

$$v = \sqrt{\frac{2(8.99 \times 10^9 \text{ N} \cdot \text{m}^2 / \text{C}^2)(2)(79)(1.60 \times 10^{-19} \text{ C})^2}{4(1.66 \times 10^{-27} \text{ kg})(3.00 \times 10^{-13} \text{ m})}} = 6.04 \times 10^6 \text{ m/s}$$ ∎

Finalize: The minimum distance in part (a) is on the order of 100 times greater than the combined radii of the particles. For part (b), the alpha particle must have more than 0.5 MeV of energy since it gets closer to the nucleus than the 455 fm found in part (a). Even so, the speed of the alpha particle in part (b) is only about 2% of the speed of light so we are justified in not using a relativistic approach. In solving this problem, we ignored the effect of the electrons around the gold nucleus that tend to "screen" the nucleus so that the alpha particle sees a reduced positive charge. If this screening effect were considered, the potential energy would be slightly reduced and, for the same initial energy, the alpha particle could get closer to the gold nucleus.

---

7. A star ending its life with a mass of four to eight times the Sun's mass is expected to collapse and then undergo a supernova event. In the remnant that is not carried away by the supernova explosion, protons and electrons combine to form a neutron star with approximately twice the mass of the Sun. Such a star can be thought of as a gigantic atomic nucleus. Assume $r = aA^{1/3}$ (Equation 30.1). If a star of mass 3.98×10^{30} kg is composed entirely of neutrons ($m_n = 1.67 \times 10^{-27}$ kg), what would its radius be?

Solution

Conceptualize: The neutron star is really held together by gravity, but it is a good approximation to think of its density as that of nuclear matter. Do you expect a radius like that of the Earth?

Categorize: We find the number of neutrons and then apply the given radius equation to find the size of the object.

Analyze: The number of nucleons in the star is

$$A = \frac{2(1.99 \times 10^{30} \text{ kg})}{1.67 \times 10^{-27} \text{ kg}} = 2.38 \times 10^{57}$$

Therefore, $r = aA^{1/3} = (1.2 \times 10^{-15} \text{ m})(2.38 \times 10^{57})^{1/3} = 16$ km.   ∎

Finalize: It is not the size of the Earth at all! The Sun itself is expected to become a *white dwarf star* after using up all its nuclear fuel in reactions fusing hydrogen to helium to carbon. Still having most of its present mass, it will then be a roughly Earth-size object made of atoms (carbon) that you can visualize as having no space between them. Its density will be in tons per teaspoonful. A star with mass greater than the Chandrasekhar limit of 1.4 times the mass of the Sun has enough gravity to crush its atoms at the end of its life. It becomes the neutron star considered here, with diameter the size of a city and a density of billions of tons per teaspoonful.

---

13. Nuclei having the same mass numbers are called *isobars*. The isotope $^{139}_{57}\text{La}$ is stable. A radioactive isobar, $^{139}_{59}\text{Pr}$, is located below the line of stable nuclei in Figure P30.13 and decays by e^+ emission. Another radioactive isobar of $^{139}_{57}\text{La}$, $^{139}_{55}\text{Cs}$, decays by e^- emission and is located above the line of stable nuclei in Figure P30.13. (a) Which of these three isobars has the highest neutron-to-proton ratio? (b) Which has the greatest binding energy per nucleon? (c) Which do you expect to be heavier, $^{139}_{59}\text{Pr}$ or $^{139}_{57}\text{Cs}$?

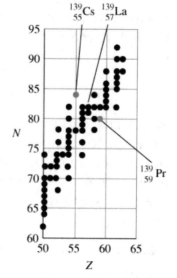

Figure P30.13

Solution

Conceptualize: This problem is about nuclear stability. An unstable nucleus is radioactive. Among unstable nuclei, a less stable nucleus has a shorter half-life.

Categorize: The neutron–proton plot of stable nuclei, Figure 30.4, suggests that special conditions have to be met for a nucleus to be stable. Figure P30.13 is an enlarged section of the overall plot.

Analyze:

(a) For $^{139}_{59}\text{Pr}$ the neutron number is $139 - 59 = 80$. For $^{139}_{55}\text{Cs}$ the neutron number is 84, so the Cs isotope has the greatest neutron-to-proton ratio. ∎

(b) Binding energy per nucleon measures stability so it is greatest for the stable nucleus, the lanthanum isotope. Note also that it has a magic number of neutrons, 82. ∎

(c) Cs-139 has 55 protons and 84 neutrons. Pr-139 has 59 protons and 80 neutrons. Because both are unstable, they are not shown on Figure 30.4 in the textbook. When we locate them on Figure P30.13 here, we see that cesium is a little farther away from the center of the zone of stable nuclei. Being less stable goes with being able to lose more energy in decay, and with having more mass. We therefore expect cesium to be heavier. ∎

Finalize: For nuclear stability there needs to be a "just right" mix of neutrons and protons. Too many protons will tear the nucleus apart with their electrostatic repulsion. Too many neutrons will leave too many unpaired with protons. We do not have a precise predictive theory on what "just right" means. We guess bits of the recipe Nature is following from the nuclei Nature stews up.

15. The radioactive isotope ^{198}Au has a half-life of 64.8 h. A sample containing this isotope has an initial activity ($t = 0$) of 40.0 μCi. Calculate the number of nuclei that decay in the time interval between $t_1 = 10.0$ h and $t_2 = 12.0$ h.

Solution

Conceptualize: One microcurie is 3.7×10^4 decays per second, so the initial decay rate of this sample is about 1.5 million decays per second or 5 billion decays per hour. The answer to the problem will be somewhat less than 10 billion decays in two hours, because the activity is always decreasing, and will be less than ten or twelve hours from now.

Categorize: We will find the number of nuclei remaining at the 10-h mark and the number at the 12-h instant. Then subtraction, not specified by any equation in the textbook, will tell us the number decaying in the time interval between.

Analyze: First, let us find λ and N_0 from the given information. The decay constant is

$$\lambda = \frac{\ln 2}{T_{1/2}} = \frac{0.693}{64.8 \text{ h}} = 0.010\ 7 \text{ h}^{-1} = 2.97 \times 10^{-6} \text{ s}^{-1}$$

The original sample size is

$$N_0 = \frac{R_0}{\lambda} = \left(\frac{40.0 \times 10^{-6} \text{ Ci}}{2.97 \times 10^{-6} \text{ s}^{-1}} \right) \left(\frac{3.70 \times 10^{10} \text{ decays/s}}{1 \text{ Ci}} \right) = 4.98 \times 10^{11} \text{ nuclei}$$

Since $N = N_0 e^{-\lambda t}$, the number of nuclei which decay between times t_2 and t_2 is

$$N_1 - N_2 = N_0 \left(e^{-\lambda t_1} - e^{-\lambda t_2} \right)$$

$$N_1 - N_2 = \left(4.98 \times 10^{11} \right) \left[e^{-\left(0.010\ 7\ \text{h}^{-1} \right)\left(10.0\ \text{h} \right)} - e^{-\left(0.010\ 7\ \text{h}^{-1} \right)\left(12.0\ \text{h} \right)} \right]$$

$$N_1 - N_2 = 9.47 \times 10^9 \text{ nuclei} \qquad\qquad \blacksquare$$

Finalize: Our estimate was good. Notice that 0.01 hours is 36 seconds, but the decay probability per time is $0.010\ 7 \text{ h}^{-1} = 2.97 \times 10^{-6} \text{ s}^{-1}$. This sort of conversion is routinely necessary. Make sure you do not get it upside down.

---

18. A freshly prepared sample of a certain radioactive isotope has an activity of 10.0 mCi. After 4.00 h, its activity is 8.00 mCi. Find (a) the decay constant and (b) the half-life. (c) How many atoms of the isotope were contained in the freshly prepared sample? (d) What is the sample's activity 30.0 h after it is prepared?

Solution

Conceptualize: More tangibly, "freshly prepared" could mean "newly isolated" or "freshly purified." Over the course of 4 hours, this isotope lost 20% of its activity, so its half-life appears to be around 10 hours, which means that its activity after 30 hours (~3 half-lives) will be about 1 mCi. The decay constant and number of atoms are not so easy to estimate.

Categorize: From the rate equation, $R = R_0 e^{-\lambda t}$, we can find the decay constant λ, which can then be used to find the half life, the original number of atoms, and the activity at any other time, t.

Analyze:

(a) The decay constant is

$$\lambda = \frac{1}{t}\ln\left(\frac{R_0}{R}\right) = \left[\frac{1}{(4.00 \text{ h})(3\ 600 \text{ s/h})}\right]\ln\left(\frac{10.0 \text{ mCi}}{8.00 \text{ mCi}}\right)$$

$$= 1.55 \times 10^{-5} \text{ s}^{-1} = 0.055\ 8 \text{ h}^{-1} \qquad\blacksquare$$

(b) The half-life is $\quad T_{1/2} = \dfrac{\ln 2}{\lambda} = \dfrac{0.693}{0.055\ 8 \text{ h}^{-1}} = 12.4 \text{ h}.$ $\qquad\blacksquare$

(c) The number of original atoms can be found if we convert the initial activity from curies into becquerels (decays per second): $1 \text{ Ci} \equiv 3.7 \times 10^{10} \text{ Bq}$.

$$R_0 = 10.0 \text{ mCi} = (10.0 \times 10^{-3} \text{ Ci})(3.70 \times 10^{10} \text{ Bq/Ci}) = 3.70 \times 10^8 \text{ Bq}$$

Since $R_0 = \lambda N_0$, the original number of nuclei is

$$N_0 = \frac{R_0}{\lambda} = \frac{3.70 \times 10^8 \text{ decays/s}}{1.55 \times 10^{-5} \text{ s}} = 2.39 \times 10^{13} \text{ atoms} \qquad\blacksquare$$

(d) The decay rate after thirty hours is

$$R = R_0 e^{-\lambda t} = (10.0 \text{ mCi})e^{-(5.58 \times 10^{-2} \text{ h}^{-1})(30.0 \text{ h})} = 1.88 \text{ mCi} \qquad\blacksquare$$

Finalize: Our estimate of the half-life was about 20% short because we did not account for the non-linearity of the decay rate. Consequently, our estimate of the final activity also fell short, but both of these calculated results are close enough to be reasonable.

The number of atoms is much less than one mole, so this appears to be a very small sample. To get a sense of how small, we can assume that the molar mass is on the order of 100 g/mol, so the sample has a mass of only

$$m \approx \frac{(2.4 \times 10^{13} \text{ atoms})(100 \text{ g/mol})}{6.02 \times 10^{23} \text{ atoms/mol}} \approx 0.004 \ \mu g$$

This sample is so small it cannot be measured by a commercial mass balance!

The problem states that this sample was "freshly prepared," from which we assumed that **all** the atoms within the sample are initially radioactive. Generally this is not the case, so that N_0 only accounts for the formerly radioactive atoms, and does not include additional atoms in the sample that were not radioactive. Realistically, then, the sample mass should be significantly greater than our estimate above.

---

25. The nucleus ${}^{15}_{8}\text{O}$ decays by electron capture. The nuclear reaction is written ${}^{15}_{8}\text{O} + e^- \rightarrow {}^{15}_{7}\text{N} + \nu$. (a) Write the process going on for a single particle within the nucleus. (b) Disregarding the daughter's recoil, determine the energy of the neutrino.

Solution

Conceptualize: The oxygen nucleus swallows one of the inner electrons. The process is a decay in the sense that the system is dropping down into a lower energy state.

Categorize: We will find the energy by comparing masses before and after.

Analyze:

(a) The reaction for one particle is $\qquad e^- + p \rightarrow n + \nu$. ∎

(b) Add 7 protons, 7 neutrons, and 7 electrons to each side to give

$${}^{15}\text{O atom} \rightarrow {}^{15}\text{N atom} + \nu$$

It is the masses of neutral atoms that we can look up in a table.

From Table A.3 (in Appendix A), $M({}^{15}\text{O}) = M({}^{15}\text{N}) + Q/c^2$:

$$|\Delta M| = 15.003\,066 - 15.000\,109 = 0.002\,957 \text{ u}$$

$$Q = (931.5 \text{ MeV/u})(0.002\,957 \text{ u}) = 2.75 \text{ MeV} \qquad ∎$$

Finalize: One proton in the parent nucleus combines with the captured electron to become a neutron. Note that charge is conserved in this process, as in any process for an isolated system. We need a reaction stated in terms of neutral atoms to use the information listed in Table A.3 (in Appendix A).

---

32. Natural gold has only one isotope, ${}^{197}_{79}\text{Au}$. If natural gold is irradiated by a flux of slow neutrons, electrons are emitted. (a) Write the reaction equation. (b) Calculate the maximum energy of the emitted electrons.

Solution

Conceptualize: The $^{197}_{79}$Au will absorb a neutron. The conservation laws for charge and mass number (more generally called baryon number) will let us identify the product nucleus . . .

Categorize: . . . and the daughter after beta decay. Comparing masses will tell us how much energy is liberated in the overall reaction.

Analyze: The $^{197}_{79}$Au absorbs a neutron $^{1}_{0}$n to become $^{198}_{79}$Au, which emits an e^{-} = $^{0}_{-1}$e to become $^{198}_{80}$Hg.

(a) For nuclei the net reaction is

$$^{197}_{79}\text{Au nucleus} + ^{1}_{0}\text{n} \rightarrow ^{198}_{80}\text{Hg nucleus} + ^{0}_{-1}\text{e} + \bar{\nu} \qquad \blacksquare$$

Note the conservation of baryon number (which you can think of as nucleon census number and call mass number in this chapter) in the superscripts: 197 + 1 = 198 + 0. Note the conservation of charge in the subscripts: 79 + 0 = 80 − 1.

(b) Adding 79 electrons to both sides gives the reaction for neutral atoms:

$$^{197}_{79}\text{Au atom} + ^{1}_{0}\text{n} \rightarrow ^{198}_{80}\text{Hg atom} + \bar{\nu}$$

From Table A.3 (in Appendix A), 196.966 569 + 1.008 665 = 197.966 769 + 0 + Q/c^2. The energy released in the reaction is

$$Q = |\Delta m| c^2$$
$$= (0.008\ 465\ \text{u})(931.5\ \text{MeV/u}) = 7.89\ \text{MeV} \qquad \blacksquare$$

Finalize: The ancient alchemists could not transmute other metals into gold by mechanical or chemical means. This nuclear reaction transmutes gold into base metal. For starting with a stable nucleus, the energy released may seem surprisingly large.

It is our choice of units—unified atomic mass units—that makes the masses of atoms so nearly numerically equal to the integer values for the mass numbers A of the atomic nuclei. But the mass numbers A and atomic numbers Z go into reasoning about *what* reaction occurs. The masses M go into different reasoning that lets us figure out the reaction energy.

35. Review. Suppose seawater exerts an average frictional drag force of 1.00×10^5 N on a nuclear-powered ship. The fuel consists of enriched uranium containing 3.40% of the fissionable isotope $^{235}_{92}\text{U}$ and the ship's reactor has an efficiency of 20.0%. If 200 MeV is released per fission event, how far can the ship travel per kilogram of fuel?

Solution

Conceptualize: Nuclear fission is much more efficient for converting mass to energy than burning fossil fuels. However, without knowing the rate of diesel fuel consumption for a comparable ship, it is difficult to estimate the nuclear fuel use rate. It seems plausible that a ship could cross the Atlantic Ocean with only a few kilograms of nuclear fuel, so a reasonable range of uranium fuel consumption might be 100 km/kg to 10 000 km/kg.

Categorize: The fuel consumption rate can be found from the energy released by the nuclear fuel and the work required to push the ship through the water. We use the particle in equilibrium model. The thrust force P exerted on the propeller by the water around it must be equal in magnitude to the backward friction (drag) force on the hull.

Analyze: One kg of enriched uranium contains 3.40% $^{235}_{92}\text{U}$, so the mass of uranium-235 is $m_{235} = 0.034\ 0(1\ 000\ \text{g}) = 34.0$ g.

In terms of number of nuclei, this is equivalent to

$$N_{235} = (34.0\ \text{g})\left(\frac{1}{235\ \text{g/mol}}\right)\left(6.02 \times 10^{23}\ \text{atoms/mol}\right)$$

$$= 8.71 \times 10^{22}\ \text{nuclei}$$

If all these nuclei fission, the energy released is equal to

$$\left(8.71 \times 10^{22}\ \text{nuclei}\right)\left(200 \times 10^6\ \text{eV/nucleus}\right)$$

$$\times \left(1.602 \times 10^{-19}\ \text{J/eV}\right) = 2.79 \times 10^{12}\ \text{J}$$

Now, for the engine,

$$\text{efficiency} = \frac{\text{work output}}{\text{heat input}} \quad \text{or} \quad e = \frac{P\Delta r \cos\theta}{Q_h}.$$

So the distance the ship can travel per kilogram of uranium fuel is

$$\Delta r = \frac{eQ_h}{P\cos(0°)} = \frac{0.200\left(2.79 \times 10^{12}\ \text{J}\right)}{1.00 \times 10^5\ \text{N}} = 5.58 \times 10^6\ \text{m} \qquad \blacksquare$$

Finalize: The ship can travel 5 580 km/kg of uranium fuel, which is on the high end of our prediction range. The distance between the ports of New York and

Southampton, United Kingdom is 5 845 km, so this ship could cross the Atlantic Ocean on just about one kilogram of uranium fuel. Compare this to the Titanic, and its sister ship the Olympic, which burned more than 5 000 tons of coal for the same voyage.

---

37. Carbon detonations are powerful nuclear reactions that temporarily tear apart the cores inside massive stars late in their lives. These blasts are produced by carbon fusion, which requires a temperature of approximately 6×10^8 K to overcome the strong Coulomb repulsion between carbon nuclei. (a) Estimate the repulsive energy barrier to fusion, using the temperature required for carbon fusion. (In other words, what is the average kinetic energy of a carbon nucleus at 6×10^8 K?) (b) Calculate the energy (in MeV) released in each of these "carbon-burning" reactions:

$$^{12}\text{C} + {}^{12}\text{C} \rightarrow {}^{20}\text{Ne} + {}^{4}\text{He}$$

$$^{12}\text{C} + {}^{12}\text{C} \rightarrow {}^{24}\text{Mg} + \gamma$$

(c) Calculate the energy in kilowatt-hours given off when 2.00 kg of carbon completely fuse according to the first reaction.

Solution

Conceptualize: The core of a massive star goes through stages late in the life of the star, getting hotter and hotter as it fuses heavier and heavier nuclei until it gets to iron. We expect less than 1 MeV as the thermal energy of the colliding nuclei and many MeV's as the energy output, amounting to a huge number of kilowatt-hours per kilogram.

Categorize: We review the energy of a particle in an ideal gas. We use Einstein's equation about loss of mass appearing as energy output.

Analyze:

(a) At 6×10^8 K, the carbon nuclei have an average energy of

$$\frac{3}{2}k_\text{B}T = \frac{3}{2}(8.62 \times 10^{-5} \text{ eV/K})(6 \times 10^8 \text{ K}) = 8 \times 10^4 \text{ eV} \qquad \blacksquare$$

(b) The energy released in the first reaction is

$$Q = E = \left[2m(^{12}\text{C}) - m(^{20}\text{Ne}) - m(^{4}\text{He}) \right]c^2$$

$$E = (24.000\ 000 - 19.992\ 440 - 4.002\ 603)(931.5) \text{ MeV} = 4.62 \text{ MeV} \qquad \blacksquare$$

In the second case, the energy released is

$$E = \left[2m\left(^{12}C \right) - m\left(^{24}Mg \right) \right]\left(931.5 \text{ MeV/u} \right)$$

$$E = \left(24.000\,000 - 23.985\,042 \right)\left(931.5 \right) \text{ MeV} = 13.9 \text{ MeV} \qquad \blacksquare$$

(c) The energy released equals the energy of reaction of the number of carbon nuclei in a 2.00-kg sample, which corresponds to

$$\Delta E = \left(\frac{2\,000 \text{ g}}{12 \text{ g/mol}} \right)\left(\frac{6.02 \times 10^{23} \text{ atoms}}{1 \text{ mol}} \right)\left(\frac{1 \text{ fusion}}{2 \text{ atoms}} \right)$$

$$\times \left(\frac{4.62 \text{ MeV}}{\text{fusion}} \right)\left(\frac{1 \text{ kWh}}{2.25 \times 10^{19} \text{ MeV}} \right)$$

$$= 10.3 \times 10^{6} \text{ kWh} \qquad \blacksquare$$

Finalize: To prove that 1 kWh is 2.25×10^{19} MeV, just review facts you should definitely know: k = 1 000, W = J/s, h = 3 600 s, M = 1 000 000, $e = 1.60 \times 10^{-19}$ C, and the really important V = J/C.

The carbon fusion reactions involve 24 nucleons, but put out much less than 24 times the energy of simple hydrogen fusion. The aging star burns through its last resources in a small fraction of the time it spent as a normal star. Tapping into carbon fusion to run a power plant is beyond the potential of the human race.

================

40. Consider the two nuclear reactions

$$\text{(I)} \quad A + B \rightarrow C + E$$

$$\text{(II)} \quad C + D \rightarrow F + G$$

(a) Show that the net disintegration energy for these two reactions ($Q_{net} = Q_I + Q_{II}$) is identical to the disintegration energy for the net reaction

$$A + B + D \rightarrow E + F + G$$

(b) One chain of reactions in the proton–proton cycle in the Sun's core is

$$^{1}_{1}H + {}^{1}_{1}H \rightarrow {}^{2}_{1}H + {}^{0}_{1}e + \nu$$

$$^{0}_{1}e + {}^{0}_{-1}e \rightarrow 2\gamma$$

$$^{1}_{1}H + {}^{2}_{1}H \rightarrow {}^{3}_{2}He + \gamma$$

$$^{1}_{1}H + {}^{3}_{2}He \rightarrow {}^{4}_{2}He + {}^{0}_{1}e + \nu$$

$$^{0}_{1}e + {}^{0}_{-1}e \rightarrow 2\gamma$$

Based on part (a), what is Q_{net} for this sequence?

Solution

Conceptualize: It seems natural that we should be able to add up reactions and use only the net reaction to determine the net energy output. We will prove this fact. Adding up the reactions in part (b) will give the net reaction 4 hydrogen-1 atoms → helium-4 atom. An example in the textbook shows that the energy output is 26.7 MeV, so we should get this value again.

Categorize: It is all based on the Einstein mass-energy relation.

Analyze: In the symbols given in the problem, the energy output of the first reaction is

$$Q_I = (M_A + M_B - M_C - M_E)c^2$$

and that of the second reaction is $Q_{II} = (M_C + M_D - M_F - M_G)c^2$.

The net energy output is then

$$Q_{net} = (M_A + M_B - M_C - M_E + M_C + M_D - M_F - M_G)c^2$$

$$= (M_A + M_B + M_D - M_E - M_F - M_G)c^2$$

(a) This value is identical to Q for the reaction A + B + D → E + F + G. Thus, any product (like "C") that is a reactant in a subsequent reaction disappears from the energy balance. ∎

(b) Adding all five reactions, we have $4\left({}^1_1H\right) + 2\left({}^0_{-1}e\right) \rightarrow {}^4_2He + 2\nu$.

Here the symbol ${}^1_1H$ represents a proton, the nucleus of a hydrogen-1 atom. So that we may use the tabulated masses of neutral atoms for calculation, we add two electrons to each side of the reaction. The four electrons on the initial side are enough to make four hydrogen atoms and the two electrons on the final side constitute, with the alpha particle, a neutral helium atom. The atomic-electronic binding energies are negligible compared to Q_{net}.

We use the arbitrary symbol "${}^1_1H$ atom" to represent a neutral atom. Then we have the reaction $4\left({}^1_1H \text{ atom}\right) \rightarrow {}^4_2He \text{ atom} + 2\nu$

and the corresponding mass-energy equation

$$4(1.007\ 825\ u) = 4.002\ 603\ u + Q_{net}/c^2$$

The energy released is

$$Q_{net} = [4(1.007\ 825\ u) - 4.002\ 603\ u](931.5\ \text{MeV/u}) = 26.7\ \text{MeV} \quad ∎$$

Finalize: Both parts worked out just right. The neutrinos created in the fusion reaction have negligible mass, but carry off some energy, so the star has to stay hot on somewhat less than 26.7 MeV per reaction.

---

51. A small building has become accidentally contaminated with radioactivity. The longest-lived material in the building is strontium-90. ($^{90}_{38}$Sr has an atomic mass 89.907 7 u, and its half-life is 29.1 yr. It is particularly dangerous because it substitutes for calcium in bones.) Assume the building initially contained 5.00 kg of this substance uniformly distributed throughout the building and the safe level is defined as less than 10.0 decays/min (which is small compared to background radiation). How long will the building be unsafe?

Solution

Conceptualize: Two fission bombs have been used in war and many have been tested. They create a mix of radioactive products, such as strontium-90, dispersed into the atmosphere. A localized blob containing five kilograms of a radioactive isotope with a short half-life will have a very high decay rate originally, but you can contain it, keep people away, and wait for the activity to go down to a low value. With five kilograms of an isotope with a very long half-life, the decay rate will be very low. We and our human and non-human ancestors have always lived with such radioactive materials in the Earth's crust. It is the medium-half-life materials, from strontium-90 to plutonium-239 (24 120 yr), that give us the most trouble. We estimate centuries.

Categorize: We can find the sample size as a number of nuclei, and then the original decay rate. Then the law of radioactive decay will tell us the time interval required for the activity to get down.

Analyze: The number of nuclei in the original sample is

$$N_0 = \frac{\text{mass present}}{\text{mass of nucleus}} = \frac{5.00 \text{ kg}}{(89.907\ 7 \text{ u})(1.66 \times 10^{-27} \text{ kg/u})}$$

$$= 3.35 \times 10^{25} \text{ nuclei}$$

The decay constant is

$$\lambda = \frac{\ln 2}{T_{1/2}} = \frac{0.693}{29.1 \text{ yr}} = 2.38 \times 10^{-2} \text{ yr}^{-1} = 4.53 \times 10^{-8} \text{ min}^{-1}$$

The original activity is

$$R_0 = \lambda N_0 = (4.53 \times 10^{-8} \text{ min}^{-1})(3.35 \times 10^{25} \text{ nuclei})$$

$$= 1.52 \times 10^{18} \text{ decays/min}$$

The law of decay then gives us

$$\frac{R}{R_0} = \frac{10.0 \text{ decays/min}}{1.52 \times 10^{18} \text{ decays/min}} = 6.59 \times 10^{-18} = e^{-\lambda t}$$

and the time interval is

$$t = \frac{-\ln(R/R_0)}{\lambda} = \frac{-\ln(6.59 \times 10^{-18})}{2.38 \times 10^{-2} \text{ yr}^{-1}} = 1\ 660 \text{ yr} \qquad \blacksquare$$

Finalize: What sort of containment can exclude all curiosity seekers, pranksters, criminals, and irresponsible politicians for millennia? The people working on this question do not have any big successes to report.

---

55. When gamma rays are incident on matter, the intensity of the gamma rays passing through the material varies with depth x as $I(x) = I_0 e^{-\mu x}$, where I_0 is the intensity of the radiation at the surface of the material (at $x = 0$) and μ is the linear absorption coefficient. For low-energy gamma rays in steel, take the absorption coefficient to be 0.720 mm^{-1}. (a) Determine the "half-thickness" for steel, that is, the thickness of steel that would absorb half the incident gamma rays. (b) In a steel mill, the thickness of sheet steel passing into a roller is measured by monitoring the intensity of gamma radiation reaching a detector below the rapidly moving metal from a small source immediately above the metal. If the thickness of the sheet changes from 0.800 mm to 0.700 mm, by what percentage does the gamma-ray intensity change?

Solution

Conceptualize: You can think of the absorption of gamma rays by steel as basically like the absorption of visible light by a filter, such as the lenses of sunglasses. Energy comes in carried by electromagnetic radiation. The portion that does not make it out the other side turns into internal energy, slightly raising the temperature of the material.

Categorize: The exponential decay in space described by the equation in the problem parallels the exponential decay in time characteristic of radioactive decay. So a certain thickness of the absorbing material will reduce the radiation intensity by half. Twice that thickness will reduce it to one-quarter of the incident intensity, and so on. The steel-mill thickness gauge described can monitor the sheet thickness in real time so that the roller pressure can be adjusted to make the thickness of the roller output more uniform.

Analyze:

(a) Let x_h represent the unknown half-thickness, according to

$$\frac{1}{2}I_0 = I_0 e^{-\mu x_h} \qquad \rightarrow \qquad 2 = e^{+\mu x_h} \qquad \rightarrow \qquad \mu x_h = \ln 2$$

so $\qquad x_h = \dfrac{\ln 2}{\mu} = \dfrac{\ln 2}{0.720/\text{mm}} = 0.963 \text{ mm.}$ $\qquad \blacksquare$

(b) The intensity reaching the detector through $x_1 = 0.800$ mm of steel is

$I_1 = I_0 e^{-\mu x_1}$. That transmitted by thickness $x_2 = 0.700$ mm is $I_2 = I_0 e^{-\mu x_2}$. The fractional change is

$$\frac{I_2 - I_1}{I_1} = \frac{I_0 e^{-\mu x_2} - I_0 e^{-\mu x_1}}{I_0 e^{-\mu x_1}} = e^{\mu(x_1 - x_2)} - 1 = e^{(0.720/\text{mm})(0.100 \text{ mm})} - 1$$

$$= e^{0.0720} - 1 = +0.074\ 7 = 7.47\%$$

As the thickness decreases, the intensity increases by 7.47%. ∎

Finalize: The fractional change in thickness has magnitude 12.5%, but the smaller change in gamma ray intensity should be easily measurable, say from a counting rate in a Geiger tube or from a current in an ionization chamber. The gamma radiation passing through the steel does no harm; the photons stopping inside the steel do not damage it, but produce only a small temperature increase. Dragging a radiation source through a pipeline can be a good way to check for cracks and faulty welds.

Note that it is impractical or impossible to set up shielding that will block *all* of the radiation from some source. The best that can be done is to reduce its intensity to be small compared to background radiation. For citizen education and in emergency situations, scientists must be careful to inform the public clearly and accurately about radiation hazards and their mitigation.

---

58. (a) Calculate the energy (in kilowatt-hours) released if 1.00 kg of ^{239}Pu undergoes complete fission and the energy released per fission event is 200 MeV. (b) Calculate the energy (in electron volts) released in the deuterium–tritium fusion reaction

$$^{2}_{1}\text{H} + ^{3}_{1}\text{H} \rightarrow ^{4}_{2}\text{He} + ^{1}_{0}\text{n}$$

(c) Calculate the energy (in kilowatt-hours) released if 1.00 kg of deuterium undergoes fusion according to this reaction. (d) **What If?** Calculate the energy (in kilowatt-hours) released by the combustion of 1.00 kg of coal if each $C + O_2 \rightarrow CO_2$ reaction yields 4.20 eV. (e) List advantages and disadvantages of each of these methods of energy generation.

Solution

Conceptualize: The problem compares one kilogram of very different fuels. We estimate that the energy from fission [in part (a)] is more than a million times the chemical energy [part (d)] and that the fusion energy [part (c)] will be more than ten times larger still.

Categorize: In parts (a), (c), and (d) we can find the correspondences by multiplying the quantity of fuel by continued conversion factors. In part (b) we will compare masses before and after the reaction to find the Q value.

Analyze:

(a) The energy release of the plutonium is

$$E = (1\ 000\ \text{g})\left(\frac{6.02 \times 10^{23}\ \text{nuclei}}{239\ \text{g}}\right)\left(\frac{200 \times 10^6\ \text{eV}}{1\ \text{nucleus}}\right)$$

$$\times \left(\frac{1.60 \times 10^{-19}\ \text{J/eV}}{3.60 \times 10^6\ \text{J/kWh}}\right)$$

$$= 2.24 \times 10^7\ \text{kWh}$$ ■

(b) We obtain the masses of the particles from Table A.3 (in Appendix A). The total mass of reactants is $m_{\text{before}} = 2.014\ 102\ \text{u} + 3.016\ 049\ \text{u} = 5.030\ 151\ \text{u}$

The mass of the products is $m_{\text{after}} = 4.002\ 603\ \text{u} + 1.008\ 665\ \text{u} = 5.011\ 268\ \text{u}$

For the change in mass we have $|\Delta m| = m_{\text{before}} - m_{\text{after}} = 0.018\ 883\ \text{u}$

The energy released is

$$\Delta E = |\Delta m| c^2 = (0.018\ 883\ \text{u})(931.494\ \text{MeV/u}) = 17.6\ \text{MeV}$$ ■

(c) One kilogram of deuterium can then release energy

$$E = \left(1\ 000\ \text{g}\ {}_1^2\text{H}\right)\left(\frac{6.02 \times 10^{23}\ \text{deuterons}}{2.014\ \text{g}\ {}_1^2\text{H}}\right)$$

$$\times \left(\frac{17.6\ \text{MeV}}{{}_1^2\text{H fusion}}\right)\left(\frac{1.60 \times 10^{-13}\ \text{J/MeV}}{3.60 \times 10^6\ \text{J/kWh}}\right)$$

$$= 2.34 \times 10^8\ \text{kWh}$$ ■

(d) Coal is essentially pure carbon. Assuming complete combustion, the energy output of one kilogram is

$$(1\ 000\ \text{g C})\left(\frac{6.02 \times 10^{23}\ \text{C atoms}}{12.0\ \text{g C}}\right)$$

$$\times \left(\frac{4.20\ \text{eV}}{\text{C atom}}\right)\left(\frac{1.60 \times 10^{-19}\ \text{J/eV}}{3.60 \times 10^6\ \text{J/kWh}}\right) = 9.36\ \text{kWh}$$ ■

(e) You likely pay the electric company to burn coal, because coal is cheap at this moment in human history. The limit on supply will inevitably drive up the price of fossil fuels. Worldwide, electrically transmitted energy from fission is an immediate option. Fission may offer advantages of reduced carbon dioxide emissions, reduced overall pollution, and even a reduction in the radiation released into the atmosphere, compared to mining and burning coal. However, the Chernobyl explosion and other accidents demonstrate that there is a continuing significant risk of catastrophic disaster. For a fair comparison to a coal-fired plant, we should think of a nuclear generating station paying for long-term disposal of its waste and paying for its own insurance, without government subsidies. In rich countries we urgently need to reduce total energy use and to employ renewable energy sources. We hope that fusion reactors will become practical in the future, because fusion might be both safer and less expensive than either coal or fission.   ■

Finalize: Our answer to part (b) agrees with the statement in the chapter text. For each kilogram of fuel, the energy output of fission is on the order of ten million times larger than that of chemical burning, and the energy output of fusion ten times larger still.

---

67. Free neutrons have a characteristic half-life of 10.4 min. What fraction of a group of free neutrons with kinetic energy 0.040 0 eV decays before traveling a distance of 10.0 km?

Solution

Conceptualize: We guess that these neutrons are moving pretty fast, so they are not in flight for very long, and perhaps one percent may decay.

Categorize: The fraction that will remain is given by the ratio N/N_0, where $N/N_0 = e^{-\lambda \Delta t}$ and Δt is the time interval it takes the neutron to travel a distance of $d = 10.0$ km. Further, we use the particle under constant speed model.

Analyze: The time of flight is given by $\Delta t = d/v$.

Since $K = \dfrac{1}{2}mv^2$, $\Delta t = \dfrac{d}{\sqrt{\dfrac{2K}{m}}} = \dfrac{10.0 \times 10^3 \text{ m}}{\sqrt{\dfrac{2(0.040\ 0 \text{ eV})\left(1.60 \times 10^{-19} \text{ J/eV}\right)}{1.67 \times 10^{-27} \text{ kg}}}} = 3.61 \text{ s.}$

The decay constant is $\lambda = \dfrac{0.693}{T_{1/2}} = \dfrac{0.693}{(10.4 \text{ min})(60 \text{ s/min})} = 1.11 \times 10^{-3} \text{ s}^{-1}.$

Therefore we have

$$\lambda \Delta t = \left(1.11 \times 10^{-3} \text{ s}\right)(3.61 \text{ s}) = 4.01 \times 10^{-3} = 0.004\ 01$$

And the fraction remaining is $\dfrac{N}{N_0} = e^{-\lambda \Delta t} = e^{-0.004\ 01} = 0.996\ 0.$

Hence, the fraction that has decayed in this time interval is

$$1 - \dfrac{N}{N_0} = 0.004\ 00 \quad \text{or} \quad 0.400\% \qquad\qquad ■$$

Finalize: We made a good guess! Neutron loss by decay is a minor concern in designing fission reactors.

Chapter 31
Particle Physics

Section 31.1   The Fundamental Forces in Nature

There are four fundamental forces in nature: **strong** (hadronic), **electromagnetic**, **weak**, and **gravitational**. A residual effect of the strong force is the nuclear force, which acts between nucleons to keep the nucleus together. The weak force is responsible for beta decay. The electromagnetic and weak forces are now considered to be manifestations of a single force called the **electroweak** force.

Section 31.2   Positrons and Other Antiparticles
Section 31.3   Mesons and the Beginning of Particle Physics

An antiparticle and a particle have the same mass, but opposite charge. Furthermore, other properties may have opposite values such as lepton number and baryon number. It is possible to produce particle-antiparticle pairs in nuclear reactions if the available energy is greater than $2mc^2$, where m is the mass of the particle.

Pair production is a process in which a gamma ray with an energy of at least 1.02 MeV interacts with a nucleus, and an electron-positron pair is created.

Pair annihilation is an event in which an electron and a positron can annihilate to produce two gamma rays, each with an energy of at least 0.511 MeV.

A **Feynman diagram** can be used to represent the interaction between two particles.

Section 31.4   Classification of Particles

We model the universe as composed of field particles and matter particles. The exchange of field particles mediates the interactions of the matter particles.

426

- **Gluons** are the field particles for the strong force.

- **Photons** are exchanged by charged particles in electromagnetic interactions.

- **W⁺, W⁻, and Z particles** mediate the weak force.

- **Gravitons** are hypothetical quantum particles of the gravitational field.

The **matter particles** fall into two broad classifications: **hadrons** and **leptons**.

Hadrons are particles that interact via all fundamental forces. They are composed of quarks. There are two types of hadrons grouped according to their masses and spins.

- **Mesons** — decay finally into electrons, positrons, neutrinos, and photons; they have spin quantum number 0 or 1.

- **Baryons** — (with the exception of the proton) decay in a manner leading to end products which include a proton; they have spin quantum number $\frac{1}{2}$ or $\frac{3}{2}$.

- **Leptons** (e^- , μ^- , τ^- , ν_e , ν_μ , ν_τ and their antiparticles) interact by the weak interaction, the electromagnetic interaction (if charged), and (presumably) the gravitational interaction, but not by the strong force. All leptons have spin quantum number $\frac{1}{2}$.

All particles have a corresponding antiparticle. A particle and its antiparticle have the same mass and spin, and have opposites of all other "charges" (electric charge, baryon number, strangeness, etc.). For a particle composed of quarks, its antiparticle is composed of the corresponding antiquarks. There are a few neutral mesons which are their own antiparticle.

Section 31.5   Conservation Laws
Section 31.6   Strange Particles and Strangeness

Conservation laws of elementary particles are based on empirical evidence:

- **Baryon number**—Whenever a nuclear reaction or decay occurs, the sum of the baryon numbers before the process must equal the sum of the baryon numbers after the process. This requirement is quantified by assigning a baryon number, B ($B = +1$ for baryons, $B = -1$ for antibaryons, $B = 0$ for all other particles).

- **Lepton number**—In this case there are three conservation laws, quantified by assigning three lepton numbers, one for each variety of lepton: electron-lepton number, L_e; muon-lepton number, L_μ; and tau-lepton number. In each case the sum of the lepton numbers before a reaction or decay must equal the sum of the lepton numbers after the reaction or decay.

- **Strangeness**—Strange particles are always produced in pairs by the strong interaction and decay very slowly, as is characteristic of the weak interaction. Strange particles are assigned a strangeness quantum number, S; quantitatively $S = \pm 1, \pm 2, \pm 3$ for strange particles and $S = 0$ for nonstrange particles. Whenever a nuclear reaction or decay occurs via the strong or electromagnetic interactions, the sum of the strangeness numbers before the process must equal the sum of the strangeness numbers after the process ($\Delta S = 0$). Decays occurring via the weak interaction are accompanied by the loss of one strange particle; the process proceeds slowly via the weak interaction and violates the conservation of strangeness.

- **Relativistic total energy**—This quantity is conserved in all reactions, but an apparent fluctuation in energy ΔE can occur in the creation of a virtual particle provided that the particle exists for a time no longer than $\Delta t = \dfrac{\hbar}{2\Delta E}$.

- **Electric charge**—This scalar quantity is conserved in all reactions.

- **Linear momentum and angular momentum**—These vector quantities are conserved in all reactions.

Integer quantum numbers for strangeness, charm, topness, and bottomness are assigned to baryons. These properties are conserved in interactions involving the strong force, but not necessarily in processes occurring via the weak interaction. These quantum numbers are accounted for in the quark model of hadrons.

Section 31.9 Quarks

Quarks and antiquarks (of six possible types or flavors) make up baryons and mesons; the identity of each particle is determined by the particular combination of quarks. *Each baryon contains three quarks and each meson contains one quark and one antiquark.* The table at right lists the charge and baryon number for each of the six quarks and antiquarks.

Quark, antiquark		Charge	Baryon number
Up:	$u, \bar{u}$	$+\frac{2}{3}e, -\frac{2}{3}e$	$\frac{1}{3}, -\frac{1}{3}$
Down:	$d, \bar{d}$	$-\frac{1}{3}e, +\frac{1}{3}e$	$\frac{1}{3}, -\frac{1}{3}$
Strange:	$s, \bar{s}$	$-\frac{1}{3}e, +\frac{1}{3}e$	$\frac{1}{3}, -\frac{1}{3}$
Charm:	$c, \bar{c}$	$+\frac{2}{3}e, -\frac{2}{3}e$	$\frac{1}{3}, -\frac{1}{3}$
Bottom:	$b, \bar{b}$	$-\frac{1}{3}e, +\frac{1}{3}e$	$\frac{1}{3}, -\frac{1}{3}$
Top:	$t, \bar{t}$	$+\frac{2}{3}e, -\frac{2}{3}e$	$\frac{1}{3}, -\frac{1}{3}$

The strong force between quarks is referred to as the **color force**, and is mediated by massless particles called **gluons**. **Color charge** (red, green, blue) is a property assigned to quarks, which allows combinations of quarks to satisfy the exclusion principle.

EQUATIONS AND CONCEPTS

Pions are in three varieties corresponding to three charge states: π^+, π^-, and π^0. Pions are very unstable; the equations at right show example decays for each type.

$$\pi^+ \rightarrow \pi^0 + e^+ + \nu_e$$
$$\pi^- \rightarrow \mu^- + \bar{\nu}_\mu$$
$$\pi^0 \rightarrow \gamma + \gamma$$

Muons are in two varieties: μ^- and μ^+ (the antiparticle). Example decay processes are shown for each type.

$$\mu^+ \rightarrow e^+ + \nu_e + \bar{\nu}_\mu$$
$$\mu^- \rightarrow e^- + \nu_\mu + \bar{\nu}_e$$

Hubble's law states a linear relationship between the velocity of a galaxy and its distance R from the Earth. The constant H is called the **Hubble parameter**.

$$v = HR \tag{31.4}$$

$$H \approx 22 \times 10^{-3} \, \text{m}/(\text{s} \cdot \text{ly})$$

REVIEW CHECKLIST

- Be aware of the four fundamental forces in nature and the corresponding field particles or quanta via which these forces are mediated.

- Understand the concepts of the antiparticle, pair production, and pair annihilation.

- Know the broad classification of particles and the characteristic properties of the several classes (relative mass value, spin, and decay mode).

ANSWER TO AN OBJECTIVE QUESTION

2. Define the average density of the solar system ρ_{ss} as the total mass of the Sun, planets, satellites, rings, asteroids, icy outliers, and comets, divided by the volume of a sphere around the Sun large enough to contain all these objects. The sphere extends about halfway to the nearest star, with a radius of approximately 2×10^{16} m, about two light-years. How does this average density of the solar system compare with the critical density ρ_c required for the Universe to stop its Hubble's-law expansion? (a) ρ_{ss} is much greater than ρ_c. (b) ρ_{ss} is approximately or precisely equal to ρ_c. (c) ρ_{ss} is much less than ρ_c. (d) It is impossible to determine.

Answer   (a) The average density of the solar system, as defined in the question, can be considered a measure of the average density of the galaxy. It is much greater than the critical density that would make gravity a strong enough force to stop the expansion of the Universe. Recall that that critical density is in turn greater than the actual average density of the Universe. It is the great gulfs between galaxies, empty as far as we know, and especially the colossal spaces between galaxy clusters, including the voids that show up on the largest scales we can map, that bring the average concentration of matter in the Universe way down.

We can estimate ρ_{ss} numerically. Most of the mass of the solar system is that in the Sun, 1.99×10^{30} kg. The volume of the sphere defined in the question is $(4/3)\pi r^3 = (4/3)\pi(2 \times 10^{16}\ \text{m})^3 = 3 \times 10^{49}\ \text{m}^3$. Then we have $\rho_{ss} = m/V = 1.99 \times 10^{30}\ \text{kg}/3 \times 10^{49}\ \text{m}^3 = 6 \times 10^{-20}\ \text{kg/m}^3$. This is ten million times larger than the critical density $3H^2/8\pi G$ computed in the chapter text. One could say that gravity in the solar system, and in the Milky Way galaxy, is such a strong force that it has stopped the Hubble expansion in our locality. Hubble's law as an observational result describes the expansion of distances between clusters of galaxies.

ANSWERS TO SELECTED CONCEPTUAL QUESTIONS

1. The Ξ^0 particle decays by the weak interaction according to the decay mode $\Xi^0 \rightarrow \Lambda^0 + \pi^0$. Would you expect this decay to be fast or slow? Explain.

Answer This decay should be slow, since decays which occur via the weak interaction typically take 10^{-10} s or longer to occur. This decay cannot occur by the strong interaction, because at the level of quarks it is uss $\rightarrow$ uds + $\bar{d}$d, changing the net strangeness quantum number.

□   □   □   □

8. Describe the properties of baryons and mesons and the important differences between them.

Answer There are two types of hadrons, called baryons and mesons. Hadrons interact primarily through the strong force and are not elementary particles, being composed of either three quarks (baryons), or a quark and an antiquark (mesons). Baryons have a non-zero baryon number with a spin of either 1/2 or 3/2. Mesons have baryon number of zero, and a spin of either 0 or 1.

□   □   □   □

9. How many quarks are in each of the following: (a) a baryon, (b) an antibaryon, (c) a meson, (d) an antimeson? (e) How do you explain that baryons have half-integral spins whereas mesons have spins of 0 or 1?

Answer (a) and (b) All baryons and antibaryons consist of three quarks. (c) and (d) All mesons and antimesons consist of two quarks. When one quantum particle with spin 1/2 combines with another particle to form a new particle, the spin of the new particle must be either 1/2 greater or 1/2 less than that of the original. Since quarks have spins of 1/2, it follows that all baryons (which consist of three quarks) must have half-integral spins, and all mesons (which consist of two quarks) must have spins of 0 or 1.

□   □   □   □

SOLUTIONS TO SELECTED END-OF-CHAPTER PROBLEMS

5. A photon with an energy $E_\gamma = 2.09$ GeV creates a proton–antiproton pair in which the proton has a kinetic energy of 95.0 MeV. What is the kinetic energy of the antiproton? *Note:* $m_p c^2 = 938.3$ MeV.

Solution

Conceptualize: An antiproton has the same mass as a proton, so it seems reasonable to expect that both particles will have similar kinetic energies.

Categorize: The total energy of each particle is the sum of its rest energy and its kinetic energy. Conservation of system energy requires that the total energy before this pair production event equal the total energy after.

Analyze: $E_\gamma = \left(E_{Rp} + K_p\right) + \left(E_{R\bar{p}} + K_{\bar{p}}\right)$

The energy of the photon is given as $E_\gamma = 2.09 \text{ GeV} = 2.09 \times 10^3 \text{ MeV}.$

From Table 31.2 or from the problem statement, we see that the rest energy of both the proton and the antiproton is

$$E_{Rp} = E_{R\bar{p}} = m_p c^2 = 938.3 \text{ MeV}$$

If the kinetic energy of the proton is observed to be 95.0 MeV, the kinetic energy of the antiproton is

$$K_{\bar{p}} = E_\gamma - E_{R\bar{p}} - E_{Rp} - K_p$$
$$= 2.09 \times 10^3 \text{ MeV} - 2(938.3 \text{ MeV}) - 95.0 \text{ MeV}$$
$$= 118 \text{ MeV} \qquad\qquad \blacksquare$$

Finalize: The kinetic energy of the antiproton is slightly (~20%) greater than the proton. The magnitude of the momentum of each product particle is less than 1/4 of the gamma ray's momentum. Thus, another particle must have been involved to satisfy system momentum conservation. It could be a pre-existing heavy nucleus with which the photon collided. This means that the nucleus must have also carried away some of the energy. For a heavy nucleus this could have been small, as little as 3 MeV for a favorable geometry. The actual value cannot be determined without further information. This extra particle also explains why the energy need not be shared equally between the proton and antiproton.

---

7.   A neutral pion at rest decays into two photons according to $\pi^0 \to \gamma + \gamma$. Find the (a) energy, (b) momentum, and (c) frequency of each photon.

Solution

Conceptualize: A pion is not a very massive particle, but we guess that it is heavy enough for these to be gamma-ray photons.

Categorize: We use the energy and momentum versions of the isolated system model.

Analyze: Since the pion is at rest and system momentum is conserved in the decay, the two gamma rays must have equal amounts of momentum in opposite directions. So they must share equally in the energy of the pion.

We have the original mass $m_{\pi^0} = 135.0$ MeV/c^2 (Table 31.2).

(a)   Therefore the energy of each photon is $E_\gamma = 67.5$ MeV $= 1.08 \times 10^{-11}$ J.   ▪

(b)   Its momentum is $p = \dfrac{E_\gamma}{c} = \dfrac{67.5 \text{ MeV}}{3.00 \times 10^8 \text{ m/s}} = 3.60 \times 10^{-20}$ kg·m/s,   ▪

(c)   and its frequency is $f = \dfrac{E_\gamma}{h} = 1.63 \times 10^{22}$ Hz.   ▪

Finalize: Gamma ray is right! This electromagnetic wave is at the top of the spectrum chart in the text's Figure 24.11. Currently [March 2012] the Fermi Space Telescope is searching for gamma rays at this energy and beyond from starburst galaxies, as compelling evidence that supernova explosions and their aftermath produce cosmic rays. The logic: In a region where many stars are being born, numerous large-mass stars are dying as supernovas. Expanding shock waves around them accelerate charged particles to energies up to the TeV range. The charged particles themselves do not go off in straight lines—they get all mixed up in direction by magnetic fields. Those that we on Earth receive as cosmic rays do not reveal their source to us. But some of the particles collide or decay within the source galaxy to produce pions, which in turn decay into photons by the process considered here. We on Earth can sample the gamma rays and (we hope) trace them straight back to the source.

---

9.   One mediator of the weak interaction is the Z^0 boson, with mass 91 GeV/c^2. Use this information to find the order of magnitude of the range of the weak interaction.

Solution

Conceptualize: Following Yukawa, we visualize the weak interaction as acting between two matter particles when they exchange a virtual Z^0 boson. The boson should not exist according to classical energy conservation, and . . .

Categorize: . . . it cannot exist for a time interval longer than that allowed by the uncertainty principle. This is the condition that tells us about the range of the force.

Analyze: The rest energy of the Z^0 boson is $E_0 = 91$ GeV. The maximum time a virtual Z^0 boson can exist is found from

$$\Delta E \Delta t \geq \tfrac{1}{2}\hbar$$

or $$\Delta t = \frac{\hbar}{2\Delta E} = \frac{1.055 \times 10^{-34} \text{ J} \cdot \text{s}}{2(91 \text{ GeV})(1.60 \times 10^{-10} \text{ J/GeV})} = 3.62 \times 10^{-27} \text{ s}$$

The maximum distance it can travel in this time interval is

$$d = c\Delta t = (3.00 \times 10^{8} \text{ m/s})(3.62 \times 10^{-27} \text{ s}) \sim 10^{-18} \text{ m}$$ ■

The distance d is an approximate value for the range of the weak interaction.

Finalize: Our answer agrees with the datum in Table 31.1.

⸻

13. Name one possible decay mode for Ω^{+}, $\overline{K}{}_{S}^{0}$, $\overline{\Lambda}{}^{0}$, and $\overline{n}$.

Solution

Conceptualize: The Ω^{+} is the antiparticle of Ω^{-} and is a baryon, the $\overline{K}{}_{S}^{0}$ is the antiparticle of the $\overline{K}{}_{S}^{0}$ kaon and is a meson, the $\overline{\Lambda}{}^{0}$ is the antiparticle of the $\overline{\Lambda}{}^{0}$ baryon, and the $\overline{n}$ is the antiparticle of the neutron.

Categorize: We use the conservation laws in Section 31.5 in the textbook, also noting that particles decay into products that are less massive than the original particle.

Analyze:

- Figure 31.10 in the textbook shows the bubble chamber photograph in which the Ω^{-} was first observed. The diagram in Figure 31.10 shows the Ω^{-} decaying into a Ξ^{0} and a negative pion. We can therefore guess that the Ω^{+} antiparticle will also undergo a similar decay process:

  $$\Omega^{+} \rightarrow \overline{\Xi}{}^{0} + \pi^{+}$$

- The $\overline{K}{}_{S}^{0}$ kaon is a meson and must decay into less massive mesons. The only mesons that are less massive than kaons are pions. Also, the $\overline{K}{}_{S}^{0}$ kaon is charge neutral, so from the conservation of charge, the total charge of the products also has to be zero. Two decay modes are possible:

  $$\overline{K}{}_{S}^{0} \rightarrow \pi^{+} + \pi^{-} \text{ (or } \pi^{0} + \pi^{0})$$

- The $\overline{\Lambda}{}^{0}$ is a baryon with baryon number –1 and neutral charge. Since baryons ultimately decay into protons, and since protons and neutrons are the only baryons less massive than the lambda particle, we guess that the one of the decay products of the $\overline{\Lambda}{}^{0}$ antiparticle is the antiproton, also with a baryon number of –1. Since the antiproton has negative charge and a baryon number of –1, the second decay product cannot be a baryon and must be a positively charged meson. Subtracting the rest masses of the $\overline{\Lambda}{}^{0}$

and the antiproton gives 1 156.6 MeV/c² − 938.3 MeV/c² = 218.3 MeV/c². The only meson with rest mass less than this value is the π^+ pion. Thus,

$$\overline{\Lambda}^0 \rightarrow \overline{p} + \pi^+$$

- The antiparticle of the neutron, the antineutron, is a baryon with baryon number −1, thus one of the decay products must be a baryon. The only baryon less massive than the neutron is the proton, so one of the decay products must be the antiparticle of the proton. Since the difference between the rest masses of the antineutron and antiproton is 939.6 MeV/c² − 938.3 MeV/c² = 1.3 MeV/c², the remaining product(s) must be one or more leptons. The antiproton has a negative charge, which necessitates a positively charged anti-lepton. Only the positron (e⁺) has low enough energy for this reaction. Finally, because of the conservation of electron lepton number, the positron (L_e = −1) must be accompanied by an electron neutrino (L_e = +1). Thus we have

$$\overline{n} \rightarrow \overline{p} + e^+ + \nu_e$$

Finalize: We used conservation of energy, baryon number, electron lepton number, and charge in finding the decay products of the four antiparticles. The elusive Ω^+ particle was first observed at the Stanford Linear Accelerator Center in 1971.

---

17. The following reactions or decays involve one or more neutrinos. In each case, supply the missing neutrino (ν_e, ν_μ, or ν_τ) or antineutrino.

(a) $\pi^- \rightarrow \mu^- + ?$ (b) $K^+ \rightarrow \mu^+ + ?$

(c) $? + p \rightarrow n + e^+$ (d) $? + n \rightarrow p + e^-$

(e) $? + n \rightarrow p + \mu^-$ (f) $\mu^- \rightarrow e^- + ? + ?$

Solution

Conceptualize: There are just six choices: an electron neutrino, a muon neutrino, a tau neutrino, or the antineutrino version of each one.

Categorize: By following the conservation of electron-lepton number, muon-lepton number, and tau-lepton number we will obtain each answer.

Analyze:

(a) The muon-lepton number in the reaction $\pi^- \rightarrow \mu^- + ?$ changes from 0 to 1 + ?, so the muon-lepton number of the final particle must be −1 and the reaction must be $\pi^- \rightarrow \mu^- + \overline{\nu}_\mu$ to give $L_\mu: 0 \rightarrow 1 - 1$.   ■

(b) Following the same logic we have $K^+ \rightarrow \mu^+ + \nu_\mu$ so that the muon lepton number will be conserved according to $L_\mu : 0 \rightarrow -1 + 1$. ∎

(c) $\bar{\nu}_e + p^+ \rightarrow n + e^+$ $\qquad L_e : -1 + 0 \rightarrow 0 - 1$ ∎

(d) $\nu_e + n \rightarrow p^+ + e^-$ $\qquad L_e : 1 + 0 \rightarrow 0 + 1$ ∎

(e) $\nu_\mu + n \rightarrow p^+ + \mu^-$ $\qquad L_\mu : 1 + 0 \rightarrow 0 + 1$ ∎

(f) $\mu^- \rightarrow e^- + \bar{\nu}_e + \nu_\mu$ $\qquad L_\mu : 1 \rightarrow 0 + 0 + 1$ and $L_e : 0 \rightarrow 1 - 1 + 0$ ∎

Finalize: The problem asked us to identify neutrinos, but we could have used the same rules to determine whether an electron, muon, or tau participated in a reaction, if the problem had told us what kind of neutrino it was.

---

21. Determine which of the following reactions can occur. For those that cannot occur, determine the conservation law (or laws) violated.

(a) $p \rightarrow \pi^+ + \pi^0$ $\qquad\qquad$ (b) $p + p \rightarrow p + p + \pi^0$

(c) $p + p \rightarrow p + \pi^+$ $\qquad\qquad$ (d) $\pi^+ \rightarrow \mu^+ + \nu_\mu$

(e) $n \rightarrow p + e^- + \bar{\nu}_e$ $\qquad\qquad$ (f) $\pi^+ \rightarrow \mu^+ + n$

Solution

Conceptualize: Conservation laws include those for a system's total energy, momentum, angular momentum, charge, baryon number, three lepton numbers, strangeness, charm, bottomness, and topness.

Categorize: The particles participating here all have zero strangeness, charm, bottomness, and topness. Their motion determines their energy, momentum, and angular momentum. So we have only conservation of charge, baryon number, and lepton numbers to check.

Analyze:

(a) $p \rightarrow \pi^+ + \pi^0$. Baryon number conservation is violated. If the reaction occurred, the baryon number would change according to $1 \rightarrow 0 + 0$. ∎

(b) $p + p \rightarrow p + p + \pi^0$ This reaction can occur. ∎

(c) $p + p \rightarrow p + \pi^+$ $\qquad$ Baryon number is violated: $1 + 1 \rightarrow 1 + 0$. ∎

(d) $\pi^+ \rightarrow \mu^+ + \nu_\mu$ $\qquad$ This reaction can occur. ∎

(e) $n \rightarrow p + e^- + \bar{v}_e$ This reaction can occur. ■

(f) $\pi^+ \rightarrow \mu^+ + n$ Violates baryon number: $0 \rightarrow 0 + 1$ and violates muon-lepton number: $0 \rightarrow -1 + 0$. Further, if the pion is not colliding with another particle, it does not have enough energy to undergo this decay. ■

Finalize: Note that reactions proceeding by the weak interaction can alter the strangeness, charm, topness, and bottomness of an isolated system.

━━━━━━━━━━━━━━━

23. Determine whether or not strangeness is conserved in the following decays and reactions.

(a) $\Lambda^0 \rightarrow p + \pi^-$ (b) $\pi^- + p \rightarrow \Lambda^0 + K^0$

(c) $\bar{p} + p \rightarrow \bar{\Lambda}^0 + \Lambda^0$ (d) $\pi^- + p \rightarrow \pi^- + \Sigma^+$

(e) $\Xi^- \rightarrow \Lambda^0 + \pi^-$ (f) $\Xi^0 \rightarrow p + \pi^-$

Solution

Conceptualize: We could think about the quark composition of each particle, but it is simplest to compare the total strangeness quantum number before and after, based on . . .

Categorize: . . . the listings for the individual particles in Table 31.2.

Analyze:

(a) For $\Lambda^0 \rightarrow p + \pi^-$ we have strangeness: $-1 \rightarrow 0 + 0$. Since -1 does not equal 0, strangeness is not conserved. ■

(b) Following the same procedure, for $\pi^- + p \rightarrow \Lambda^0 + K^0$ we have strangeness: $0 + 0 \rightarrow -1 + 1$. This reduces to $0 = 0$ so strangeness is conserved. ■

(c) $\bar{p} + p \rightarrow \bar{\Lambda}^0 + \Lambda^0$ gives strangeness: $0 + 0 \rightarrow +1 - 1$. Since $0 = 0$, strangeness is conserved. ■

(d) $\pi^- + p \rightarrow \pi^- + \Sigma^+$ gives strangeness: $0 + 0 \rightarrow 0 - 1$. Since 0 does not equal -1, strangeness is not conserved. ■

(e) $\Xi^- \rightarrow \Lambda^0 + \pi^-$ gives strangeness: $-2 \rightarrow -1 + 0$. Since -2 does not equal -1, strangeness is not conserved. ■

(f) $\Xi^0 \rightarrow p + \pi^-$ gives strangeness: $-2 \rightarrow 0 + 0$. Since -2 does not equal 0, strangeness is not conserved. ■

Finalize: Note again that reactions proceeding by the weak interaction can alter the strangeness of an isolated system.

═══════════════

29. If a $\bar{K}_S^0$ meson at rest decays in 0.900×10^{-10} s, how far does a $\bar{K}_S^0$ meson travel if it is moving at $0.960c$?

Solution

Conceptualize: The motion of the $\bar{K}_S^0$ particle is relativistic. Just like the astronaut who leaves for a distant star and returns to find his family long gone, the kaon appears to us to have a longer lifetime.

Categorize: We identify the effect of time dilation and use the particle under constant speed model.

Analyze: The time-dilated lifetime is

$$t = \gamma t_0 = \frac{0.900 \times 10^{-10} \text{ s}}{\sqrt{1 - v^2/c^2}} = \frac{0.900 \times 10^{-10} \text{ s}}{\sqrt{1 - (0.960)^2}} = 3.21 \times 10^{-10} \text{ s}$$

During this time interval, we see the kaon travel at $0.960c$. It travels for a distance of

$$d = vt = (0.960)(3.00 \times 10^8 \text{ m/s})(3.21 \times 10^{-10} \text{ s})$$
$$= 0.092\ 6 \text{ m} = 9.26 \text{ cm} \qquad \blacksquare$$

Finalize: This distance can fit into a particle detector, such as the bubble-chamber photographs in the text's Figures 31.7 and 31.10.

═══════════════

37. Analyze each of the following reactions in terms of constituent quarks and show that each type of quark is conserved.

(a) $\pi^+ + p \rightarrow K^+ + \Sigma^+$

(b) $K^- + p \rightarrow K^+ + K^0 + \Omega^-$

(c) Determine the quarks in the final particle for this reaction:

$$p + p \rightarrow K^0 + p + \pi^+ + ?$$

(d) In the reaction in part (c), identify the mystery particle.

Solution

Conceptualize: We should see the same quarks differently arranged on the two sides of each reaction.

Categorize: We look up the quark constituents of the particles in Tables 31.4 and 31.5.

Analyze:

(a) The reaction in terms of quarks reads $\bar{d}u + uud \rightarrow u\bar{s} + uus$, which nets 3u, 0d, 0s before and after. ∎

(b) Following the same procedure gives $\bar{u}s + uud \rightarrow u\bar{s} + d\bar{s} + sss$, which shows conservation at 1u, 1d, 1s before and after. ∎

(c) In terms of quarks, the reaction can be written as $uud + uud \rightarrow d\bar{s} + uud + u\bar{d} + ?$, which requires the mystery particle to have quark composition uds to balance the quarks on the two sides of the reaction. ∎

(d) A uds is either a Λ^0 or a Σ^0. ∎

Finalize: With larger mass, the Σ^0 can be thought of as an excited state of the Λ^0, with the quarks inside it moving around faster.

---

54. The energy flux carried by neutrinos from the Sun is estimated to be on the order of 0.400 W/m² at Earth's surface. Estimate the fractional mass loss of the Sun over 10^9 yr due to the emission of neutrinos. The mass of the Sun is 1.989×10^{30} kg. The Earth–Sun distance is 1.496×10^{11} m.

Solution

Conceptualize: Our Sun is estimated to have a life span of about 10 billion years, so in this problem, we are examining the radiation of neutrinos over a considerable fraction of the Sun's life. However, the mass carried away by the neutrinos is a very small fraction of the total mass involved in the Sun's nuclear fusion process, so even over this long time interval, the mass of the Sun may not change significantly (probably by less than 1%).

Categorize: The change in mass of the Sun can be found from the energy flux received by the Earth and Einstein's famous equation, $E = mc^2$. We will also use the inverse square law and the definition of power.

Analyze: Since the neutrino flux from the Sun reaching the Earth is 0.400 W/m², the total energy emitted per second by the Sun in neutrinos in all directions is that which would irradiate the surface of a great sphere around it, with the Earth's orbit as its equator.

$$\left(0.400 \text{ W/m}^2\right)\left(4\pi r^2\right) = \left(0.400 \text{ W/m}^2\right)\left[4\pi\left(1.496 \times 10^{11} \text{ m}\right)^2\right]$$
$$= 1.12 \times 10^{23} \text{ W}$$

In a period of 10^9 yr, the Sun emits a total energy of $E = P\Delta t$.

$$E = (1.12 \times 10^{23} \text{ J/s})(10^9 \text{ yr})(3.156 \times 10^7 \text{ s/yr}) = 3.55 \times 10^{39} \text{ J}$$

carried by neutrinos. This energy corresponds to an annihilated mass according to

$$E = m_\nu c^2 = 3.55 \times 10^{39} \text{ J} \quad \text{or} \quad m_\nu = 3.94 \times 10^{22} \text{ kg}.$$

Since the Sun has a mass of 1.989×10^{30} kg, this corresponds to a loss of only about 1 part in 50 400 000 of the Sun's mass over 10^9 yr in the form of neutrinos. ∎

Finalize: It appears that the neutrino flux changes the mass of the Sun by so little that it would be difficult to measure the difference in mass, even over its lifetime!

---

61. Assume that the half-life of free neutrons is 614 s. What fraction of a group of free thermal neutrons with kinetic energy 0.040 0 eV will decay before traveling a distance of 10.0 km?

Solution

Conceptualize: We guess that these neutrons are moving pretty fast, so they are not in flight for very long, and perhaps one percent may decay.

Categorize: The fraction that will remain is given by the ratio N/N_0, where $N/N_0 = e^{-\lambda \Delta t}$ and Δt is the time interval it takes the neutron to travel a distance of $d = 10.0$ km. Further, we use the particle under constant speed model.

Analyze: The time of flight is given by $\Delta t = d/v$.

Since $K = \dfrac{1}{2}mv^2$, $\quad \Delta t = \dfrac{d}{\sqrt{\dfrac{2K}{m}}} = \dfrac{10.0 \times 10^3 \text{ m}}{\sqrt{\dfrac{2(0.040\ 0 \text{ eV})(1.60 \times 10^{-19} \text{ J/eV})}{1.67 \times 10^{-27} \text{ kg}}}} = 3.61 \text{ s}.$

The decay constant is $\lambda = \dfrac{\ln 2}{T_{1/2}} = \dfrac{0.693}{614 \text{ s}} = 1.13 \times 10^{-3} \text{ s}^{-1}.$

Therefore we have

$$\lambda \Delta t = (1.13 \times 10^{-3} \text{ s})(3.61 \text{ s}) = 4.08 \times 10^{-3} = 0.004\ 08$$

And the fraction remaining is $\quad \dfrac{N}{N_0} = e^{-\lambda \Delta t} = e^{-0.004\ 08} = 0.995\ 9.$

Hence, the fraction that has decayed in this time interval is

$$1 - \dfrac{N}{N_0} = 0.004\ 07 \quad \text{or} \quad 0.407\%. \qquad ∎$$

Finalize: We made a good guess! Neutron loss by decay is a minor concern in designing fission reactors.

===

64. A Σ^0 particle at rest decays according to $\Sigma^0 = \Lambda^0 + \gamma$. Find the gamma-ray energy.

Solution

Conceptualize: The Λ^0 and the Σ^0 have the same quark composition. You can think of the more massive Σ^0 as an excited state of the Λ^0, with the u, d, and s quarks in it moving around more rapidly. From Table 31.2, the rest energies are $E_{0\Sigma} = m_\Sigma c^2 = 1\,192.5\,\text{MeV}$ and $E_{0\Lambda} = m_\Lambda c^2 = 1\,115.6\,\text{MeV}$. The photon has zero rest energy. Subtracting gives the Q value of the reaction as 76.9 MeV. This energy must be shared between the lambda particle and the photon. The photon will have most of the decay energy but not all of it, because the lambda particle must move in the opposite direction to the photon with enough speed to have an equal amount of momentum.

Categorize: We use the energy and momentum versions of the isolated system model. The lambda particle will be moving at less than one-tenth of the speed of light, so we can use the classical relation $K = \frac{1}{2}mv^2 = \frac{p^2}{2m}$ for it. The only relativistic equations we need are the identification of the Q value as $m_\Sigma c^2 - m_\Lambda c^2$ and $E = pc$ for the photon.

Analyze: For $\Sigma^0 \rightarrow \Lambda^0 + \gamma$ conservation of system energy in the decay requires

$$E_{0\Sigma} = \left(E_{0\Lambda} + K_\Lambda\right) + E_\gamma \quad \text{or} \quad m_\Sigma c^2 = \left(m_\Lambda c^2 + \frac{p_\Lambda^2}{2m_\Lambda}\right) + E_\gamma$$

or $\qquad\qquad m_\Sigma c^2 = \left(m_\Lambda c^2 + \frac{p_\Lambda^2 c^2}{2m_\Lambda c^2}\right) + E_\gamma.$

System momentum conservation gives $\left|p_\Lambda\right| = \left|p_\gamma\right|$, so the last result may be written as

$$m_\Sigma c^2 = m_\Lambda c^2 + \frac{p_\gamma^2 c^2}{2m_\Lambda c^2} + E_\gamma$$

Recognizing that $p_\gamma c = E_\gamma$, we now have $\qquad Q = \frac{E_\gamma^2}{2m_\Lambda c^2} + E_\gamma.$

Solving this quadratic equation gives

$$E_\gamma = m_\Lambda c^2 \left(\sqrt{1 + \frac{2Q}{m_\Lambda c^2}} - 1 \right) = (1\ 115.6\ \text{MeV})\left(\sqrt{1 + \frac{2(76.9)}{1\ 115.6}} - 1 \right)$$

$$E_\gamma = 74.4\ \text{MeV}$$ ∎

Finalize: The kinetic energy of the lambda particle is only about $Q - E_\gamma = 2.48$ MeV, so its speed is

$$v = \sqrt{\frac{2K}{m}} = c\sqrt{\frac{2K}{mc^2}} = 3.00 \times 10^8\ \frac{\text{m}}{\text{s}} \sqrt{\frac{2(2.48)}{1\ 115.6}} = 0.066\ 7c = 2.00 \times 10^7\ \frac{\text{m}}{\text{s}}$$

This is less than $0.1c$, confirming our assumption that it moves nonrelativistically.

━━━━━━━━

65. Determine the kinetic energies of the proton and pion resulting from the decay of a Λ^0 at rest:

$$\Lambda^0 \to p + \pi^-$$

Solution

Conceptualize: We obtain the rest energy of each of the particles from Table 31.2. Their difference gives us the kinetic energy of the products of the reaction.

Categorize: We use the relativistic energy version of the isolated system model.

Analyze:

From Table 31.2, $m_\Lambda c^2 = 1\ 115.6$ MeV, $m_p c^2 = 938.3$ MeV, and $m_\pi c^2 = 139.6$ MeV.

Since the Λ^0 is at rest, the difference between its rest energy of the and the rest energies of the proton and the pion is the sum of the kinetic energies of the proton and the pion.

$$K_p + K_\pi = 1\ 115.6\ \text{MeV} - 938.3\ \text{MeV} - 139.6\ \text{MeV} = 37.7\ \text{MeV}$$

Now, since $p_p = p_\pi = p$, applying conservation of relativistic energy to the decay process, we have

$$\left[\sqrt{(938.3)^2 + p^2 c^2} - 938.3 \right] + \left[\sqrt{(139.6)^2 + p^2 c^2} - 139.6 \right] = 37.7\ \text{MeV}$$

Solving yields

$$p_\pi c = p_p c = 100.4\ \text{MeV}$$

Then,

$$K_p = \sqrt{\left(m_p c^2\right)^2 + (100.4)^2} - m_p c^2 = \boxed{5.35 \text{ MeV}}$$

$$K_\pi = \sqrt{(139.6)^2 + (100.4)^2} - 139.6 = \boxed{32.3 \text{ MeV}}$$

Finalize: The kinetic energies of the products is much smaller than their rest energies. The pion is the lightest known meson, which is why its kinetic energy is much greater than that of the proton.

Solar System Data

Body	Mass (kg)	Mean Radius (m)	Period (s)	Mean Distance from the Sun (m)
Mercury	3.30×10^{23}	2.44×10^6	7.60×10^6	5.79×10^{10}
Venus	4.87×10^{24}	6.05×10^6	1.94×10^7	1.08×10^{11}
Earth	5.97×10^{24}	6.37×10^6	3.156×10^7	1.496×10^{11}
Mars	6.42×10^{23}	3.39×10^6	5.94×10^7	2.28×10^{11}
Jupiter	1.90×10^{27}	6.99×10^7	3.74×10^8	7.78×10^{11}
Saturn	5.68×10^{26}	5.82×10^7	9.29×10^8	1.43×10^{12}
Uranus	8.68×10^{25}	2.54×10^7	2.65×10^9	2.87×10^{12}
Neptune	1.02×10^{26}	2.46×10^7	5.18×10^9	4.50×10^{12}
Pluto[a]	1.25×10^{22}	1.20×10^6	7.82×10^9	5.91×10^{12}
Moon	7.35×10^{22}	1.74×10^6	—	—
Sun	1.989×10^{30}	6.96×10^8	—	—

[a]In August 2006, the International Astronomical Union adopted a definition of a planet that separates Pluto from the other eight planets. Pluto is now defined as a "dwarf planet" (like the asteroid Ceres).

Physical Data Often Used

Average Earth–Moon distance	3.84×10^8 m
Average Earth–Sun distance	1.496×10^{11} m
Average radius of the Earth	6.37×10^6 m
Density of air (20°C and 1 atm)	1.20 kg/m³
Density of air (0°C and 1 atm)	1.29 kg/m³
Density of water (20°C and 1 atm)	1.00×10^3 kg/m³
Free-fall acceleration	9.80 m/s²
Mass of the Earth	5.97×10^{24} kg
Mass of the Moon	7.35×10^{22} kg
Mass of the Sun	1.99×10^{30} kg
Standard atmospheric pressure	1.013×10^5 Pa

Note: These values are the ones used in the text.

Some Prefixes for Powers of Ten

Power	Prefix	Abbreviation	Power	Prefix	Abbreviation
10^{-24}	yocto	y	10^1	deka	da
10^{-21}	zepto	z	10^2	hecto	h
10^{-18}	atto	a	10^3	kilo	k
10^{-15}	femto	f	10^6	mega	M
10^{-12}	pico	p	10^9	giga	G
10^{-9}	nano	n	10^{12}	tera	T
10^{-6}	micro	μ	10^{15}	peta	P
10^{-3}	milli	m	10^{18}	exa	E
10^{-2}	centi	c	10^{21}	zetta	Z
10^{-1}	deci	d	10^{24}	yotta	Y